# Topics in Almost Hermitian Geometry and Related Fields

# Topics in Almost Hermitian Geometry and Related Fields

## PROCEEDINGS IN HONOR OF PROFESSOR K SEKIGAWA'S 60TH BIRTHDAY

Niigata, Japan 1 – 3 November 2004


Editors

**Yasuo Matsushita**
University of Shiga Prefecture, Japan

**Eduardo García-Río**
Universidade de Santiago de Compostela, Spain

**Hideya Hashimoto**
Meijyo University, Japan

**Takashi Koda**
Toyoma University, Japan

**Takashi Oguro**
Tokyo Denki University, Japan



World Scientific

NEW JERSEY • LONDON • SINGAPORE • BEIJING • SHANGHAI • HONG KONG • TAIPEI • CHENNAI

*Published by*

World Scientific Publishing Co. Pte. Ltd.
5 Toh Tuck Link, Singapore 596224
*USA office:* 27 Warren Street, Suite 401-402, Hackensack, NJ 07601
*UK office:* 57 Shelton Street, Covent Garden, London WC2H 9HE

**British Library Cataloguing-in-Publication Data**
A catalogue record for this book is available from the British Library.

**TOPICS IN ALMOST HERMITIAN GEOMETRY AND RELATED FIELDS**
**Proceedings in Honor of Professor K Sekigawa's 60th Birthday**



ISBN 981-256-417-9

Printed in Singapore by B & JO Enterprise

# PREFACE

The Conference "Topics in Almost Hermitian Geometry and the Related Fields" was held during November 1–3, 2004 in Niigata University in honor of Professor Kouei Sekigawa in his 60th years. This volume is the Proceedings of the Conference which includes not only works talked at the Conference but also contributions by authors who could not attend there. Thanks to Professor Sekiagwa's wide variety of mathematical activity and due to his warmhearted personal character, many mathematicians, who have discussed with him, who have been taught and supervised by him, have made their effort to complete a single issue, "Sekigawa's Volume". The articles in the present volume are concerned with various topics in differential geometry and mathematical physics, with almost Hermitian geometry as a core subject. The organizers of the Conference are also working as the editors of the present volume, and would like to express our cordial thanks to all the contributors for their kind support in various aspects, and greatly acknowledge a grant from Japan Society for the Promotion of Sciences. This volume will be an important step to the future development of Sekigawa's school and also of differential geometry. The organizers and editors, together with all contributors, would like to dedicate the present volume to Professor Kouei Sekigawa.

April 2005

YM., E.G-R., HH., TK., and TO.

Niigata, November 2, 2004

# CONTENTS

# CONTRIBUTORS

**Mohamed Tahar Kadaoui Abbassi:** Département des Mathématiques, Faculté des sciences Dhar El Mahraz, Université Sidi Mohamed Ben Abdallah, B.P. 1796, Fès-Atlas, Fès, Morocco.

**Toshiaki Adachi:** Department of Mathematics, Nagoya Institute of Technology, Gokiso, Nagoya 466-8555, Japan.

**Cornelia-Livia Bejan:** Seminar Matematic, Universitatea "A. I. Cuza", Iasi 700506, Romania.

**Tran Quoc Binh:** Department of Mathematics and Informatics, Debrecen University, H-4010 Debrecen, P.O. Box 12, Hungary.

**Eric Boeckx:** Department of Mathematics, Katholieke Universiteit Leuven, Celestijnenlaan 200B, 3001 Leuven, Belgium.

**Jong Taek Cho:** Department of Mathematics, Chonnam National University, CNU The Institute of Basic Sciences, Kwangju 500-757, Korea.

**Sun Hyang Chun:** Department of Mathematics, Graduate School Chonnam National University, Kwangju 500-757, Korea.

**Iago García-Ramírez:** Department of Geometry and Topology, Faculty of Mathematics, University of Santiago de Compostela, 15782 Santiago de Compostela, Spain.

**Eduardo García-Río:** Department of Geometry and Topology, Faculty of Mathematics, University of Santiago de Compostela, 15782 Santiago de Compostela, Spain.

**Peter B Gilkey:** Mathematics Department, University of Oregon, Eugene Or 97403 USA.

**Mitsuhiro Itoh:** Institute of Mathematics, University of Tsukuba 305-8571, Japan.

**Włodzimierz Jelonek:** Institute of Mathematics, Technical University of Cracow, Warszawska 24, 31-155 Kraków, Poland.

**Oldřich Kowalski:** Faculty of Mathematics and physics, Charles University, Sokolovská 83, 186 75 Prague 8, Czech Republic.

**Tatsuya Kawabe:** Department of Mathematical Science, Graduate School of Science and Technology, Niigata University, Ikarashi, Niigata 950-2181, Japan.

**U-Hang Ki:** Department of Mathematics, Kyungpook National University, Daegu 702-701, Korea.

**Sadahiro Maeda:** Department of Mathematics, Shimane University, Matsue, Shimane 690-8504, Japan.

**Mancho Manev:** University of Plovdiv, Faculty of Mathematics and Informatics, 236 Bulgaria blvd., Plovdiv 4003, Bulgaria.

**Katsuya Mashimo:** Department of Mathematics, Tokyo University of Agriculture and Technology, Koganei, Tokyo 184-8588, Japan.

**Yasuo Matsushita:** Section of Mathematics, School of Engineering, University of Shiga Prefecture, Hikone 522-8533, Japan.

**Takashi Nitta:** Department of Education, Mie University, Kamihama, Tsu 514-8507, Japan.

**Setsuo Nagai:** Department of Education, Toyama University, Toyamashi 930-8555, Japan.

**Stana Nikčević:** Mathematical Institute, SANU, Knez Mihailova 35, p.p. 367, 11001 Belgrade, Serbia and Montenegro.

**Tomoko Okada:** Graduate school of Mathematics, Nagoya University, Chikusa-ku, Nagoya 464-8602, Japan.

**Takashi Okayasu:** Faculty of Education, Ibaraki University, 2-1-1 Bunkyo, Mito, Ibaraki 310-8512, Japan.

**Takuji Sato:** Graduate School of Natural Science and Technology, Kanazawa University, Kanazawa 920-1192, Japan.

**Young Jin Suh:** Kyungpook National University, Department of Mathematics, Taegu 702-701, Korea.

**Ryoichi Takagi:** Department of Mathematics and Informatics, Chiba University, Chibashi 263-8522, Japan.

**Tadashi Taniguchi:** Sendai National College of Technology, Aobaku, Sendai 989-3128, Japan.

**Kazumi Tsukada:** Department of Mathematics, Ochanomizu University, 2-1-1 Otsuka, Bunkyo-ku, Tokyo 112-8610, Japan.

**Hajime Urakawa:** Division of Mathematics, Graduate School of Information Sciences, Tohoku University, Aoba 09, Sendai 980-8579, Japan.

**Lieven Vanhecke:** Department of Mathematics, Katholieke Universiteit Leuven, Celestijnenlaan 200B, 3001 Leuven, Belgium.

# ON $G$-NATURAL METRICS WITH CONSTANT SCALAR CURVATURE ON UNIT TANGENT SPHERE BUNDLES

MOHAMED TAHAR KADAOUI ABBASSI*

*Département des Mathématiques,*
*Faculté des sciences Dhar El Mahraz,*
*Université Sidi Mohamed Ben Abdallah,*
*B.P. 1796, Fès-Atlas, Fès, Morocco.*
*E-mail: mtk_abbassi@Yahoo.fr*

OLDŘICH KOWALSKI†

*Faculty of Mathematics and physics,*
*Charles University,*
*Sokolovská 83, 186 75 Prague 8, Czech Republic.*
*E-mail: kowalski@karlin.mff.cuni.cz*

Let $(M, g)$ be a Riemannian manifold, $TM$ its tangent bundle and $T_1M$ its unit tangent sphere bundle. The family of all Riemannian $g$-natural metrics $G$ on $TM$ have been determined by the second author and M. Sekizawa. It depends on 6 arbitrary functions of the norm of a vector $u \in TM$. In the present paper, we study the induced metrics $\tilde{G}$ on $T_1M$ corresponding to the subfamily of all Riemannian $g$-natural metrics of the form $G = a.g^s + b.g^h + c.g^v$, where $a$, $b$, $c$ are constants. We calculate the Levi-Civita connection and the standard sorts of curvatures for the above family of metrics. Then, in the three-dimensional case, we derive necessary and sufficient conditions for such metrics to have constant scalar curvature (in terms of the curvature of the underlying manifold). Hereby we generalize some earlier results by E. Boeckx and L. Vanhecke, where the *curvature homogeneity* of the base manifold plays a crucial role.

*Keywords*: unit tangent (sphere) bundle.

## Introduction and main results

The major object for the study in Riemannian geometry is the Riemannian

2000 *Mathematics Subject Classification.* Primary 53C07; Secondary 53B20, 53A55, 53C20, 53C25.

*The first author would like to express his sincere thanks to the second author for his hospitality at the occasion of his invitation to the Charles University in Prague.

†The second author was supported by the grant GA ČR 201/05/2707 and the project MSM 0021620839.

curvature. Imposing some conditions (or constraints) to the curvature, the geometers have studied interesting classes of Riemannian manifolds whose Riemannian curvatures satisfy the prescribed condition. For example, if we require that, for each pair of points $p$ and $q$ in $M$, there exists a linear isometry $F : M_p \to M_q$ between the corresponding tangent spaces such that $F^* R_q = R_p$, where $R$ is the Riemannian curvature of $(M, g)$, then we obtain the class of *curvature homogeneous spaces* [41]. Locally homogeneous spaces are trivial examples of such spaces, but there exist much more examples which are not locally homogeneous. The first examples of a *complete* curvature homogeneous space which is not locally homogeneous were given by K. Sekigawa in [38], [39]. In the three-dimensional case, the curvature homogeneity is equivalent to the constancy of the Ricci roots. In the case when the constant Ricci eigenvalues are prescribed and just two of them are distinct (but arbitrary), and if we do not require completeness, then all corresponding spaces are parametrized (up to local isometries) by two arbitrary functions of one variable and they are not locally homogeneous, in general (see [23] and [27]). A survey on curvature homogeneous spaces can be found in [12]. S. Ivanov and I. Petrova have shown that, in dimension 3, curvature homogeneous spaces with two zero Ricci roots and one non-zero constant Ricci root are, besides spaces of constant curvature, the only three-dimensional Riemannian manifolds on which the skew-symmetric curvature operator has globally constant eigenvalues [19].

Recently, it appeared that the curvature homogeneous spaces of the Ivanov-Petrova type can be characterized by the constancy of the scalar curvature of the corresponding unit tangent bundle equipped with the induced Sasaki metric. Namely, E. Boeckx and L. Vanhecke have proved the following:

**Theorem 1.** ([14]) *Let $(M^3, g)$ be a connected 3-dimensional Riemannian manifold. Then the unit tangent sphere bundle $T_1M^3$ with the Sasaki metric is a space of constant scalar curvature if and only if $(M^3, g)$ has constant curvature or $(M^3, g)$ is a curvature homogeneous space with constant Ricci roots $\rho_1 = \rho_2 = 0 \neq \rho_3$.*

As mentioned in the Theorem above, Boeckx and Vanhecke endowed the unit tangent sphere bundle with the induced Sasaki metric. The obvious reason was that the Riemannian geometry of the *tangent bundle $TM$* of a Riemannian manifold $(M, g)$ has been traditionally related to the Sasaki metric $g^s$. Here the constancy of the scalar curvature occurs only in a trivial situation. Namely, E. Musso and F. Tricerri proved in 1986 the following result: if the scalar curvature of $(TM, g^s)$ is constant then $(M, g)$ is flat [32]. Inspired by a paper of J. Cheeger and D. Gromoll, they also defined a new $g$-natural metric $g_{CG}$ on $TM$ which they called the Cheeger-Gromoll metric. M. Sekizawa [40] has shown that the scalar curvature of $(TM, g_{CG})$ is never constant if the original metric on the base manifold has constant sectional curvature (see also [18]). Furthermore, the first author and M. Sarih have proved that $(TM, g_{CG})$ is never a space of constant sectional curvature (cf. [3]). In particular, $g_{CG}$ is never flat.

V. Oproiu and his collaborators constructed a rather sophisticated family of Riemannian metrics on the tangent bundles of Riemannian manifolds which possess interesting geometrical properties (cf. [33]–[36]). In particular, the scalar curvature of $TM$ can be constant also for a *non-flat* $(M, g)$ with constant sectional curvature. This family was found by the use of the computer program, and other examples with this property are not known. The first author and M. Sarih proved in [6] that the considered metrics by Oproiu form a particular subclass of what they called (cf. [4]) *g-natural metrics* on the tangent bundle $TM$, initially classified by the second author and M. Sekizawa [26] (see also [21] for another presentation of the results and [1] for the improvement of the classification Theorems). Riemannian $g$-natural metrics on $TM$ have been studied systematically by the first author and M. Sarih (cf. [1], [4]–[7]).

Now back to the tangent sphere bundles. The second author and M. Sekizawa studied the geometry of tangent sphere bundles with *arbitrary radii* endowed with the induced Sasaki metric (cf. [28]), showing that this geometry is essentially different for the small radii and for the big radii. As concerns the geometry of $T_1M$, equipped with the metric $\tilde{g}_{CG}$ induced by the Cheeger-Gromoll metric $g_{CG}$, it was mentioned in [28] that $(T_1M, \tilde{g}_{CG})$ is isometric to the tangent sphere bundle $T_{1/\sqrt{2}}M$ of radius $1/\sqrt{2}$, equipped with the metric induced by the Sasaki metric. Generalizing this particular observation, we shall prove the following:

**Theorem 2.** *Let $(M, g)$ be a Riemannian manifold and $\tilde{g^s}$ be the induced Sasaki metric on the tangent sphere bundle $T_rM$ of an arbitrary radius $r > 0$. Then, for the Riemannian g-natural metric $G = r^2.g^s + (1 - r^2).g^v$ on $TM$, $(T_rM, \tilde{g^s})$ is isometric to $(T_1M, \tilde{G})$ , where $\tilde{G}$ is the restriction of $G$ to $T_1M$.*

**Remark 1.** Combining Theorem 2 with the results of [28], we can prove easily that Riemannian metrics of the form $a.g^s + b.g^h + c.g^v$ on the unit tangent bundle $T_1M$ change essentially their curvature properties with the change of the parameters.

In this paper, we shall consider a broader family of metrics on $TM$, which can be called "classical metrics", namely all metrics of the form $G = a.g^s + b.g^h + c.g^v$, with $a > 0$ and $\alpha := a(a+c) - b^2 > 0$, where $g^s$, $g^h$ and $g^v$ are the Sasaki lift, the horizontal lift and the vertical lift to $TM$ of the metric $g$, respectively, and we shall study their restrictions $\tilde{G}$ to the unit tangent sphere bundle $T_1M$. The inequalities above just express the property that the corresponding metric $G$ is properly Riemannian — see [7].

**Remark 2.** The choice of $g$-natural metrics on tangent sphere bundles induced by those of the form $a.g^s + b.g^h + c.g^v$ is not very restrictive when compared with the general case. Indeed, it is easy to check (see [2] for a proof) that any Riemannian metric on any tangent sphere bundle induced from any arbitrary Riemannian $g$-natural metric on $TM$ is a constant linear combination of the three classical lifts of $g$ and the vertical lift of the $F$-metric $k$ on $TM$ defined by

$$k(u; X, Y) = g(u, X)g(u, Y), \quad \text{for all} \quad (u, X, Y) \in TM \oplus TM \oplus TM.$$

(See [26], for example, for the definition of an $F$-metric and more details).

Now, the main purpose of this paper is to give, for any Riemannian 3-manifold $(M^3, g)$ and any arbitrarily chosen Riemannian metric of the form $G = a.g^s + b.g^h + c.g^v$ on $TM^3$, the necessary and sufficient conditions for the unit tangent sphere bundle $T_1M^3$, endowed with the metric $\tilde{G}$ induced by $G$, to be of constant scalar curvature.

At first, if we replace the Sasaki metric by any Riemannian metric of the form $a.g^s + c.g^v$ on $T_1M^3$, then we obtain:

**Theorem 3.** *Let $(M^3, g)$ be a connected 3-dimensional Riemannian manifold and $G = a.g^s + c.g^v$ be a Riemannian $g$-natural metric on $TM^3$, i.e., such that $a > 0$ and $a + c > 0$. Then we have*

a) *$(T_1M^3, \tilde{G})$ cannot have a constant scalar curvature $\widetilde{S} > \frac{11}{a}$.*
b) *$(T_1M^3, \tilde{G})$ is of constant scalar curvature $\frac{3}{a} < \widetilde{S} \leq \frac{11}{a}$ if and only if $(M^3, g)$ has constant curvature.*
c) *$(T_1M^3, \tilde{G})$ is of constant scalar curvature $\widetilde{S} \leq \frac{3}{a}$ if and only if $(M^3, g)$ has constant curvature or $(M^3, g)$ is a curvature homogeneous space with constant Ricci roots $\rho_1 = \rho_2 = 0 \neq \rho_3$.*

**Remark 3.** If we take $a = 1$ and $c = 0$ in Theorem 3, then the metric $G$ is nothing but the Sasaki metric. Consequently, Theorem 3, with $a = 1$ and $c = 0$, can be considered as a slight improvement of Theorem 1.

**Corollary 1.** *Let real numbers $s$ and $u \neq 0$ be given. Then, on the unit tangent sphere bundle of each curvature homogeneous 3-space with constant Ricci roots $\rho_1 = \rho_2 = 0$ and $\rho_3 = u$, there is a one-parameter family $S_u$ of Riemannian g-natural metrics with constant scalar curvatures equal to $s$.*

For the Riemannian $g$-natural metrics of the form $a.g^s + b.g^h + c.g^v$ with $b \neq 0$ on $TM^3$, we shall prove the two following results:

**Theorem 4.** *Let $(M^3, g)$ be a connected 3-dimensional Riemannian manifold and $G = a.g^s + b.g^h + c.g^v$ be such that $b \neq 0$ and $a(a+c) \neq 5b^2$. Then $(T_1M^3, \tilde{G})$ is a space of constant scalar curvature if and only if $(M^3, g)$ is a space of constant curvature.*

**Theorem 5.** *Let $(M^3, g)$ be a connected 3-dimensional Riemannian manifold and $G = a.g^s + b.g^h + c.g^v$ be such that $a(a+c) = 5b^2$. Then we have*

*a) $(T_1M^3, \tilde{G})$ cannot have a constant scalar curvature $\widetilde{S} > \frac{251627}{256000.a}$.*
*b) $(T_1M^3, \tilde{G})$ is of constant scalar curvature $-\frac{12315}{2816.a} < \widetilde{S} \leq \frac{251627}{256000.a}$ if and only if $(M^3, g)$ is a space of constant curvature.*
*c) $(T_1M^3, \tilde{G})$ is of constant scalar curvature $\widetilde{S} \leq -\frac{12315}{2816.a}$ if and only if $(M^3, g)$ is a space of constant curvature or $(M^3, g)$ is a curvature homogeneous space with constant Ricci roots $\rho_1 = \rho_2 \neq \rho_3$, where*

$$\rho_3 = \frac{20}{9}\rho_1 + \frac{277(a+c)}{45a}. \tag{0.1}$$

Part c) of Theorem 5 has the following corollaries:

**Corollary 2.** *For any connected 3-dimensional curvature homogeneous space $(M^3, g)$ with constant Ricci roots $\rho_1 = \rho_2 \neq \rho_3$ such that $\rho_3 > \frac{20}{9}\rho_1$ and $\rho_1$ arbitrary, there is a 1-parameter family $S'$ of g-natural metrics on $T_1M^3$ with constant negative scalar curvatures.*

**Corollary 3.** *Let $s$, $v$, $w$ be real numbers such that $s < 0$ and $w > \frac{20}{9}v$. Then on the unit tangent sphere bundle of each curvature homogeneous 3-space with constant Ricci roots $\rho_1 = \rho_2 = v$ and $\rho_3 = w$, there is a Riemannian g-natural metric with constant scalar curvature equal to $s$.*

**Remark 4.** In the case when all the eigenvalues are negative and we have $\rho_1 < \rho_3$ (which is still a particular case occurring in Corollary 2), then there is no locally homogeneous space with constant Ricci roots $\rho_1 = \rho_2 \neq \rho_3$ (cf. [25]).

**Remark 5.** It is an easy task to show that, for a two-dimensional Riemannian manifold $(M^2, g)$ and any arbitrary Riemannian metric of the form $G = a.g^s + b.g^h + c.g^v$ on $TM^2$, the unit tangent sphere bundle $T_1M^2$, with the metric induced by $G$, is of constant scalar curvature if and only if $(M^2, g)$ is of constant sectional curvature.

## 1. Basic formulas on tangent bundles

Let $\nabla$ be the Levi-Civita connection of $g$. Then the tangent space of $TM$ at any point $(x,u) \in TM$ splits into the horizontal and vertical subspaces with respect to $\nabla$:

$$(TM)_{(x,u)} = H_{(x,u)} \oplus V_{(x,u)}.$$

If $(x,u) \in TM$ is given then, for any vector $X \in M_x$, there exists a unique vector $X^h \in H_{(x,u)}$ such that $p_* X^h = X$, where $p : TM \to M$ is the natural projection. We call $X^h$ the *horizontal lift* of $X$ to the point $(x,u) \in TM$. The *vertical lift* of a vector $X \in M_x$ to $(x,u) \in TM$ is a vector $X^v \in V_{(x,u)}$ such that $X^v(df) = Xf$, for all functions $f$ on $M$. Here we consider 1-forms $df$ on $M$ as functions on $TM$ (i.e., $(df)(x,u) = uf$). Note that the map $X \to X^h$ is an isomorphism between the vector spaces $M_x$ and $H_{(x,u)}$. Similarly, the map $X \to X^v$ is an isomorphism between the vector spaces $M_x$ and $V_{(x,u)}$. Obviously, each tangent vector $\tilde{Z} \in (TM)_{(x,u)}$ can be written in the form $\tilde{Z} = X^h + Y^v$, where $X, Y \in M_x$ are uniquely determined vectors.

In an obvious way we can define horizontal and vertical lifts of vector fields on $M$. These are uniquely defined vector fields on $TM$. Each system of local coordinates $\{(U; x^i, i = 1, \ldots, m)\}$ in $M$ induces on $TM$ a system of local coordinates $\{(p^{-1}(U); x^i, u^i, i = 1, \ldots, m)\}$. Let $X = \sum_i X^i \left(\frac{\partial}{\partial x^i}\right)_x$ be the local expression in $\{(U; x^i, i = 1, \ldots, m)\}$ of a vector $X$ in $M_x$, $x \in M$. Then, the horizontal lift $X^h$ and the vertical lift $X^v$ of $X$ to $(x,u) \in TM$ are given, with respect to the induced coordinates, by:

$$X^h = \sum X^i \left(\frac{\partial}{\partial x^i}\right)_{(x,u)} - \sum \Gamma^i_{jk} u^j X^k \left(\frac{\partial}{\partial u^i}\right)_{(x,u)}, \quad \text{and} \tag{1.1}$$

$$X^v = \sum X^i \left(\frac{\partial}{\partial u^i}\right)_{(x,u)}, \tag{1.2}$$

where $(\Gamma^i_{jk})$ denote the Christoffel's symbols of $g$.

The canonical vertical vector field on $TM$ is a vector field $\mathcal{U}$ defined, in terms of local coordinates, by $\mathcal{U} = \sum_i u^i \partial/\partial u^i$. Here $\mathcal{U}$ does not depend on the choice of local coordinates and it is defined globally on $TM$. For a vector $u = \sum_i u^i(\partial/\partial x^i)_x \in M_x$, we see that $u^v_{(x,u)} = \sum_i u^i(\partial/\partial x^i)^v_{(x,u)} = \mathcal{U}_{(x,u)}$ and $u^h_{(x,u)} = \sum_i u^i(\partial/\partial x^i)^h_{(x,u)} = \xi_{(x,u)}$, where $\xi$ is the geodesic flow on $TM$.

The Riemannian curvature $R$ of $g$ is defined by

$$R(X,Y) = [\nabla_X, \nabla_Y] - \nabla_{[X,Y]}. \tag{1.3}$$

Now, the three classical constructions of metrics on tangent bundles are given as follows:

(*a*) The *Sasaki metric* $g^s$ is a positive definite Riuemannian metric on $TM$ derived from $g$ as follows:

$$\begin{cases} g^s_{(x,u)}(X^h,Y^h) = g_x(X,Y), & g^s_{(x,u)}(X^h,Y^v) = 0, \\ g^s_{(x,u)}(X^v,Y^h) = 0, & g^s_{(x,u)}(X^v,Y^v) = g_x(X,Y), \end{cases}$$

for all $X$, $Y \in M_x$.

(*b*) The *horizontal lift* $g^h$ of $g$ is a pseudo-Riemannian metric on $TM$ with signature $(m,m)$ which is given by:

$$\begin{cases} g^h_{(x,u)}(X^h,Y^h) = 0, & g^h_{(x,u)}(X^h,Y^v) = g_x(X,Y), \\ g^h_{(x,u)}(X^v,Y^h) = g_x(X,Y), & g^h_{(x,u)}(X^v,Y^v) = 0, \end{cases}$$

for all $X$, $Y \in M_x$.

(*c*) The *vertical lift* $g^v$ of $g$ is a degenerate metric of rank $m$ on $TM$ which is given by:

$$\begin{cases} g^v_{(x,u)}(X^h,Y^h) = g_x(X,Y), & g^v_{(x,u)}(X^h,Y^v) = 0, \\ g^v_{(x,u)}(X^v,Y^h) = 0, & g^v_{(x,u)}(X^v,Y^v) = 0, \end{cases}$$

for all $X$, $Y \in M_x$.

Combining linearly these classical metrics, we obtain the family $\{G = a.g^s + b.g^h + c.g^v$, $a$, $b$ and $c$ are constants$\}$ of metrics on $TM$, and whose elements belong to the broader class of $g$-natural metrics on $TM$ (cf. [4] for the definition of $g$-natural metrics and [21] for the general definition of naturality). We shall present here some basic results on the metrics of the form $G = a.g^s + b.g^h + c.g^v$ on $TM$, which had been studied in [7]. First, such a metric is Riemannian if and only if the inequalities $a > 0$ and $\alpha = a(a+c) - b^2 > 0$ hold. The Levi-Civita connection and the

Riemannian curvature tensor of a such Riemannian $g$-natural metric are given by (cf. [7]):

**Proposition 1.** *Let $(M, g)$ be a Riemannian manifold, $\nabla$ its Levi-Civita connection and $R$ its curvature tensor. Let $G = a.g^s + b.g^h + c.g^v$, $a$, $b$ and $c$ are constants satisfying $a > 0$ and $a(a+c) - b^2 > 0$. Then the Levi-Civita connection $\bar{\nabla}$ of $(TM, G)$ is characterized by*

$$(i) \qquad (\bar{\nabla}_{X^h} Y^h)_{(x,u)} = \{(\nabla_X Y)_x - \frac{ab}{2\alpha}[R(X_x, u)Y_x + R(Y_x, u)X_x]\}^h - \frac{1}{\alpha}\{b^2 R(X_x, u)Y_x + \frac{a(a+c)}{2} R(X_x, Y_x)u\}^v,$$

$$(ii) \qquad (\bar{\nabla}_{X^h} Y^v)_{(x,u)} = -\frac{a^2}{2\alpha}\{R(Y_x, u)X_x\}^h + \{(\nabla_X Y)_x + \frac{ab}{2\alpha} R(Y_x, u)X_x\}^v,$$

$$(iii) \qquad (\bar{\nabla}_{X^v} Y^h)_{(x,u)} = -\frac{a^2}{2\alpha}\{R(X_x, u)Y_x\}^h + \frac{ab}{2\alpha}\{R(X_x, u)Y_x\}^v,$$

$$(iv) \qquad (\bar{\nabla}_{X^v} Y^v)_{(x,u)} = 0$$

*for all vector fields $X$, $Y$ on $M$ and $(x, u) \in TM$.*

**Proposition 2.** *Let $(M, g)$ be a Riemannian manifold and $G = a.g^s + b.g^h + c.g^v$, $a$, $b$ and $c$ are constants satisfying $a > 0$ and $a(a+c) - b^2 > 0$. Denote by $\nabla$ and $R$ the Levi-Civita connection and the Riemannian curvature tensor of $(M, g)$, respectively. If we denote by $\bar{R}$ the Riemannian curvature tensor of $(TM, G)$, then:*

$$(1.4) \qquad \bar{R}(X^h, Y^h)Z^h$$
$$= \{R(X,Y)Z + \frac{ab}{2\alpha}[2(\nabla_u R)(X,Y)Z - (\nabla_Z R)(X,Y)u]$$
$$+ \frac{a^2}{4\alpha}[R(R(Y,Z)u, u)X - R(R(X,Z)u, u)Y]$$
$$+ \frac{a^2 b^2}{4\alpha^2}[R(X,u)R(Y,u)Z + R(X,u)R(Z,u)Y$$
$$- R(Y,u)R(X,u)Z - R(Y,u)R(Z,u)X)]$$
$$- \frac{a^2}{2\alpha} R(R(X,Y)u, u)Z\}^h + \{-\frac{b^2}{\alpha}(\nabla_u R)(X,Y)Z$$
$$+ \frac{a(a+c)}{2\alpha}(\nabla_Z R)(X,Y)u + \frac{ab^3}{2\alpha^2}[R(R(Y,u)Z, X)u$$
$$- R(X,u)R(Z,u)Y - R(R(X,u)Z, Y)u + R(Y,u)R(Z,u)X]$$

$$
\begin{aligned}
&+ \frac{a^2 b\,(a+c)}{4\alpha^2}\,[R(X,R(Y,u)Z)u + R(X,R(Z,u)Y)u \\
&- R(Y,R(X,u)Z)u - R(Y,R(Z,u)X)u - R(R(Y,Z)u,u)X \\
&+ R(R(X,Z)u,u)Y] + \frac{a\,b}{2\alpha}\,R(R(X,Y)u,u)Z\}^v,
\end{aligned}
$$

$$
\begin{aligned}
(1.5)\quad &\bar{R}(X^h,Y^v)Z^h \\
&= \{-\frac{a^2}{2\alpha}\,(\nabla_X R)(Y,u)Z + \frac{a^3 b}{4\alpha^2}\,[R(X,u)R(Y,u)Z \\
&- R(Y,u)R(Z,u)X - R(Y,u)R(X,u)Z] \\
&+ \frac{a\,b}{2\alpha}\,[R(X,Y)Z + R(Z,Y)X]\}^h \\
&+ \{\frac{a\,b}{2\alpha}\,(\nabla_X R)(Y,u)Z - \frac{a^2 b^2}{4\alpha^2}\,[R(X,u)R(Y,u)Z \\
&- R(Y,u)R(Z,u)X - R(Y,u)R(X,u)Z] + \frac{a^2}{4\alpha}\,R(X,R(Y,u)Z)u \\
&- \frac{b^2}{\alpha}\,R(X,Y)Z + \frac{a\,(a+c)}{2\alpha}\,R(X,Z)Y\}^v,
\end{aligned}
$$

$$
(1.6)\qquad \bar{R}(X^v,Y^v)Z^v = 0,
$$

*for all $x \in M$ and $X$, $Y$, $Z \in M_x$.*

Let us remark that the curvature terms of the form $\bar{R}(X^h,Y^v)Z^v$ are not needed in the sequel.

## 2. Riemannian metrics on $T_1M$ induced from Riemannian $g$-natural metrics on $TM$ of the form $a.g^s + b.g^h + c.g^v$

Let $r$ be a positive number. Then the *tangent sphere bundle of radius $r$* over a Riemannian manifold $(M,g)$ is the hypersurface $T_rM = \{\,(x,u) \in TM \mid g_x(u,u) = r^2\,\}$. The tangent space of $T_rM$, at a point $(x,u) \in T_rM$, is given by

$$
(2.1)\qquad (T_rM)_{(x,u)} = \{\,X^h + Y^v \mid X \in M_x,\ Y \in \{u\}^\perp \subset M_x\,\}.
$$

When $r = 1$, $T_1M$ is called *the unit tangent (sphere) bundle.*

Let $G = a.g^s + b.g^h + c.g^v$ be a Riemannian $g$-natural metric on $TM$, i.e., such that $a > 0$ and $a(a+c) - b^2 > 0$, and let $\tilde{G}$ be the metric on $T_1M$ induced by $G$. Remark, at first, that the canonical vector field $\mathcal{U}$ on $TM$ is not necessarily normal to $T_1M$ with respect to $G$ (if $b \neq 0$), but it is transverse to $T_1M$. Now, we shall determine a unit vector field $N^G$ on $TM$, which is normal to $T_1M$. For this, we put $N^G_{(x,u)} = X^h + \lambda\xi_{(x,u)} + Y^v + \mu\mathcal{U}_{(x,u)}$, where $(x,u) \in TM$, $X$, $Y \in u^\perp \subset M_x$, $\xi$ is the geodesic flow vector field on $TM$ and $\lambda$ and $\mu$ are real numbers. Using (2.1) we see that $N^G$ is normal to $T_1M$ if and only if

$$\text{(2.2)} \qquad \begin{cases} 0 = G_{(x,u)}(N^G_{(x,u)}, \xi_{(x,u)}) = \lambda(a+c) + \mu b, \\ 0 = G_{(x,u)}(N^G_{(x,u)}, Z^h_{(x,u)}) = (a+c)g_x(X,Z) + bg_x(Y,Z), \\ 0 = G_{(x,u)}(N^G_{(x,u)}, Z^v_{(x,u)}) = bg_x(X,Z) + ag_x(Y,Z), \end{cases}$$

for all $(x,u) \in T_1M$ and $Z \in u^\perp \subset M_x$. The two last equations of (2.2) are equivalent, by virtue of $a(a+c) - b^2 \neq 0$, to the identity $g_x(X,Z) = g_x(Y,Z) = 0$, for all $Z \in u^\perp$. It follows that $X = Y = 0$, since $X$, $Y \in u^\perp$. Taking into account the first identity of (2.2), we deduce that $N^G$ is normal to $T_1M$ if and only if $N^G = \mu(-\frac{b}{a+c}\xi + \mathcal{U})$, for some real $\mu$. On the other hand, to have $N^G$ unitary, it is necessary that $\mu$ satisfy the equality $\mu^2 = \frac{a+c}{a(a+c)-b^2}$. Choosing $\mu = \sqrt{\frac{a+c}{a(a+c)-b^2}}$ and recalling the notation $\alpha = a(a+c) - b^2$, we deduce that

$$\text{(2.3)} \qquad N^G = \frac{1}{\sqrt{(a+c)\alpha}}[-b\xi + (a+c)\mathcal{U}]$$

is a unit vector field on $TM$ which is normal to $T_1M$.

Now, we define the $G$-tangential lift $X^{t_G}$ of a vector $X \in M_x$ to $(x,u) \in T_1M$ by

$$\text{(2.4)} \qquad \begin{aligned} X^{t_G} &= X^v - G_{(x,u)}(X^v, N^G_{(x,u)})\, N^G_{(x,u)} \\ &= X^v - \sqrt{\frac{\alpha}{a+c}} g_x(X,u)\, N^G_{(x,u)}. \end{aligned}$$

In an obvious way we can define the $G$-tangential lift of vector fields on $M$.

**Remark 6.** If $b = 0$ (which implies that $\alpha = a(a+c)$), then the unit vector field $N_G$ normal to $T_1M$ is nothing but the canonical vector field $\mathcal{U}$ on $TM$, and the $G$-tangential lift of a vector $X \in M_x$ to $(x,u) \in T_1M$ is exactly the classical tangential lift

$$\text{(2.5)} \qquad X^t = X^v - g_x(X,u)\,\mathcal{U}_{(x,u)}$$

with respect to the Sasaki metric. For this reason we shall always write briefly $X^t$ instead of $X^{t_G}$ if there is no risk of confusion.

**Remark 7.** The $G$-tangential lift $X^t$ of a vector $X \in M_x$ to a point $(x,u) \in T_1M$ is the tangential component of the vertical lift $X^v$ of $X$ to $(x,u)$, with respect to the orthogonal decomposition

$$(TM)_{(x,u)} = (T_1M)_{(x,u)} \oplus \langle N^G_{(x,u)} \rangle. \tag{2.6}$$

Remark also that, with respect to the same decomposition, the tangential component of the horizontal lift $X^h$ of $X$ to $(x,u)$ is exactly $X^h$, since $X^h \in (T_1M)_{(x,u)}$. For this reason, the terminology of tangential lift is used here to describe the tangential component of the vertical lift of vectors.

Now, the tangent space $(T_1M)_{(x,u)}$ of $T_1M$ at $(x,u)$ is spanned by vectors of the form $X^h$ and $Y^t$, where $X$, $Y \in M_x$. Using this fact, the Riemannian metric $\tilde{G}$ on $T_1M$, induced from $G$, is completely determined by the identities

$$\begin{cases} \tilde{G}_{(x,u)}(X^h, Y^h) = (a+c)g_x(X,Y), \\ \tilde{G}_{(x,u)}(X^h, Y^t) \;= bg_x(X,Y), \\ \tilde{G}_{(x,u)}(X^t, Y^t) \;\;= ag_x(X,Y) - \frac{\alpha}{a+c} g_x(X,u)g_x(Y,u), \end{cases} \tag{2.7}$$

for all $(x,u) \in T_1M$ and $X$, $Y \in M_x$.

**Proposition 3.** *The Levi-Civita connection $\tilde{\nabla}$ associated to $\tilde{G}$ is given, at $(x,u) \in T_1M$, by*

$$\begin{aligned}
&(\tilde{\nabla}_{X^h} Y^h)_{(x,u)} \\
&= \{(\nabla_X Y)_x - \frac{ab}{2\alpha}\,[R(X_x,u)Y_x + R(Y_x,u)X_x]\}^h \\
&\quad - \frac{1}{\alpha}\,\{b^2\,R(X_x,u)Y_x + \frac{a\,(a+c)}{2}\,R(X_x,Y_x)u\}^t, \\
&(\tilde{\nabla}_{X^h} Y^t)_{(x,u)} \\
&= -\frac{a}{2\alpha(a+c)}\,\{a(a+c)\,R(Y_x,u)X_x + b^2\,g(Y_x,u)R(X_x,u)u\}^h \\
&\quad + \{(\nabla_X Y)_x + \frac{b}{2\alpha(a+c)}\,[a(a+c)R(Y_x,u)X_x \\
&\qquad - (2b^2 + a(a+c)\,g(Y_x,u)R(X_x,u)u]\}^t, \\
&(\tilde{\nabla}_{X^t} Y^h)_{(x,u)} \\
&= \{\frac{b}{a+c}\,g(X_x,u)\nabla_u Y \\
&\qquad - \frac{a}{2\alpha(a+c)}\,[a(a+c)\,R(X_x,u)Y_x + b^2\,g(X_x,u)R(Y_x,u)u]\}^h
\end{aligned}$$

$$+\frac{ab}{2\alpha}\{R(X_x,u)Y_x+g(X_x,u)R(Y_x,u)u\}^t,$$

$$(\tilde{\nabla}_{X^t}Y^t)_{(x,u)}$$
$$=\frac{b}{a+c}\{g(Y_x,u)X_x-\frac{a^2}{2\alpha}[g(X_x,u)R(Y_x,u)u+g(Y_x,u)R(X_x,u)u]\}^h$$
$$+\{-g(Y_x,u)X_x+\frac{b}{a+c}g(X_x,u)\nabla_u Y$$
$$+\frac{ab^2}{2\alpha(a+c)}[g(X_x,u)R(Y_x,u)u+g(Y_x,u)R(X_x,u)u]\}^t,$$

*for all* $X$, $Y\in\mathfrak{X}(M)$.

**Proof.** At first, we shall give the shape operator $\widetilde{S}_G$ of $T_1M$ in $(TM,G)$ derived from the unit normal vector field $N^G$ along $T_1M$. By definition, $\widetilde{S}_G Z$ is the tangential component $(-\bar{\nabla}_Z N^G)^t$ of $-\bar{\nabla}_Z N^G$, with respect to the pointwise decomposition (2.6), for all $Z\in T(T_1M)$. Then, using Proposition 1, we obtain

$$\widetilde{S}_G X^h=-\frac{b}{2\alpha\sqrt{(a+c)\alpha}}\{ab.\,[R(X,u)u]^h+(2b^2+a(a+c)).\,[R(X,u)u]^t\},\tag{2.8}$$

$$\widetilde{S}_G X^t=-\frac{1}{\sqrt{(a+c)\alpha}}\{b[-X+\frac{a^2}{2\alpha}R(X,u)u]^h+[(a+c)X-\frac{ab^2}{2\alpha}R(X,u)u]^t\},\tag{2.9}$$

for all $(x,u)\in T_1M$ and $X\in M_x$.

On the other hand, as $(T_1M,\tilde{G})$ is a Riemannian hypersurface of $(TM,G)$, its Levi-Civita connection $\tilde{\nabla}$ is given by the identity

$$\tilde{\nabla}_V W=\bar{\nabla}_V W-G(\bar{\nabla}_V W,N^G)N^G,\tag{2.10}$$

for all $V$, $W\in\mathfrak{X}(T_1M)$.

Now, the first and the third identities follow from Proposition 1 and from the identity (2.10). The second and the fourth identities follow from Proposition 1 and from the identities (2.10), (2.8) and (2.9). □

**Convention 1.** Because $u^t=0$, for $(x,u)\in T_1M$ and hence the tangent space $(T_1M)_{(x,u)}$ coincides with the set

$$\{\,X^h+Y^t\mid X\in M_x,\ Y\in\{u\}^\perp\subset M_x\,\},\tag{2.11}$$

the operation of $G$-tangential lift from $M_x$ to a point $(x,u) \in T_1M$ will be always applied only to vectors of $M_x$ which are orthogonal to $u$.

Now, we shall give three components of the Riemannian curvature tensor of $(T_1M, \tilde{G})$ (which will be needed later):

**Proposition 4.** *Let $(M,g)$ be a Riemannian manifold and $G = a.g^s + b.g^h + c.g^v$, $a$, $b$ and $c$ are constants satisfying $a > 0$ and $a(a+c) - b^2 > 0$. Denote by $\nabla$ and $R$ the Levi-Civita connection and the Riemannian curvature tensor of $(M,g)$, respectively. If we denote by $\tilde{R}$ the Riemannian curvature tensor of $(T_1M, \tilde{G})$, then:*

$$
\begin{aligned}
(2.12)\quad & \tilde{R}(X^h, Y^h)Z^h \\
&= \{R(X,Y)Z + \frac{ab}{2\alpha}\,[2(\nabla_u R)(X,Y)Z - (\nabla_Z R)(X,Y)u] \\
&+ \frac{a^2}{4\alpha}\,[R(R(Y,Z)u,u)X - R(R(X,Z)u,u)Y] \\
&+ \frac{a^2b^2}{4\alpha^2}\,[R(X,u)R(Y,u)Z + R(X,u)R(Z,u)Y \\
&- R(Y,u)R(X,u)Z - R(Y,u)R(Z,u)X)] \\
&- \frac{a^2}{2\alpha}\,R(R(X,Y)u,u)Z + \frac{ab^4}{2\alpha^2(a+c)}[g(R(X,u)u,Z)R(Y,u)u \\
&- g(R(Y,u)u,Z)R(X,u)u]\}^h + \{-\frac{b^2}{\alpha}\,(\nabla_u R)(X,Y)Z \\
&+ \frac{a\,(a+c)}{2\alpha}\,(\nabla_Z R)(X,Y)u + \frac{a\,b^3}{2\alpha^2}\,[R(R(Y,u)Z,X)u \\
&- R(X,u)R(Z,u)Y - R(R(X,u)Z,Y)u + R(Y,u)R(Z,u)X] \\
&+ \frac{a^2b\,(a+c)}{4\alpha^2}\,[R(X,R(Y,u)Z)u + R(X,R(Z,u)Y)u \\
&- R(Y,R(X,u)Z)u - R(Y,R(Z,u)X)u - R(R(Y,Z)u,u)X \\
&+ R(R(X,Z)u,u)Y] + \frac{a\,b}{2\alpha}\,R(R(X,Y)u,u)Z \\
&+ \frac{b^3(2b^2 + a(a+c)}{2\alpha^2(a+c)}[g(R(X,u)u,Z)R(Y,u)u \\
&- g(R(Y,u)u,Z)R(X,u)u]\}^t,
\end{aligned}
$$

$$
\begin{aligned}
(2.13)\quad & \tilde{R}(X^h, Y^t)Z^h \\
&= \{-\frac{b^3}{\alpha(a+c)}g(R(X,u)u,Z)Y - \frac{a^2}{2\alpha}\,(\nabla_X R)(Y,u)Z
\end{aligned}
$$

$$
\begin{aligned}
&+ \frac{a^3 b}{4\alpha^2}\,[R(X,u)R(Y,u)Z - R(Y,u)R(Z,u)X \\
&- R(Y,u)R(X,u)Z] + \frac{a\,b}{2\alpha}\,[R(X,Y)Z + R(Z,Y)X] \\
&+ \frac{a^2 b^3}{4\alpha^2(a+c)}[2g(R(X,u)u,Z)R(Y,u)u \\
&+ g(R(Y,u)u,Z)R(X,u)u]\}^h + \{\frac{b^2}{\alpha} g(R(X,u)u,Z)Y \\
&+ \frac{a\,b}{2\alpha}\,(\nabla_X R)(Y,u)Z - \frac{a^2 b^2}{4\alpha^2}\,[R(X,u)R(Y,u)Z \\
&- R(Y,u)R(Z,u)X - R(Y,u)R(X,u)Z] + \frac{a^2}{4\alpha}\,R(X,R(Y,u)Z)u \\
&- \frac{b^2}{\alpha}\,R(X,Y)Z + \frac{a\,(a+c)}{2\alpha}\,R(X,Z)Y \\
&+ \frac{ab^2}{4\alpha^2(a+c)}[-2b^2 g(R(X,u)u,Z)R(Y,u)u \\
&+ (2b^2 + a(a+c))g(R(Y,u)u,Z)R(X,u)u]\}^t,
\end{aligned}
$$

$$
\begin{aligned}
(2.14)\quad & \tilde{R}(X^t,Y^t)Z^t \\
&= -\frac{b}{a+c}\{g(Y,Z)X - g(X,Z)Y \\
&- \frac{a^2}{2\alpha}[g(Y,Z)R(X,u)u - g(X,Z)R(Y,u)u]\}^h \\
&+ \{g(Y,Z)X - g(X,Z)Y \\
&- \frac{ab^2}{2\alpha(a+c)}[g(Y,Z)R(X,u)u - g(X,Z)R(Y,u)u]\}^t,
\end{aligned}
$$

*for all $x \in M$, $(x,u) \in T_1M$ and all arbitrary vectors $X$, $Y$ and $Z \in M_x$ satisfying Convention 1.*

**Proof.** Let $\theta : \mathfrak{X}(T_1M)\times\mathfrak{X}(T_1M) \to C^\infty(T_1M)$ be the second fundamental form of $T_1M$ (as a hypersurface immersed in $TM$), associated to the normal vector field $N^G$ on $T_1M$, i.e., for all vector fields $V$ and $W$ on $T_1M$, we have $\bar{\nabla}_V W = \tilde{\nabla}_V W + \theta(V,W).N^G$. This means that $\theta(V,W) = G(\bar{\nabla}_V W, N^G)$, for all $V$, $W \in \mathfrak{X}(T_1M)$. We deduce, then, easily from Proposition 1 the following identities

$$(2.15)\qquad \theta(X^h,Y^h)_{(x,u)} = \frac{b^2}{\sqrt{\alpha(a+c)}} g(R(X_x,u)u,Y_x),$$

$$(2.16)\qquad \theta(X^h,Y^t)_{(x,u)} = -\frac{ab}{2\sqrt{\alpha(a+c)}} g(R(X_x,u)u,Y_x),$$

$$\theta(X^t, Y^h)_{(x,u)} = -\frac{ab}{2\sqrt{\alpha(a+c)}} g(R(X_x, u)u, Y_x), \tag{2.17}$$

$$\theta(X^t, Y^t)_{(x,u)} = \sqrt{\frac{\alpha}{a+c}}[-g(X_x, Y_x) + g(X_x, u)g(Y_x, u)], \tag{2.18}$$

for all vector fields $X$ and $Y$ on $M$ and $(x, u) \in T_1M$. Denoting by $\bar{R}$ and $\tilde{R}$ the Riemannian curvature tensors of $(TM, G)$ and $(T_1M, \tilde{G})$, respectively, we deduce from the Gauss equation for hypersurfaces that the tangential component $(\bar{R}(V, W)Z)^t$ of $\bar{R}(V, W)Z$, with respect to the pointwise decomposition (2.6), satisfy the equation

$$\tilde{R}(V, W)Z = (\bar{R}(V, W)Z)^t - \theta(V, Z).\widetilde{S}_G W + \theta(W, Z).\widetilde{S}_G V, \tag{2.19}$$

for all $(x, u) \in T_1M$ and $V$, $W$ and $Z$ in $(T_1M)_{(x,u)}$, where $\widetilde{S}_G$ is the shape operator of $T_1M$ in $(TM, G)$ derived from the unit normal vector field $N^G$ along $T_1M$.

Now, applying (2.19) to various classical lifts to $T_1M$ of vectors on $M$, and using Proposition 2 and the identities (2.8), (2.9) and (2.15)-(2.18), we obtain the identities (2.12)-(2.14). □

Now, we shall calculate the Ricci tensor $\widetilde{\text{Ric}}$ and the scalar curvature $\widetilde{S}$ of $(T_1M, \tilde{G})$. Let $(x, u)$ be a point of $T_1M$ and $\{E_1, \dots, E_m\}$ be an orthonormal basis for $M_x$ such that $E_m = u$. Then putting

$$\begin{aligned} F_i &= \frac{1}{\sqrt{a+c}}.E_i^h; \quad i = 1, \dots, m; \quad \text{and} \\ F_{m+j} &= \frac{b}{\sqrt{(a+c)\,\alpha}}.E_j^h - \frac{\sqrt{a+c}}{\sqrt{\alpha}}.E_j^t; \quad j = 1, \dots, m-1, \end{aligned} \tag{2.20}$$

we get an orthonormal basis $\{F_1, \dots, E_{2m-1}\}$ for the tangent space $(T_1M)_{(x,u)}$.

**Proposition 5.** *If we denote by Ric the Ricci tensor of $(M, g)$, then the Ricci tensor $\widetilde{\text{Ric}}$ of $(T_1M, \tilde{G})$ is given by:*

$$\begin{aligned} &\widetilde{\text{Ric}}(X^h, Y^h) \\ &= (1 - \frac{b^2}{\alpha}) \text{Ric}(X, Y) + \frac{ab}{2\alpha} [2(\nabla_u \text{Ric})(X, Y) \\ &- (\nabla_Y \text{Ric})(u, X) - (\nabla_X \text{Ric})(u, Y)] \\ &- \frac{a^2}{2\alpha} \sum_{i=1}^{m} g(R(u, E_i)X, R(u, E_i)Y) \end{aligned} \tag{2.21}$$

$$
\begin{aligned}
&+ \Big(\frac{a^2b^2}{4\alpha^2} - \frac{a^2b^4}{4\alpha^3}\Big) \sum_{i=1}^{m} g(R(u,X)E_i, R(u,Y)E_i) \\
&+ \frac{a^2b^4}{4\alpha^3} \sum_{i=1}^{m} [g(R(u,E_i)X, R(u,Y)E_i) \\
&+ g(R(u,E_i)Y, R(u,X)E_i)] \\
&- \frac{b^3}{\alpha(a+c)} g((\nabla_u R)(X,u)u, Y) \\
&- (m-2)\frac{b^2}{\alpha} g(R(X,u)u,Y) + \frac{ab^2}{2\alpha^3(a+c)}[a^2(a+c)^2 \\
&- 5a(a+c)b^2 - 2b^4] g(R(X,u)u, R(Y,u)u) \\
&+ 2\frac{ab^6}{\alpha^3(a+c)} g(R(X,u)u,Y)\mathrm{Ric}(u,u),
\end{aligned}
$$

$$
\begin{aligned}
&\widetilde{\mathrm{Ric}}(X^h, Y^t) \\
&= -\frac{ab}{2\alpha}\,\mathrm{Ric}(X,Y) - \frac{ab^2}{2\alpha(a+c)} g((\nabla_u R)(X,u)u,Y) \\
&+ \frac{a^2}{2\alpha}\,[(\nabla_u \mathrm{Ric})(X,Y) - (\nabla_Y \mathrm{Ric})(u,X)] \\
&+ \frac{a^3b}{4\alpha^2} \sum_{i=1}^{m} g(R(u,X)E_i, R(u,Y)E_i) \\
&+ \frac{b}{2\alpha(a+c)}[(m+1)a(a+c) - 4b^2] g(R(X,u)u,Y) \\
&+ \frac{a^2b^3}{2\alpha^3(a+c)}[a(a+c) + 5b^2] g(R(X,u)u, R(Y,u)u) \\
&- \frac{a^2b^3}{\alpha^3(a+c)}[a(a+c) + 3b^2] g(R(X,u)u,Y)\mathrm{Ric}(u,u),
\end{aligned} \tag{2.22}
$$

$$
\begin{aligned}
&\widetilde{\mathrm{Ric}}(X^t, Y^t) \\
&= \frac{a^4}{4\alpha^2} \sum_{i=1}^{m} g(R(u,X)E_i, R(u,Y)E_i) \\
&+ (m-2)g(X,Y) - 2\frac{ab^2}{\alpha(a+c)} g(X,Y)\mathrm{Ric}(u,u) \\
&+ \frac{ab^2}{2\alpha^2(a+c)}[2\alpha - b^2] g(R(X,u)u,Y) \\
&- \frac{a^3b^2}{2\alpha^3(a+c)}[a(a+c) + b^2] g(R(X,u)u, R(Y,u)u),
\end{aligned} \tag{2.23}
$$

*for all $x \in M$, $(x,u) \in T_1M$ and all arbitrary vectors $X$, $Y$ and $Z \in M_x$ satisfying Convention 1.*

**Proof.** Using the basis (2.20) of $(T_1M)_{(x,u)}$, we can write

$$\begin{aligned}
&\widetilde{\mathrm{Ric}}(X^h, Y^h)\\
&= \sum_{I=1}^{2m-1} \tilde{G}(\tilde{R}(X^h, F_I)F_I, Y^h)\\
&= \frac{a}{\alpha}\sum_{i=1}^{m} \tilde{G}(\tilde{R}(X^h, E_i^h)E_i^h, Y^h) - \frac{b^2}{\alpha(a+c)}\tilde{G}(\tilde{R}(X^h, E_m^h)E_m^h, Y^h)\\
&\quad - \frac{b}{\alpha}\sum_{i=1}^{m-1}[\tilde{G}(\tilde{R}(X^h, E_i^h)E_i^t, Y^h) + \tilde{G}(\tilde{R}(X^h, E_i^t)E_i^h, Y^h)]\\
&\quad + \frac{a+c}{\alpha}\sum_{i=1}^{m-1} \tilde{G}(\tilde{R}(X^h, E_i^t)E_i^t, Y^h),
\end{aligned}$$

where we have used the identity $1 + \frac{b^2}{\alpha} = \frac{a(a+c)}{\alpha}$. Using the three formulas (2.12)-(2.14) of the Riemannian curvature tensor $\tilde{R}$ and the identities of (2.7), we find

$$\begin{aligned}
&\tilde{G}(\tilde{R}(X^h, E_i^h)E_i^h, Y^h)\\
&= (a+c)\, g(R(X, E_i)E_i, Y) + b\, g((\nabla_u R)(X, E_i)E_i, Y)\\
&\quad + (-\frac{3a}{4} + \frac{ab^4}{4\alpha^2})\, g(R(X, E_i)u, R(Y, E_i)u)\\
&\quad - \frac{ab^2}{2\alpha}\, g(R(X,u)Y + R(Y,u)X, R(E_i, u)E_i)\\
&\quad + \frac{ab^2(\alpha - b^2)}{4\alpha^2}\, [g(R(X,u)E_i, R(Y,u)E_i) + g(R(E_i,u)X, R(E_i,u)Y)]\\
&\quad + \frac{a^2b^2(a+c)}{4\alpha^2}\, [g(R(E_i,u)X, R(Y,u)E_i) + g(R(E_i,u)Y, R(X,u)E_i)]\\
&\quad + \frac{b^4(a(a+c)+b^2)}{\alpha^2(a+c)}[g(R(X,u)u, E_i)g(R(Y,u)u, E_i)\\
&\quad - g(R(X,u)u, Y)g(R(E_i,u)u, E_i)],\\
&\tilde{G}(\tilde{R}(X^h, E_m^h)E_m^h, Y^h)\\
&= (a+c)\, g(R(X,u)u, Y) + b\, g((\nabla_u R)(X,u)u, Y)\\
&\quad + \frac{a}{4}(\frac{b^2}{\alpha} - 3) g(R(X,u)u, R(Y,u)u),\\
&\tilde{G}(\tilde{R}(X^h, E_i^h)E_i^t, Y^h) = -G(\tilde{R}(X^h, E_i^h)Y^h, E_i^t)
\end{aligned}$$

$$
\begin{aligned}
&= b\, g(R(X,E_i)E_i,Y) + \frac{a}{2}\, g((\nabla_Y R)(X,E_i)E_i,u) \\
&\quad + \frac{a^2 b}{4\alpha}\,[g(R(E_i,u)X, R(E_i,u)Y + R(Y,u)E_i) \\
&\quad - g(R(X,u)Y + R(Y,u)X, R(E_i,u)E_i)] \\
&\quad + \frac{ab^3(a(a+c)+3b^2)}{2\alpha^2(a+c)}[g(R(X,u)u,E_i)g(R(Y,u)u,E_i) \\
&\quad - g(R(X,u)u,Y)g(R(E_i,u)u,E_i)], \\
&\tilde{G}(\tilde{R}(X^h,E_i^t)E_i^t,Y^h) = -\tilde{G}(\tilde{R}(X^h,E_i^t)Y^h,E_i^t) \\
&= -\frac{b^2}{a+c}\, g(R(X,u)u,Y) + \frac{a^3}{4\alpha}\, g(R(E_i,u)X, R(E_i,u)Y) \\
&\quad - \frac{a^2 b^2(a(a+c)+3b^2)}{4\alpha^2(a+c)} g(R(X,u)u,E_i)g(R(Y,u)u,E_i).
\end{aligned}
$$

Now, since $\{E_1,\dots,E_m\}$ is an orthonormal basis of $M_x$, we have

$$
g(Z,W) = \sum_{j=1}^{m} g(Z,E_j)g(W,E_j), \tag{2.24}
$$

for all $Z$, $W \in M_x$. Putting $Z = R(X,E_i)u$ and $W = R(Y,E_i)u$ in (2.24), where $X$, $Y \in M_x$, we obtain

$$
\begin{aligned}
&\sum_{i=1}^{m} g(R(X,E_i)u, R(Y,E_i)u) \\
&= \sum_{i,j=1}^{m} g(R(X,E_i)u,E_j)g(R(Y,E_i)u,E_j) \\
&= \sum_{i,j=1}^{m} g(R(u,E_j)X,E_i)g(R(u,E_j)Y,E_i) \\
&= \sum_{j=1}^{m} g(R(u,E_j)X, R(u,E_j)Y).
\end{aligned} \tag{2.25}
$$

A straightforward calculation, using (2.25) and the formula above, gives then (2.21). The identities (2.22) and (2.23) can be obtained in a similar way. □

Now, we are in position to give the scalar curvature $\widetilde{S}$ of $(T_1M,\tilde{G})$:

**Proposition 6.** *If we denote by $S$ the scalar curvature of $(M,g)$, then the* *scalar curvature* $\widetilde{S}$ *of* $(T_1M, \tilde{G})$ *is given by:*

$$\widetilde{S}_{(x,u)} = \frac{a}{\alpha}.S_x + (m-1)(m-2)\frac{a+c}{\alpha} - \frac{a^3}{4\alpha^2}.\xi_x(u,u) + d.H(u) + e.\mathrm{Ric}_x(u,u) + f.[\mathrm{Ric}_x(u,u)]^2, \tag{2.26}$$

*for all* $(x,u) \in T_1M$, *where*

- $d$, $e$ *and* $f$ *are the constants given by*

$$\begin{cases} d = \dfrac{a^2b^2}{2\alpha^4(a+c)}[a(a+c)+b^2][a(a+c)-11b^2], \\ e = -4(m-1)\dfrac{ab^2}{\alpha^2} + \dfrac{b^4}{2\alpha^3(a+c)}[12\alpha - a(a+c)], \\ f = 2\dfrac{a^2b^4}{\alpha^4(a+c)}[a(a+c)+4b^2], \end{cases}$$

- $\xi$ *is, as in [9], the symmetric bilinear differential form defined by*

$$\xi_x(u,v) = \sum_{i,j=1}^{m} g_x(R(u,E_i)E_j, R(v,E_i)E_j), \tag{2.27}$$

*for all* $x \in M$ *and* $u$, $v \in M_x$,

- $H$ *is the function on* $TM$ *given by*

$$H(u) = \mathrm{trace}(R_u \circ R_u), \tag{2.28}$$

*for all* $u \in TM$, $R_u$ *being the Jacobi operator defined by* $R_u(v) = R(u,v)u$, *for all* $v \in M_{p_M(u)}$.

**Proof.** With the notations of (2.20), the scalar curvature $\widetilde{S}$ is, by definition, given by

$$\begin{aligned}\widetilde{S} &= \sum_{I=1}^{2m-1} \widetilde{\mathrm{Ric}}(F_I, F_I) \\ &= \frac{a}{\alpha}\sum_{i=1}^{m} \widetilde{\mathrm{Ric}}(E_i^h, E_i^h) - \frac{2b}{\alpha}\sum_{i=1}^{m} \widetilde{\mathrm{Ric}}(E_i^h, E_i^t) + \frac{a+c}{\alpha}\sum_{i=1}^{m} \widetilde{\mathrm{Ric}}(E_i^t, E_i^t) \\ &\quad - \frac{b^2}{\alpha(a+c)}\widetilde{\mathrm{Ric}}(u^h, u^h) + \frac{2b}{\alpha}\widetilde{\mathrm{Ric}}(u^h, u^t) - \frac{a+c}{\alpha}\widetilde{\mathrm{Ric}}(u^t, u^t).\end{aligned}$$

Using the identities (2.21)–(2.23), and the equality

$$\begin{aligned}&2g(R(E_i,u)E_j, R(E_j,u)E_i) \\ &= \|R(E_i,u)E_j\|^2 + \|R(E_j,u)E_i\|^2 - \|R(E_i,u)E_j - R(E_j,u)E_i\|^2 \\ &= \|R(E_i,u)E_j\|^2 + \|R(E_j,u)E_i\|^2 - \|R(E_i,E_j)u\|^2.\end{aligned}$$

we obtain obviously (2.26), which completes the proof of Proposition 6. □

A surprising observation is that the formula (2.26) for $\widetilde{S}$ does not involve the covariant derivatives $\nabla R$ and $\nabla\mathrm{Ric}$, which were still present in the formulas (2.21), (2.22) for the Ricci tensor $\widetilde{\mathrm{Ric}}$.

**Corollary 4.** *If $(M,g)$ is a space of constant sectional curvature $\kappa$, then the scalar curvature $\widetilde{S}$ of $(T_1M,\tilde{G})$ is a constant equal to*

$$\begin{aligned}\widetilde{S} &= (m-1)\{(m-2)\frac{a+c}{\alpha}\\ &+\kappa[m\frac{a}{\alpha}-4(m-1)\frac{ab^2}{\alpha^2}+\frac{b^4}{2\alpha^3(a+c)}(12\alpha-a(a+c))]\\ &+\kappa^2[-\frac{a^2}{2\alpha(a+c)}+2(m-2)\frac{a^2b^4}{\alpha^3(a+c)}+10(m-1)\frac{a^2b^6}{\alpha^4(a+c)}]\}.\end{aligned}\tag{2.29}$$

## 3. Proofs of the main results

**Proof of Theorem 2.** Let $\psi$ be the diffeomorphism from $T_1M$ onto $T_rM$ defined by $\psi(x,u)=(x,r.u)$, for all $(x,u)\in T_1M$. We shall construct a Riemannian $g$-natural metric of the form $a.g^s+c.g^v$ on $TM$ such that $\psi:(T_1M,\tilde{G})\to(T_rM,\tilde{g^s})$ is an isometry. By virtue of Remark 6, the tangential lift, with respect to the $g$-natural metric of the desired form, coincides with the classical tangential lift (with respect to the Sasaki metric) given by the identity (2.5).

Let $(x,u)\in T_1M$ and $Z=(x^i,u^i,X^i,Y^i)\in(T_1M)_{(x,u)}$. It is easy to see that $(\psi_*)(Z)=(x^i,r.u^i,X^i,r.Y^i)$ which is an element of $(T_rM)_{(x,r.u)}$. In particular, a simple calculation, using coordinates, gives $(\psi_*)(X^h_{(x,u)})=X^h_{(x,r.u)}$ and $(\psi_*)(X^t_{(x,u)})=r.X^t_{(x,r.u)}$. It follows that

$$\begin{aligned}\tilde{g^s}_{(x,r.u)}((\psi_*)(X^h_{(x,u)}),(\psi_*)(Y^h_{(x,u)})) &= \tilde{g^s}_{(x,r.u)}(X^h_{(x,u)},Y^h_{(x,u)})=g_x(X,Y),\\ \tilde{g^s}_{(x,r.u)}((\psi_*)(X^h_{(x,u)}),(\psi_*)(Y^t_{(x,u)})) &= r.\tilde{g^s}_{(x,r.u)}(X^h_{(x,u)},Y^t_{(x,u)})=0,\\ \tilde{g^s}_{(x,r.u)}((\psi_*)(X^t_{(x,u)}),(\psi_*)(Y^t_{(x,u)})) &= r^2.\tilde{g^s}_{(x,r.u)}(X^t_{(x,u)},Y^t_{(x,u)})\\ &= r^2.g_x(X,Y),\end{aligned}$$

for all $X$, $Y\in M_x$, $x\in M$.

We deduce then that $\psi^*\tilde{g^s}=\tilde{G}$, where $\tilde{G}$ is the metric induced on $T_1M$ by the metric $G=r^2.g^s+(1-r^2).g^v$ given on $TM$. □

Now, in order to generalize the case of Sasaki metric treated in [14], we shall determine all three-dimensional Riemannian manifolds $(M^3,g)$ whose unit tangent sphere bundles $(T_1M^3, \tilde{G})$ have constant scalar curvature, $G$ being an arbitrary Riemannian $g$-natural metric of the form $a.g^s + b.g^h + c.g^v$.

Generally, the curvature tensor $R$ of a Riemannian three-dimensional manifold $(M^3, g)$ is of the form (see, e.g., [10]):

$$R = \mathrm{Ric} \owedge g - \frac{S}{4} g \owedge g, \tag{3.1}$$

where $\owedge$ is the Kulkarni-Nomizu product of symmetric 2-tensors defined as follows:

$$\begin{aligned}(h \owedge k)(X,Y,Z,V) &= h(X,Z)k(Y,V) + h(Y,V)k(X,Z) \\ &\quad - h(X,V)k(Y,Z) - h(Y,Z)k(X,V).\end{aligned}$$

A straightforward calculation, using (3.1), gives

$$\xi(u,u) = (2|\mathrm{Ric}|^2 - S^2) + 2S\mathrm{Ric}(u,u) - 2\sum_{i=1}^{3}[\mathrm{Ric}(u,E_i)]^2,$$

$$H(u) = |\mathrm{Ric}|^2 - \frac{1}{2}S^2 + S\mathrm{Ric}(u,u) + [\mathrm{Ric}(u,u)]^2 - 2\sum_{i=1}^{3}[\mathrm{Ric}(u,E_i)]^2,$$

for all $u \in T_1M^3$. Hence (2.26) becomes

$$\begin{aligned}\widetilde{S}_{(x,u)} &= 2\frac{a+c}{\alpha} + (\frac{a}{\alpha} - \frac{p}{2}.S_x).S_x + p.|\mathrm{Ric}_x|^2 \\ &\quad + (q + p.S_x).\mathrm{Ric}_x(u,u) + k.[\mathrm{Ric}_x(u,u)]^2 \\ &\quad + l.\sum_{i=1}^{3}[\mathrm{Ric}_x(u,E_i)]^2,\end{aligned} \tag{3.2}$$

for all $(x,u) \in T_1M$, where the constants $q$, $p$, $k$ and $l$ are given by

$$\begin{cases} q = -\dfrac{b^2}{2\alpha^3(a+c)}[16a^2(a+c)^2 - 27a(a+c)b^2 + 12b^4], \\ p = -\dfrac{a^2}{2\alpha^4(a+c)}[\alpha^3 + 4b^4(2a(a+c)+3b^2)], \\ k = \dfrac{a^2b^2}{2\alpha^3(a+c)}(a(a+c) - 5b^2), \\ l = \dfrac{a^2}{\alpha^4(a+c)}[\alpha^3 + 4b^4(2a(a+c)+3b^2)] - \dfrac{a^3}{2\alpha^2}. \end{cases}$$

We deduce then the following Lemma:

**Lemma 1.** *The scalar curvature $\widetilde{S}$ is constant on $T_1M^3$ if and only if the following conditions are satisfied:*

*i) for each $x \in M$, the function on $M^3_x$ defined by*

$$u \mapsto (q + p.S_x).\mathrm{Ric}_x(u,u) + k.[\mathrm{Ric}_x(u,u)]^2 + l.\sum_{i=1}^{3}[\mathrm{Ric}_x(u,E_i)]^2$$

*is a constant on the unit sphere of $M^3_x$, equal to*

$$\check{S}_x := \frac{1}{3}\{(q + p.S_x).S_x + k.\sum_{i=1}^{3}[\mathrm{Ric}_x(E_i,E_i)]^2 + l.|\mathrm{Ric}_x|^2\};$$

*ii) the function $(\frac{a}{\alpha} - \frac{p}{2}.S).S + p.|\mathrm{Ric}|^2 + \check{S}$ is constant on $M^3$, equal to $\widetilde{S} - 2\frac{a+c}{\alpha}$.*

Now, we denote by $\rho_1$, $\rho_2$ and $\rho_3$ the Ricci roots and we suppose that, for all $x \in M^3$, the orthonormal basis $\{E_1, E_2, E_3\}$ of $M^3_x$ consists of eigenvectors of the Ricci tensor $\mathrm{Ric}_x$ at $x$, i.e., $\mathrm{Ric}_x(E_i, E_j) = (\rho_i)_x\delta_{ij}$, $i, j = 1, 2, 3$. If we express any $u \in M^3_x \cap T_1M^3$ in the basis $\{E_1, E_2, E_3\}$ as $u = u^1.E_1 + u^2.E_2 + u^3.E_3$, then we have $(u^1)^2 + (u^2)^2 + (u^3)^2 = 1$. If we calculate the function given in the first term of $i$), Lemma 1, using the coordinate expression for the vector $u$ and then eliminate the last coordinate $u^3$, we obtain the following function $F_x : [0,1] \times [0,1] \to \mathbb{R}$:

$$\begin{aligned}
F_x(u^1,u^2) := {} & k[((\rho_1)_x - (\rho_3)_x)^2(u^1)^4 + ((\rho_2)_x - (\rho_3)_x)^2(u^2)^4 \\
& + 2((\rho_1)_x - (\rho_3)_x)((\rho_2)_x - (\rho_3)_x)(u^1)^2(u^2)^2] \\
& + [(q + pS_x)((\rho_1)_x - (\rho_3)_x) + l((\rho_1)_x^2 - (\rho_3)_x^2) \\
& + 2k(\rho_3)_x((\rho_1)_x - (\rho_3)_x)](u^1)^2 \\
& + [(q + pS_x)((\rho_2)_x - (\rho_3)_x) + l((\rho_2)_x^2 - (\rho_3)_x^2) \\
& + 2k(\rho_3)_x((\rho_2)_x - (\rho_3)_x)](u^2)^2 \\
& + (q + pS_x)(\rho_3)_x + (k + l)(\rho_3)_x^2,
\end{aligned} \tag{3.3}$$

for all $(u^1, u^2) \in [0,1] \times [0,1]$. Thus the condition $i$) of Lemma 1 is equivalent to the fact that

(3.4) $\quad F_x$ is constant on $[0,1] \times [0,1]$ and equal to $\check{S}_x$, where

$$\check{S}_x := \frac{1}{3}\{(q + pS_x).S_x + (k + l)((\rho_1)_x^2 + (\rho_2)_x^2 + (\rho_3)_x^2)\}. \tag{3.5}$$

Note that $S_x$ can be also expressed in terms of $(\rho_i)_x$ as $S_x = (\rho_1)_x + (\rho_2)_x + (\rho_3)_x$.

**Proof of Theorem 3.** If $b = 0$, then $q = k = 0$, $p = -\frac{a}{2(a+c)^2}$ and $l = \frac{a}{2(a+c)^2}$. It follows that

$$F_x(u^1, u^2) = -\frac{a}{2(a+c)^2}[(\rho_2)_x.((\rho_1)_x - (\rho_3)_x)(u^1)^2 + (\rho_1)_x.((\rho_2)_x - (\rho_3)_x)(u^2)^2 + (\rho_3)_x.((\rho_1)_x + (\rho_2)_x)],$$

for all $(u^1, u^2) \in [0,1]\times[0,1]$. We suppose now that $(T_1M^3, \tilde{G})$ is of constant scalar curvature. Then condition (3.4) implies that $(\rho_2)_x.((\rho_1)_x - (\rho_3)_x) = (\rho_1)_x.((\rho_2)_x - (\rho_3)_x) = 0$. If we suppose that $(\rho_1)_x \neq (\rho_3)_x$, then $(\rho_2)_x = 0$, and hence $(\rho_1)_x.(\rho_3)_x = 0$. Hence, either $(\rho_2)_x = (\rho_3)_x = 0$, or $\rho_1)_x = (\rho_2)_x = 0$. If $(\rho_1)_x = (\rho_3)_x$, then $(\rho_1)_x.((\rho_2)_x - (\rho_1)_x) = 0$. We deduce that either $(\rho_1)_x = (\rho_3)_x = 0$, or $(\rho_1)_x = (\rho_2)_x = (\rho_3)_x$. It follows that, at any point of $M^3$, either two Ricci roots are zero, or all Ricci roots are the same.

Now, we show exactly as in the proof of Theorem 1 in [14], using condition *ii)* of Lemma 1, that all Ricci roots must be globally constant on $M^3$. Consequently, either $(M^3, g)$ has constant curvature, or $(M^3, g)$ is a curvature homogeneous space with Ricci roots $\rho_1 = \rho_2 = 0 \neq \rho_3$.

If we suppose that $(M^3, g)$ has constant curvature $\kappa$, then we have, by virtue of (2.29),

$$\widetilde{S} = -\frac{a}{(a+c)^2}\kappa^2 + \frac{6}{a+c}\kappa + \frac{2}{a}, \tag{3.6}$$

which has a maximal value $\widetilde{S} = \frac{11}{a}$ (corresponding to $\kappa = \frac{3(a+c)}{a}$). On the other hand, if $(M^3, g)$ is a curvature homogeneous space with Ricci roots $\rho_1 = \rho_2 = 0 \neq \rho_3$, then condition *ii)* of Lemma 1 implies that

$$\widetilde{S} = -\frac{a}{4(a+c)^2}\rho_3^2 + \frac{1}{a+c}\rho_3 + \frac{2}{a}, \tag{3.7}$$

which has a maximal value $\widetilde{S} = \frac{3}{a}$ (corresponding to $\rho_3 = \frac{2(a+c)}{a}$).

Conversely, if $(M^3, g)$ has constant curvature or is a curvature homogeneous space with Ricci roots $\rho_1 = \rho_2 = 0 \neq \rho_3$, then we deduce from the identities (3.6) and (3.7), respectively, that $(T_1M^3, \tilde{G})$ is of constant scalar curvature. □

**Proof of Corollary 1.** Let $(M^3, g)$ be a curvature homogeneous 3-space with constant Ricci roots $\rho_1 = \rho_2 = 0$ and $\rho_3 = u$. For any $a > 0$ such that $as \leq 3$, we are looking for a constant $c$ with $a + c > 0$ such that the metric on $T_1M^3$ induced from the metric $G = a.g^s + c.g^v$ has constant scalar curvature $s$. For this, suppose that $(T_1M^3, \tilde{G})$ has constant scalar curvature $s$. Then it follows from (3.7) the identity

$$(2 - as)(a+c)^2 + au(a+c) - \frac{a^2u^2}{4} = 0. \tag{3.8}$$

If $as = 2$, then the solution of (3.8) is given by $a + c = \frac{au}{4}$. Since $a + c$ must be positive, then $u$ must also be positive.

If $as \neq 2$, then there are two solutions of (3.8) given by

$$a + c = \frac{au}{2(1+\sqrt{3-as})} \tag{3.9}$$

$$a + c = \frac{au}{2(1-\sqrt{3-as})}, \tag{3.10}$$

and which must be positive. We deduce that

- If $u > 0$, then (3.9) is always positive, and (3.10) is positive only if $2 < as \leq 3$. Hence the required family $\mathcal{S}_u$ is that of metrics $a.g^s + c.g^v$, such that $(a, c)$ runs through the set $\{(a, \frac{au}{4} - a),\ as = 2\} \cup \{(a, \frac{au}{2(1+\sqrt{3-as})} - a),\ a > 0\} \cup \{(a, \frac{au}{2(1-\sqrt{3-as})} - a),\ a > 0,\ 2 < as \leq 3\}$.
- If $u < 0$, then (3.9) is always negative, and (3.10) is positive only if $as < 2$. Hence the required family $\mathcal{S}_u$ is that of metrics $a.g^s + c.g^v$, such that $(a, c)$ runs through the set $\{(a, \frac{au}{2(1-\sqrt{3-as})} - a),\ a > 0,\ as < 2\}$. □

**Proof of Theorem 4.** Suppose that $b \neq 0$ and $a(a+c) \neq 5b^2$. This means that $k \neq 0$. If we suppose that $(T_1M^3, \tilde{G})$ is of constant scalar curvature, then we obtain the condition (3.4). Now the part of degree 4 of the polynomial (3.3) is nontrivial and it must be zero. It follows that $(\rho_1)_x = (\rho_2)_x = (\rho_3)_x$, for all $x \in M$. We deduce that $(M^3, g)$ has constant sectional curvature. The converse of the Theorem follows from Corollary 4 at the end of section 2. □

**Proof of Theorem 5.** Suppose that $a(a+c) = 5b^2$, which implies that $k = 0$. In this case, $F_x$ can be expressed as follows:

$$\begin{aligned} F_x(u^1, u^2) := {} & ((\rho_1)_x - (\rho_3)_x)[(q + pS_x) + l((\rho_1)_x + (\rho_3)_x)](u^1)^2 \\ & + ((\rho_2)_x - (\rho_3)_x)[(q + pS_x) + l((\rho_2)_x + (\rho_3)_x)](u^2)^2 \\ & + (\rho_3)_x[(q + pS_x) + l(\rho_3)_x], \end{aligned}$$

for all $(u^1, u^2) \in [0,1] \times [0,1]$. Suppose that $(T_1M^3, \tilde{G})$ is of constant scalar curvature, and that $(\rho_1)_x \neq (\rho_3)_x$ and $(\rho_2)_x \neq (\rho_3)_x$. The condition (3.4) implies that the function $F_x(u^1, u^2)$ is constant and hence the coefficients of $(u^1)^2$ and $(u^2)^2$ are zero, i.e.,

$$\begin{cases} (q + pS_x) + l((\rho_1)_x + (\rho_3)_x) = 0 \\ (q + pS_x) + l((\rho_2)_x + (\rho_3)_x) = 0. \end{cases}$$

This gives, by virtue of $l \neq 0$, $(\rho_1)_x = (\rho_2)_x$. So, *at any point of* $M^3$, *at least two Ricci roots are the same*, and $F_x(u^1, u^2)$ reduces to

$$F_x(u^1, u^2) = (\rho_3)_x[(q + pS_x) + l(\rho_3)_x], \tag{3.11}$$

for all $(u^1, u^2) \in [0, 1] \times [0, 1]$.

The case when all Ricci roots are the same at $x \in M^3$ corresponds to the identity $\mathrm{Ric}_x = (\rho_1)_x.g_x$. Suppose now that, at $x \in M^3$, $(\rho_1)_x = (\rho_2)_x \neq (\rho_3)_x$. It is easy to check that (3.11) and (3.5) reduce, by virtue of $a(a + c) = 5b^2$, to

$$\begin{aligned} F_x(u^1, u^2) = {} & [-\frac{277}{128(a + c)} - \frac{145a}{128(a + c)^2}(2(\rho_1)_x + (\rho_3)_x)](\rho_3)_x \\ & + \frac{95a}{64(a + c)^2}(\rho_3)_x^2, \end{aligned}$$

$$\begin{aligned} \breve{S}_x = {} & \frac{1}{3}\{[-\frac{277}{128(a + c)} - \frac{145a}{128(a + c)^2}(2(\rho_1)_x + (\rho_3)_x)](2(\rho_1)_x + (\rho_3)_x) \\ & + \frac{95a}{64(a + c)^2}(2(\rho_1)_x^2 + (\rho_3)_x^2)\}, \end{aligned}$$

and condition (3.4) implies that

$$((\rho_1)_x - (\rho_3)_x)[-277 + \frac{a}{a + c}(-100(\rho_1)_x + 45(\rho_3)_x)] = 0.$$

Since $(\rho_1)_x \neq (\rho_3)_x$, then (0.1) holds at $x$.

So, we have proved that at any point $x \in M^3$, either all Ricci roots are the same, or two of them are equal and the third is related to them by the identity (0.1).

Now, as in the proof of Theorem 1 in [14], let $U$ be the dense open subset on $M^3$ on which the multiplicity of the Ricci roots is locally constant. If on a connected component of $U$ there is only one root with multiplicity three, then this component has constant curvature. On the other hand, if on a connected component of $U$ we have a root $\rho_1$ with multiplicity two, then substituting from the identities $a(a + c) = 5b^2$ and (0.1) into condition *ii)* of Lemma 1, we obtain by a long calculation the following condition

$$275\frac{a}{a + c}\,\rho_1^2 + 1402\,\rho_1 \text{ is a constant on } M^3, \tag{3.12}$$

$$\text{equal to } \frac{a + c}{400.a}(103680.a\widetilde{S} + 1168181),$$

which implies that $\rho_1$ must be constant on the connected component (hence $\rho_3$ is also constant on this component by virtue of (0.1)). In particular, the Ricci roots are always locally constant, hence globally constant and $U = M^3$.

If we suppose that $(M^3, g)$ has constant curvature $\kappa$, then we have, by virtue of (2.29),

$$\widetilde{S} = -\frac{125a}{32(a+c)^2}\kappa^2 + \frac{203}{64(a+c)}\kappa + \frac{5}{2a}, \tag{3.13}$$

which has a maximal value $\widetilde{S} = \frac{251627}{256000.a}$ (corresponding to $\kappa = \frac{203(a+c)}{1000.a}$). On the other hand, if $(M^3, g)$ is a curvature homogeneous space with Ricci roots $\rho_1 = \rho_2 \neq \rho_3$, then condition (3.12) implies that

$$\widetilde{S} = -\frac{5}{1296.a(a+c)^2}[275.a^2\rho_1^2 + 1402.a(a+c)\rho_1 + \frac{1168181}{400}(a+c)^2], \tag{3.14}$$

which has a maximal value $\widetilde{S} = -\frac{12315}{2816.a}$ (corresponding to $\rho_1 = \frac{1402(a+c)}{275.a}$).

Conversely, if $(M^3, g)$ has constant curvature or is a curvature homogeneous space with Ricci roots $\rho_1 = \rho_2 \neq \rho_3$, then we deduce from the identities (3.13) and (3.14), respectively, that $(T_1M^3, \tilde{G})$ is of constant scalar curvature. □

**Proof of Corollary 2.** It suffices to choose the subfamily $S'$ of Riemannian $g$-natural metrics $G = a.g^s + b.g^h + c.g^v$ on $T_1M^3$, such that $a > 0$, $c = \frac{a}{277}(45\rho_3 - 100\rho_1 - 277)$ and $5b^2 = a(a+c)$, and to apply Theorem 5. □

**Proof of Corollary 3.** Let $(M^3, g)$ be a curvature homogeneous 3-space with constant Ricci roots $\rho_1 = \rho_2 = v$ and $\rho_3 = w$. The required Riemannian $g$-natural metric $G = a.g^s + b.g^h + c.g^v$ on $TM^3$ should be chosen such that $a(a+c) = 5b^2$. In this case, if $(T_1M^3, \tilde{G})$ has constant scalar curvature $s$, then we have, by virtue of Theorem 5, the following

$$s \leq -\frac{12315}{2816.a}, \tag{3.15}$$

$$w = \frac{20}{9}v + \frac{277.(a+c)}{45.a}. \tag{3.16}$$

Furthermore, we have by (3.12), the identity

$$s = -\frac{5}{1296.a(a+c)^2}[275.a^2v^2 + 1402.a(a+c)v + \frac{1168181}{400}(a+c)^2]. \tag{3.17}$$

Substituting from (3.16) into (3.17), we obtain

$$a = -\frac{(277)^2 \times 5}{1296.sz^2}[275.v^2 + \frac{1402}{277}.vz + \frac{1168181}{400 \times (277)^2}z^2], \tag{3.18}$$

where $z = 45.w - 100.v$. Consider $a$ as a quadratic function in $\frac{v}{z}$, then $a$ has a minimal value equal to $-\frac{12315}{2816.s}$, corresponding to the value $v = -\frac{701.z}{275\times 277}$. This is in accordance with the necessary inequality (3.15). Recall that $s$ was negative and $a$ positive.

Now, the Riemannian $g$-natural metric $G = a.g^s + b.g^h + c.g^v$, with the constants $a$, $b$ and $c$ given by (3.18), (3.16) and $a(a+c) = 5b^2$, has constant scalar curvature $s$. This completes the proof of Corollary 3. □

## References


1. K.M.T. Abbassi, *Note on the classification Theorems of g-natural metrics on the tangent bundle of a Riemannian manifold* $(M, g)$, Comment. Math. Univ. Carolinae. **45** (4) (2004), 591–596.
2. K.M.T. Abbassi and O. Kowalski, *On g-natural metrics on unit tangent sphere bundles of Riemannian manifolds*, preprint.
3. K.M.T. Abbassi and M. Sarih, *Killing vector fields on tangent bundles with Cheeger-Gromoll metric*, Tsukuba J. Math. **27** (2) (2003), 295–306.
4. K.M.T. Abbassi and M. Sarih, *On natural metrics on tangent bundles of Riemannian manifolds*, to appear in Arch. Math. (Brno).
5. K.M.T. Abbassi and M. Sarih, *The Levi-Civita connection of Riemannian natural metrics on the tangent bundle of an oriented Riemannian manifold*, preprint.
6. K.M.T. Abbassi and M. Sarih, *On some hereditary properties of Riemannian g-natural metrics on tangent bundles of Riemannian manifolds*, Diff. Geometry and Appl., **22** (1) (2005), 19–47.
7. K.M.T. Abbassi and M. Sarih, *On Riemannian g-natural metrics of the form* $a.g^s + b.g^h + c.g^v$ *on the tangent bundle of a Riemannian manifold* $(M, g)$, to appear in Mediter. J. Math.
8. J. Berndt F. Tricerri and L. Vanhecke, *Generalized Heisenberg groups and Damek-Ricci Harmonic spaces*, Lecture Notes in Math., Vol. 1598, Springer 1995.
9. A. Besse, *Manifolds all of whose geodesics are closed*, Ergebnisse der Math., vol. **10**, Springer-Verlag, Berlin. Heidelberg, 1987.
10. A. Besse, *Einstein manifolds*, Ergebnisse der Math. und ihrer grenzgebiete, vol. 93, Springer-Verlag, Berlin, 1978.
11. D.E. Blair, When is the tangent sphere bundle locally symmetric? *Geometry and Topology*, World Scientific Publ. Co., Singapore, 1989, 15–30.
12. E. Boeckx,O. Kowalski and L. Vanhecke, *Riemannian manifolds of conullity two*, (Monograph). World Scientific Publishers, Singapore, 1996, 300 pages.
13. E. Boeckx, D. Peronne and L. Vanhecke, *Unit tangent sphere bundles and two-point homogeneous spaces*, Periodica Math. Hung. **36** (1998), 79–95.
14. E. Boeckx and L. Vanhecke, *Unit tangent sphere bundles with constant scalar curvature*, Czechoslovak Math. J. **53** (125)(221) (2001), 523–544.
15. E. Boeckx and L. Vanhecke, *Curvature homogeneous unit tangent sphere bundles*, Publ. Math. Debrecen **53** (1998), 389–413.

16. A.A. Borisenko and A.L. Yampol'skii, *Riemannian geometry of fiber bundles*, Russian Math. surveys **46** (6) (1991), 55–106.
17. P. Dombrowski, *On the geometry of the tangent bundle*, J. Reine Angew. Math. **210** (1962), 73–82.
18. S. Gudmundsson and E. Kappos, *On the geometry of the tangent bundle with the Cheeger-Gromoll metric*, Tokyo J. Math. **14** (2) (2002), 407–717.
19. S. Ivanov and I. Petrova, *Riemannian manifolds in which the skew-symmetric curvature operator has pointwise constant eigenvalues*, Geom. Dedicata **70** (1998), 269–282.
20. S. Kobayashi and K. Nomizu, Foundations of differential geometry, Intersci. pub. New York (I, 1963 and II, 1967).
21. I. Kolář, P.W. Michor and J. Slovák, Natural operations in differential geometry, Springer-Verlag, Berlin, 1993.
22. O. Kowalski, *Curvature of the induced Riemannian metric of the tangent bundle of Riemannian manifold*, J. Reine Angew. Math. **250** (1971), 124–129.
23. O. Kowalski, *A classification of Riemannian 3-manifolds with constant principal Ricci curvatures $\rho_1 = \rho_2 \neq \rho_3$*, Nagoya Math. J. **132** (1993), 1–36.
24. O. Kowalski, *An explicit classification of 3-dimensional Riemannian spaces satisfying $R(X,Y).R = 0$*, Czechoslovak Math. J. **46** (1996), 427–474.
25. O. Kowalski and S.Ž. Nikčević, *On Ricci eigenvalues of locally homogeneous Riemannian manifolds*, Geom. Dedicata **62** (1996), 65–72.
26. O. Kowalski and M. Sekizawa, *Natural transformations of Riemannian metrics on manifolds to metrics on tangent bundles —a classification—*, Bull. Tokyo Gakugei Univ. (4) **40** (1988), 1–29.
27. O. Kowalski and M. Sekizawa, *Local isometry classes of Riemannian 3-manifolds with constant Ricci eigenvalues $\rho_1 = \rho_2 \neq \rho_3 > 0$*, Arch. Math. (Brno) **32** (1996), 137–145.
28. O. Kowalski and M. Sekizawa, *On tangent sphere bundles with small or large constant radius*, Ann. Global Anal. Geom. **18** (2000), 207–219.
29. O. Kowalski and M. Sekizawa, *On the scalar curvature of tangent sphere bundles with arbitrary constant radius*, Bull. Greek Math. Soc. **44** (2000), 17–30.
30. O. Kowalski and M. Sekizawa, *On Riemannian manifolds whose tangent sphere bundles can have nonnegative sectional curvature*, Univ. Iagellonicae Acta Math. **40** (2002), 245–256.
31. O. Kowalski, M. Sekizawa and Z. Vlášek, *Can tangent sphere bundles over Riemannian manifolds have strictly positive sectional curvature?*, Global Differential Geometry: The mathematical Legacy of Alfred Gray (M. Fernandez and J.A. Wolf), Contemporary Math., vol. 288, Amer. Math. Soc., 2001, pp. 110–118.
32. E. Musso and F. Tricerri, *Riemannian metrics on tangent bundles*, Ann. Math. Pura Appl. (4) **150** (1988), 1–20.
33. V. Oproiu, *Some new geometric structures on the tangent bundle*, Publ. Math. Debrecen **55** (1999), 261–281.
34. V. Oproiu, *A locally symmetric Kähler Einstein structure on the tangent*

*bundle of a space form*, Beiträge Zur Algebra und Geometrie/Contributions to Algebra and Geometry, **40** (1999), 363–372.
35. V. Oproiu, *A Kähler Einstein structure on the tangent bundle of a space form*, Int. J. Math. Math. Sci. **25** (2001), 183–195.
36. V. Oproiu and N. Papaghiuc, *Locally symmetric space structures on the tangent bundle*, (in Differential Geometry and applications, Satellite conf. of ICM in Berlin, Aug. 10–14, 1998, Brno) Masaryk Univ. in Brno 1999, 99–109.
37. S. Sasaki, *On the differential geometry of tangent bundles of Riemannian manifolds*, Tôhoku Math. J. (I, **10** (1958) 338–354; II, **14** (1962) 146–155).
38. K. Sekigawa, *On some 3-dimensional Riemannian manifolds*, Hokkaido Math. J. **2** (1973), 259–270.
39. K. Sekigawa, *On the Riemannian manifolds of the form* $B \times_f F$, Kodai Math. Sem. Rep. **2** (1975), 343–347.
40. M. Sekizawa, *Curvatures of tangent bundles with Cheeger-Gromoll metric*, Tokyo J. Math. **14** (2) (1991), 407–417.
41. I.M. Singer, *Infinitesimally homogeneous spaces*, Comm. Pure Appl. Math. **13** (1960), 685–697.
42. Z.I. Szabó, *Structure theorems on Riemannian manifolds satisfying* $R(X, Y)$ $.R = 0$, *I, Local version*, J. Diff. Geom. **17** (1982), 531–582.
43. K. Yano and S. Ishihara, Tangent and cotangent bundles, differential geometry, Marcel Dekker Inc. New York, 1973.

# MAGNETIC MEAN OPERATORS ON A KÄHLER MANIFOLD

TOSHIAKI ADACHI*

*Department of Mathematics, Nagoya Institute of Technology*
*Gokiso, Nagoya, 466-8555, JAPAN*
*E-mail: adachi@nitech.ac.jp*

We define magnetic mean operators which are generators of random walks under action of magnetic fields. We calculate their formal Taylor expansions and show that the influence of magnetic fields first appears on the third term in general and on the fourth term for Kähler magnetic fields.

*Keywords*: Magnetic fields, trajectories, magnetic mean operators, magnetic random walk.

## 1. Introduction

Let $(M, \langle\ ,\ \rangle)$ be a complete Riemannian manifold. For a magnetic field $\mathbb{B}$ on $M$, which is a closed 2-form on $M$, we define a skew symmetric operator $\Omega_{\mathbb{B}} : TM \to TM$ on the tangent bundle $TM$ of $M$ by $\mathbb{B}(u, v) = \langle u, \Omega_{\mathbb{B}}(v)\rangle$ for all tangent vectors $u, v \in T_xM$ at an arbitrary point $x \in M$. A smooth curve $\gamma : \mathbb{R} \to M$ parameterized by its arc-length is called a *trajectory* for $\mathbb{B}$ if it satisfies the equation $\nabla_{\dot\gamma}\dot\gamma = \Omega_{\mathbb{B}}(\dot\gamma)$. When $\mathbb{B}$ is a trivial magnetic field, we see $\Omega_{\mathbb{B}} = 0$ and trajectories are geodesics. A magnetic field $\mathbb{B}$ is called *uniform* if $\Omega_{\mathbb{B}}$ is parallel. Kähler magnetic fields, which are constant multiples of the Kähler form on a Kähler manifold, are typical examples of uniform magnetic fields. In his papers [1,2] the author studied trajectories for Kähler magnetic fields from a classical Riemannian geometric point of view.

It is known that there is a correspondence between geodesics and the Laplace operators. For example, Sunada [9] studied a spherical mean oper-

*The author is partially supported by Grant-in-Aid for Scientific Research (C) (No. 14540075), Ministry of Education, Science, Sports Culture and Technology.

ator $\mathcal{M}_r$ on $\mathrm{L}^2(M)$, which takes averages of functions of $M$ on each geodesic sphere of radius $r$ with respect to the density of unit tangent sphere. He showed it is a self-adjoint operator and is a generator of geodesic random walks. He also pointed out the correspondence of this operator to the Laplace operator by giving the Taylor expansion of $\mathcal{M}_r$ with respect to $r$.

Another mean-value operator was studied by Gray and Willmore [6]. In order to generalize the mean-value property for harmonic functions on a Euclidean space, they considered the operator which is obtained by taking averages of functions on each geodesic sphere of radius $r$ with respect to the area element. By giving power series expansions for such mean-value operators with respect to $r$, they extended Pizzetti's formula for mean-values on a Euclidean space to Riemannian manifolds. Their mean-value operators are not generally self-adjoint and are coincide with spherical mean operators on a harmonic Riemannian space.

In this paper, we focus our attention on generating operators for "magnetic random walks". As a generalization of spherical mean operators, we define a magnetic mean operator which takes averages of functions on each magnetic sphere of radius $r$. This should show how the base space is warped by a magnetic field. Along the lines in [6] and [9] we give a formal expansion of magnetic mean operators, especially for the case of Kähler magnetic fields, and study how the effect of magnetic fields appear on these operators.

The author is grateful to Professor Norio Ejiri for his valuable advice.

## 2. Magnetic mean operators

Let $\mathbb{B}$ be a magnetic field on a complete Riemannian manifold $M$, and $\tau : UM \to M$ denote the unit tangent bundle of $M$. For a unit tangent vector $v \in UM$ we denote by $\gamma_v$ a trajectory for $\mathbb{B}$ with initial vector $\dot{\gamma}(0) = v$. The magnetic exponential map $\mathbb{B}\mathrm{exp}_x : T_xM \to M$ at a point $x \in M$ is given by

$$\mathbb{B}\mathrm{exp}_x(u) = \begin{cases} \gamma_{u/\|u\|}(\|u\|), & \text{when } u \neq 0, \\ x, & \text{when } u = 0, \end{cases}$$

and the magnetic flow $\mathbb{B}\varphi_t$ on the unit tangent bundle is given by $\mathbb{B}\varphi_t(v) = \dot{\gamma}_v(t)$. For a trivial magnetic field, magnetic exponential maps and the magnetic flow are exponential maps and the geodesic flow.

For a number $r$, we define a magnetic mean operator $\mathcal{M}_{\mathbb{B},r}$ on the set

$C^0(M)$ of all continuous functions on $M$ by

$$\mathcal{M}_{\mathbb{B},r} f(x) = \int_{U_x M} f(\mathbb{B}\exp_x(rv)) dS_x(v),$$

where $dS_x$ denotes the normalized canonical density of the unit sphere $U_x M$. When $\mathbb{B}$ is trivial, it is the usual spherical mean operator $\mathcal{M}_r$ defined by Sunada [9], which is given as $\mathcal{M}_r f(x) = \int_{U_x M} f(\exp_x(rv)) dS_x(v)$ by use of the exponential map $\exp_x$.

It is clear that $\mathcal{M}_{\mathbb{B},0}$ is the identity $I$. For a magnetic filed $\mathbb{B}$ we can consider another magnetic field $-\mathbb{B}$. Since $\Omega_{-\mathbb{B}} = -\Omega_{\mathbb{B}}$, we see for each trajectory $\gamma$ for $\mathbb{B}$ the curve $\sigma(t) = \gamma(-t)$ obtained by reversing the orientation is a trajectory for $-\mathbb{B}$. Thus we have $\mathcal{M}_{\mathbb{B},-r} = \mathcal{M}_{-\mathbb{B},r}$. But we should note that $\mathcal{M}_{-\mathbb{B},r}$ does not necessarily coincide with $\mathcal{M}_{\mathbb{B},r}$. If we define the fiber integral $\tau_* : C^0(UM) \to C^0(M)$ by

$$\tau_* h(x) = \int_{U_x M} h(v) dS_x(v),$$

we see that $\mathcal{M}_{\mathbb{B},r} = \tau_* \circ \mathbb{B}\varphi_r^* \circ \tau^*$ with the pull backs $\mathbb{B}\varphi_r^*$ and $\tau^*$ by $\mathbb{B}\varphi_r$ and $\tau$. Since Liouville's theorem guarantees that the magnetic flow $\mathbb{B}\varphi_t$ preserves the density $dS$ on $UM$ given by

$$\int_{UM} h(v) dS(v) = \int_M dvol(x) \int_{U_x M} h(v) dS_x(v),$$

we see

$$\begin{aligned} \|\mathcal{M}_{\mathbb{B},r} f\|_{\mathrm{L}^2}^2 &= \int_{UM} \mathbb{B}\varphi_r^* \left(\tau^* f \cdot \tau^* f\right)(v)\, dS(v) \\ &= \int_{UM} (\tau^* f)^2(v) dS(v) = \|f\|_{\mathrm{L}^2}^2. \end{aligned}$$

Hence $\mathcal{M}_{\mathbb{B},r}$ can be extended on $\mathrm{L}^2(M)$. Moreover, we find

$$\begin{aligned} \langle \mathcal{M}_{\mathbb{B},r} f, g \rangle_{\mathrm{L}^2} &= \int_M dvol(x) \int_{U_x M} \left(\mathbb{B}\varphi_r^* \circ \tau^*\right) f(v) \cdot \tau^* g(v)\, dS_x(v) \\ &= \int_{UM} \mathbb{B}\varphi_r^* \left(\tau^* f \cdot \left(\mathbb{B}\varphi_{-r}^* \circ \tau^*\right) g\right)(v)\, dS(v) \\ &= \int_{UM} \tau^* f(v) \cdot \left(\mathbb{B}\varphi_{-r}^* \circ \tau^*\right) g(v)\, dS(v) \\ &= \langle f, \mathcal{M}_{\mathbb{B},-r} g \rangle_{\mathrm{L}^2} = \langle f, \mathcal{M}_{-\mathbb{B},r} g \rangle_{\mathrm{L}^2}. \end{aligned}$$

Hence $\mathcal{M}_{\mathbb{B},r}$ and $\mathcal{M}_{-\mathbb{B},r}$ are adjoint operators on $\mathrm{L}^2(M)$ each other. When $M$ is of finite volume, as $\mathcal{M}_{-\mathbb{B},r} 1 = 1$ we see

$$\int_M \mathcal{M}_{\mathbb{B},r} f(x) dvol(x) = \int_M f(x) dvol(x)$$

for every $f \in \mathrm{L}^2(M)$.

## 3. Expansion of magnetic means with respect to radius

In his paper [9] Sunada gave a formal Taylor expansion of spherical means $\mathcal{M}_r$ with respect to the radius $r$.

**Proposition 3.1.** *Given a smooth function $f$ on a Riemannian manifold $M$ of dimension $m$, we can write for each $k$ that*

$$\mathcal{M}_r f = f + \frac{r^2}{2!}P_2 f + \frac{r^4}{4!}P_4 f + \cdots + \frac{r^{2k}}{(2k)!}P_{2k} f + o\big(r^{2(k+1)}\big),$$

*where $P_k f$ is given by*

$$P_k f(x) = \sum_{\alpha_1,\cdots,\alpha_k} \Big(\int_{U_xM} v^{\alpha_1}v^{\alpha_2}\cdots v^{\alpha_k}\,dS_x(v)\Big)\Big(\nabla_{\partial_{\alpha_1}}\nabla_{\partial_{\alpha_2}}\cdots\nabla_{\partial_{\alpha_k}} f\Big)(x)$$

*with a local coordinate $(x^1,\ldots,x^m,v^1,\ldots,v^m)$ of the tangent bundle $TM$, and $o\big(r^{2(k+1)}\big)$ is a smooth function satisfying $\lim_{r\downarrow 0}\frac{1}{r^{2k}}o\big(r^{2(k+1)}\big)=0$.*

We here note that all the odd terms $P_{2k+1}$ vanish because they correspond to integrals of homogeneous polynomials of odd degrees on unit tangent spaces. He also pointed out that the operator $P_{2k}$ is a formally self-adjoint and elliptic, has the same principal symbol as of the operator

$$\frac{\Gamma(m/2)(2k)!}{2^{2k}k!\Gamma(\frac{n-2}{2}+k+1)}(-\Delta)^k,$$

and satisfies

$$\begin{aligned} P_2 f &= -\frac{1}{m}\Delta f,\\ P_4 f &= \frac{1}{m(m+2)}\big(3\Delta^2 f + \langle\nabla f,\nabla\tau\rangle - 2\langle\nabla^2 f,\rho\rangle\big) \end{aligned}$$

for a smooth function $f$ by use of the Ricci curvature $\rho$ and the scalar curvature $\tau$ (c.f. [6]).

We here observe the corresponding expansion of magnetic mean operators. By direct computation we see for each smooth function $f$ on $M$ that

$$\begin{aligned} \frac{d}{dr}\Big|_{r=0}\mathcal{M}_{\mathbb{B},r}f(x) &= \int_{U_xM}\langle\nabla f, v\rangle\,dS_x(v) = 0,\\ \frac{d^2}{dr^2}\Big|_{r=0}\mathcal{M}_{\mathbb{B},r}f(x) &= \int_{U_xM}\big\{\langle\nabla_v(\nabla f),v\rangle + \langle\nabla f,\Omega_{\mathbb{B}}(v)\rangle\big\}\,dS_x(v)\\ &= \int_{U_xM}\langle\nabla_v(\nabla f),v\rangle dS_x(v) = -\frac{1}{m}\Delta f(x). \end{aligned}$$

In order to write down $\frac{d^k}{dr^k}\Big|_{r=0}\mathcal{M}_{\mathbb{B},r}f$ clearly, we use a tensor $T_\ell$ of $(\ell,0)$ type given by $T_1 = \nabla f$ and $T_{\ell+1} = \nabla T_\ell$. That is, for vector fields $X_1,\dots,X_{\ell+1}$ we have

$$\begin{aligned}
&T_{\ell+1}(X_1,\dots,X_{\ell+1})\\
&\quad = X_{\ell+1}\big(T_\ell(X_1,\dots,X_\ell)\big) - \sum_{i=1}^{\ell} T_\ell(X_1,\dots,\nabla_{X_{\ell+1}}X_i,\dots,X_\ell).
\end{aligned}$$

As $T_\ell$'s are tensors, we see

$$\begin{aligned}
&\dot\gamma_v\Big(\dot\gamma_v(\nabla_{\dot\gamma_v}f)\Big)\Big|_{t=0} = \dot\gamma_v\Big(T_2(\dot\gamma_v,\dot\gamma_v) + T_1(\nabla_{\dot\gamma_v}\dot\gamma_v)\Big)\Big|_{t=0}\\
&= \Big\{T_3(\dot\gamma_v,\dot\gamma_v,\dot\gamma_v) + T_2(\Omega_{\mathbb{B}}(\dot\gamma_v),\dot\gamma_v) + T_2(\dot\gamma_v,\Omega_{\mathbb{B}}(\dot\gamma_v))\Big\}\Big|_{t=0}\\
&\qquad + \Big\{T_2(\Omega_{\mathbb{B}}(\dot\gamma_v),\dot\gamma_v) + T_1((\nabla_{\dot\gamma_v}\Omega_{\mathbb{B}})(\dot\gamma_v)) + T_1(\Omega_{\mathbb{B}}^2(\dot\gamma_v))\Big\}\Big|_{t=0}\\
&= T_3(v,v,v) + 2T_2(\Omega_{\mathbb{B}}(v),v) + T_2(v,\Omega_{\mathbb{B}}(v))\\
&\qquad + T_1((\nabla_v\Omega_{\mathbb{B}})(v)) + T_1(\Omega_{\mathbb{B}}^2(v)).
\end{aligned}$$

Hence we find

$$\frac{d^3}{dr^3}\Big|_{r=0}\mathcal{M}_{\mathbb{B},r}f(x) = \int_{U_xM}\{3\langle\nabla_v(\nabla f),\Omega_{\mathbb{B}}(v)\rangle + \langle\nabla f,(\nabla_v\Omega_{\mathbb{B}})(v)\rangle\}\,dS_x(v).$$

Thus we see the following:

**Proposition 3.2.** *For each smooth function $f$ on an $m$-dimensional Riemannian manifold $M$ and each $k$, we have*

$$\mathcal{M}_{\mathbb{B},r}f = f - \frac{r^2}{2!m}\Delta f + \frac{r^3}{3!}Q_3 f + \cdots + \frac{r^k}{k!}Q_k f + o(r^{k+1}).$$

*The operator $Q_3$ does not vanish in general even if $\mathbb{B}$ is a uniform magnetic field.*

We here restrict ourselves on Kähler magnetic fields on a Kähler manifold $(M,J)$. For a Kähler magnetic field $\mathbb{B}_\kappa = \kappa\mathbb{B}_J$, which is a constant multiple of the Kähler form $\mathbb{B}_J$, we have $\Omega_{\mathbb{B}_\kappa} = \kappa J$, hence it is a uniform magnetic field. For a uniform magnetic field $\mathbb{B}$ we have

$$\begin{aligned}
&\dot\gamma_v\Big(\dot\gamma_v(\dot\gamma_v(\nabla_{\dot\gamma_v}f))\Big)\Big|_{t=0}\\
&\quad = T_4(v,v,v,v) + 3T_3(\Omega_{\mathbb{B}}(v),v,v) + 2T_3(v,\Omega_{\mathbb{B}}(v),v) + T_3(v,v,\Omega_{\mathbb{B}}(v))\\
&\qquad + 4T_2(\Omega_{\mathbb{B}}^2(v),v) + 3T_2(\Omega_{\mathbb{B}}(v),\Omega_{\mathbb{B}}(v)) + T_1(\Omega_{\mathbb{B}}^3(v)),
\end{aligned}$$

and

$$\frac{d^4}{dr^4}\bigg|_{r=0} \mathcal{M}_{\mathbb{B}_\kappa,r} f(x)$$
$$= \int_{U_xM} \big\{T_4(v,v,v,v) + 4T_2(\Omega_{\mathbb{B}}^2(v),v) + 3T_2(\Omega_{\mathbb{B}}(v),\Omega_{\mathbb{B}}(v))\big\} dS_x(v).$$

Since $\langle \nabla_{Jv}(\nabla f), J(Jv)\rangle = -\langle \nabla_v(\nabla f), Jv\rangle$, we obtain for a Kähler magnetic field $\mathbb{B}_\kappa$

$$\frac{d^3}{dr^3}\bigg|_{r=0} \mathcal{M}_{\mathbb{B}_\kappa,r} f(x) = 0,$$
$$\frac{d^4}{dr^4}\bigg|_{r=0} \mathcal{M}_{\mathbb{B}_\kappa,r} f(x) = \int_{U_xM} \big\{T_4(v,v,v,v) - \kappa^2 T_2(v,v)\big\} dS_x(v).$$

We hence obtain the following expression on magnetic means for Kähler magnetic fields.

**Theorem 3.1.** *Given a smooth function $f$ on a Kähler manifold $M$ of complex dimension $n$, we can write for each $k$ that*

$$\mathcal{M}_{\mathbb{B}_\kappa,r} f = f + \frac{r^2}{2!}Q_2 f + \frac{r^4}{4!}Q_4 f + \frac{r^5}{5!}Q_5 f + \cdots + \frac{r^k}{k!}Q_k f + o\big(r^{k+1}\big),$$

*where*

$$Q_2 f = -\frac{1}{2n}\Delta f,$$
$$Q_4 f = \frac{1}{4n(n+1)}\big(3\Delta^2 f - \langle \nabla f, \nabla \tau\rangle - 2\langle \nabla^2 f, \rho\rangle\big) + \frac{\kappa^2}{2n}\Delta f.$$

*The operator $Q_5$ does not vanish in general.*

In [4] Maeda, Ogiue and the author gave a characterization of parallel embeddings of a Kähler manifold into a Euclidean space in terms of all trajectories for $\mathbb{B}_{\pm\kappa}$, which was an extension of Sakamoto's characterization in terms of geodesics ([8]). Recently, Maeda and Tanabe [7] generalized it in terms of "Kähler Frenet curves". Although it is trivial, we see by Propositions 3.1 and 3.2 that the principal term of the difference operator $2n(\mathcal{M}_{\mathbb{B},r} - I)$ is formally $-r^2\Delta$ and does not depend on $\mathbb{B}$. This suggests us that some results by using local properties of geodesics might be generalized to results by using local properties of trajectories for generalized Kähler magnetic fields $\pm g\mathbb{B}_J$ with a function $g$ on a Kähler manifold. Theorem 3.1 also suggests us the difference $\mathcal{M}_{\mathbb{B}_\kappa,r} - \mathcal{M}_{\mathbb{B}_\nu,r}$ might show some phenomena of trajectories for Kähler magnetic fields.

We here study magnetic means on a complex space form $M^n(c;\mathbb{C})$ of constant holomorphic sectional curvature $c$. As a consequence of the results in [1] we have the following:

**Proposition 3.3.** *On $M^n(c;\mathbb{C})$, the length $r$ of trajectory segment for a nontrivial Kähler magnetic field $\mathbb{B}_\kappa$ and the distance $\ell_\kappa(r)$ between its origin and terminus have the following relationship:*

1) *When $c > 0$ and $r < \pi/\sqrt{\kappa^2+c}$, they satisfy*

$$\frac{1}{\sqrt{c}}\sin\frac{\sqrt{c}}{2}\ell_\kappa(r) = \frac{1}{\sqrt{\kappa^2+c}}\sin\frac{1}{2}\sqrt{\kappa^2+c}\,r.$$

2) *When $c = 0$ and $r < \pi/|\kappa|$, they satisfy $\ell_\kappa(r) = \frac{2}{|\kappa|}\sin\frac{|\kappa|}{2}r$.*

3) *When $c < 0$ and $|\kappa| < \sqrt{|c|}$, they satisfy*

$$\frac{1}{\sqrt{|c|}}\sinh\frac{\sqrt{|c|}}{2}\ell_\kappa(r) = \frac{1}{\sqrt{|c|-\kappa^2}}\sinh\frac{1}{2}\sqrt{|c|-\kappa^2}\,r.$$

4) *When $c < 0$ and $\kappa = \pm\sqrt{|c|}$, they satisfy $\frac{1}{\sqrt{|c|}}\sinh\frac{\sqrt{|c|}}{2}\ell_\kappa(r) = r$.*

5) *When $c < 0, |\kappa| > \sqrt{|c|}$ and $r < \pi/\sqrt{\kappa^2+c}$, they satisfy*

$$\frac{1}{\sqrt{|c|}}\sinh\frac{\sqrt{|c|}}{2}\ell_\kappa(r) = \frac{1}{\sqrt{\kappa^2+c}}\sin\frac{1}{2}\sqrt{\kappa^2+c}\,r.$$

*In particular, a magnetic mean operator $\mathcal{M}_{\mathbb{B}_\kappa,r}$ coincides with a spherical mean $\mathcal{M}_{\ell_\kappa(r)}$.*

Thus on a complex space form magnetic mean operators are essentially the same. Coming back to the expansions of Kähler magnetic means $\mathcal{M}_{\mathbb{B}_\kappa,r}$, we see every odd term vanishes.

**Theorem 3.2.** *On a complex space form $M^n(c;\mathbb{C})$ we have the following:*

(1) *$\mathcal{M}_{\mathbb{B}_\kappa,r} = \mathcal{M}_{\mathbb{B}_{-\kappa},r}$. In particular, Kähler magnetic means are self-adjoint on $\mathrm{L}^2(M^n(c;\mathbb{C}))$.*

(2) *For each smooth function $f$ and positive integer $k$ we see*

$$\mathcal{M}_{\mathbb{B}_\kappa,r}f = f + \frac{r^2}{2!}Q_2 f + \frac{r^4}{4!}Q_4 f + \cdots + \frac{r^{2k}}{(2k)!}Q_{2k}f + o\big(r^{2(k+1)}\big),$$

*where $Q_2 f = -\frac{1}{2n}\Delta f$, $Q_4 f = \frac{3}{4n(n+1)}\Delta^2 f + \frac{\kappa^2+c}{2n}\Delta f$.*

**Remark 3.1.** For a complex space form $M^n(c;\mathbb{C})$ and for $\kappa$ with $\kappa^2+c>0$, we consider a positive number $r$ $(<\pi/\sqrt{\kappa^2+c})$. We see

$$\begin{aligned}
&\mathcal{M}_{\mathbb{B}_\kappa, r+(2i\pi/\sqrt{\kappa^2+c})} = \mathcal{M}_{\mathbb{B}_\kappa, r},\\
&\mathcal{M}_{\mathbb{B}_\kappa, r+((2i+1)\pi/\sqrt{\kappa^2+c})} = \mathcal{M}_{\mathbb{B}_\kappa, r-(\pi/\sqrt{\kappa^2+c})},\\
&\mathcal{M}_{\mathbb{B}_\kappa, 2i\pi/\sqrt{\kappa^2+c}} = I,
\end{aligned}$$

for every positive integer $i$.

We are interested in the self-adjoint property of Kähler magnetic means. If this property holds for sufficiently small $r$, it should show some symmetry of the base Kähler manifold.

Here we make mention of magnetic spherical means. For a magnetic filed $\mathbb{B}$ on $M$ with global vector potential $\mathbb{A}$, which is a 1-form of $d\mathbb{A}=\mathbb{B}$, we define a magnetic spherical mean $\mathcal{L}_{\mathbb{A},r}$ on $\mathrm{L}^2(M)$ by

$$\begin{aligned}
\mathcal{L}_{\mathbb{A},r}f(x) &= \int_{U_xM} a_{\mathbb{A},r}(v) f(\mathbb{B}\mathrm{exp}_x(rv))dS_x(v),\\
a_{\mathbb{A},r}(v) &= \exp\Big(-\sqrt{-1}\int_0^r \mathbb{A}(\dot\gamma_v(t))dt\Big).
\end{aligned}$$

We denote by $H_{\mathbb{A}}$ the Schrödinger operator associated with $\mathbb{A}$ which is defined by

$$H_{\mathbb{A}} = (d-\sqrt{-1}\mathbb{A})^*(d-\sqrt{-1}\mathbb{A}).$$

If we locally express $\mathbb{A}=\sum A_j dx^j$, it is represented as

$$-(1/\sqrt{G})\sum(\partial_j-\sqrt{-1}A_j)g^{ij}\sqrt{G}(\partial_i-\sqrt{-1}A_i),$$

where $G=\det(g_{ij})$ and $(g^{ij})$ denotes the inverse of $(g_{ij})$. By direct computation, we see

$$\frac{d}{dr}\Big|_{r=0}\mathcal{L}_{\mathbb{A},r}f(x) = \int_{U_xM}\left\{-\sqrt{-1}\mathbb{A}(v)f(x)+\langle\nabla f, v\rangle\right\}dS_x(v) = 0,$$

$$\frac{d^2}{dr^2}\Big|_{r=0}\mathcal{L}_{\mathbb{A},r}f(x) = \int_{U_xM}\Big\{-\sqrt{-1}\Big((\nabla_v\mathbb{A})(v)+\mathbb{A}(\Omega_{\mathbb{B}}(v))\Big)f(x) - \mathbb{A}(v)^2 f(x) - 2\sqrt{-1}\,\mathbb{A}(v)\langle\nabla f, v\rangle + \langle\nabla_v(\nabla f), v\rangle + \langle\nabla f, \Omega_{\mathbb{B}}(v)\rangle\Big\}\, dS_x(v)$$

$$= \int_{U_xM}\{\langle\nabla_v(\nabla f), v\rangle - \mathbb{A}(v)^2 f(x) - \sqrt{-1}((\nabla_v\mathbb{A})(v) + 2\mathbb{A}(v)\langle\nabla f, v\rangle)\}\, dS_x(v)$$

$$= -\frac{1}{m}H_{\mathbb{A}}f(x).$$

Similarly we have

$$\frac{d^3}{dr^3}\Big|_{r=0}\mathcal{L}_{\mathbb{A},r}f(x)$$

$$= \int_{U_xM}\Big\{-\sqrt{-1}\Big(2(\nabla_v\mathbb{A})(\Omega_{\mathbb{B}}(v)) + (\nabla_{\Omega_{\mathbb{B}}(v)}\mathbb{A})(v)\Big)f(x) - \Big(3\mathbb{A}(v)\mathbb{A}(\Omega_{\mathbb{B}}(v)) + \sqrt{-1}\,\mathbb{A}\big((\nabla_v\Omega_{\mathbb{B}})(v)\big)\Big)f(x) - 3\sqrt{-1}\Big(\mathbb{A}(\Omega_{\mathbb{B}}(v))\nabla_v f + \mathbb{A}(v)\nabla_{\Omega_{\mathbb{B}}(v)}f\Big)\Big\}dS_x(v)$$

$$+ \frac{d^3}{dr^3}\Big|_{r=0}\mathcal{M}_{\mathbb{B},r}f(x).$$

Thus we obtain the following:

**Theorem 3.3.** *For each smooth function $f$ on an $m$-dimensional Riemannian manifold $M$ and each positive integer $k$, we have*

$$\mathcal{L}_{\mathbb{A},r}f = f - \frac{r^2}{2!m}H_{\mathbb{A}}f + \frac{r^3}{3!}R_3 f + \cdots + \frac{r^k}{k!}R_k f + o(r^{k+1}).$$

*The operator $R_3$ does not vanish in general even if $\mathbb{B}$ is a uniform magnetic field.*

The reader should compare this theorem to Proposition 3.2. From the view point of quantization, magnetic spherical means are important. Still, from Riemannian geometric point of view, Kähler magnetic fields and magnetic means may have some significance (c.f. [3]).

## 4. Magnetic trajectory chains

In the last stage we show our magnetic means are generators for random walks under an action of magnetic field by following [9]. We call a smooth curve $\sigma : [a, a+r] \to M$ a trajectory segment of length $r$ for $\mathbb{B}$ if it is a restriction of a trajectory for $\mathbb{B}$, that is, there is a trajectory $\gamma$ for $\mathbb{B}$ with $\sigma(t+a) = \gamma(t)$ for all $0 \le t \le r$. A continuous curve $\rho : [0, jr] \to M$ is called a $j$-steps chain of trajectory segments of length $r$ for $\mathbb{B}$ if each $\rho|_{[(i-1)r, ir]}$ $(i = 1, 2, \dots, j)$ is a trajectory segment of length $r$ for $\mathbb{B}$. Clearly the set $\mathcal{C}_j(r; \mathbb{B})$ of all $j$-steps chains of trajectory segment of length $r$ for $\mathbb{B}$ on $M$ is naturally identified by parallel translations with the fiber bundle over $M$ whose fiber at $x \in M$ is the $j$-product $U_xM \times \cdots \times U_xM$. Thus $\mathcal{C}_j(r; \mathbb{B})$ is equipped with a measure. By the same way the set $\mathcal{C}_\infty(r; \mathbb{B})$ of all (infinite steps) chains of trajectory segment of length $r$ for $\mathbb{B}$ on $M$ is equipped with a measure. If we define a shift operator $\mathcal{T} : \mathcal{C}_\infty(r; \mathbb{B}) \longrightarrow \mathcal{C}_\infty(r; \mathbb{B})$ by $\mathcal{T}(\rho)(t) = \rho(t+r)$, the ergodicity of the dynamical system $(\mathcal{C}_\infty(r; \mathbb{B}), \mathcal{T})$ is equivalent to the condition that $\mathcal{M}_{\mathbb{B},r} f = f$ for some $f \in \mathrm{L}^2(M)$ implies $f$ is constant.

We now consider a magnetic random walk on a complex space form $M^n(c, \mathbb{C})$. For real numbers $\kappa$ and $c$ we take a set $S(\kappa, c)$ given by

$$S(\kappa, c) = \begin{cases} \{\, i\pi/\sqrt{c} \mid i = 0, \pm 1, \pm 2, \cdots \}, & \text{when } c > 0 \text{ and } \kappa = 0, \\ \emptyset, & \text{when } \kappa^2 + c \le 0, \\ \{\, 2i\pi/\sqrt{\kappa^2 + c} \mid i = 0, \pm 1, \pm 2, \cdots \}, & \text{otherwize.} \end{cases}$$

Proposition 3.3 also guarantees that for every pair of points $p, q \in M^n(c, \mathbb{C})$ there is a finite steps chain of trajectory segments of length $r$ for $\mathbb{B}_\kappa$ joining $p$ and $q$ when $r \notin S(\kappa, c)$. Therefore on a compact quotient $N = M^n(c, \mathbb{C})/\Gamma$, for $r \notin S(\kappa, c)$, there is an infinite steps chain $\rho : [0, \infty) \longrightarrow N$ of trajectory segments of length $r$ for $\mathbb{B}_\kappa$ such that the set $\{\, \rho(ri) \mid i = 1, 2, \dots \}$ of foot points of $\rho$ is a dense set in $N$. More precisely, Proposition 3.3 and the results in [9] shows the following:

**Proposition 4.1.** *Magnetic mean $\mathcal{M}_{\mathbb{B}_\kappa, r}$ for a Kähler magnetic field on a complex projective space $\mathbb{C}P^n(c)$ is ergodic if and only if $r \notin S(\kappa, c)$.*

**Proposition 4.2.** *Let $N = M^n(c, \mathbb{C})/\Gamma$ be a compact quotient. When $r \notin S(\kappa, c)$, almost all infinite step chains of trajectory segments $\rho$ of length $r$ for $\mathbb{B}_\kappa$ with respect the measure are uniformly distributed on $N$ in the following sense: For each measurable set $K$ in $N$ it satisfies*

$$\lim_{k \to \infty} \frac{1}{k} \sum_{i=0}^{k} \chi_K(\rho(i)) = \mathrm{vol}(K)/\mathrm{vol}(N),$$

*where $\chi_K$ denotes the defining function.*

## References


1. T. Adachi, *Kähler magnetic flows on a manifold of constant holomorphic sectional curvature*, Tokyo J. Math. 18(1995), 473–483.
2. ———, *Kähler magnetic fields on a complex hyperbolic space*, A Report on KOREA-JAPAN Joint Workshop in Mathematics 2000, Differential Geometry and its Related Topics, 9–20, (2001).
3. ———, *A comparison theorem on crescents for Kähler magnetic fields*, preprint.
4. T. Adachi, S. Maeda and K. Ogiue, *Extrinsic shape of circles and standard imbeddings of projective spaces*, Manuscripta Math. 93(1997), 267–272.
5. A. Gray, *Mean-values in Riemannian manifolds*, Proceedings of the IV International colloquium on differential geometry.
6. A. Gray and T.J. Willmore, *Mean-value theorems for Riemannian manifolds*, Proc. Royal Soc. Edinburgh Sect. A 92(1982), 343–364.
7. S. Maeda and H. Tanabe, *Totally geodesic immersions of Kähler manifolds and Kähler Frenet curves*, preprint.
8. K. Sakamoto, *Planar geodesic immersions*, Tôhoku Math. J. 29(1977) 25–56.
9. T. Sunada, *Spherical means and geodesic chains on a Riemannian manifold*, Trans. A. M. S. 267(1981), 483–501.
10. T.J. Willmore, *An extension of Pizzetti's Formula to Riemannian manifolds*, Astérisque 80, 53–56.

# YANG-MILLS FIELDS ANALOGUE OF BIHARMONIC MAPS

CORNELIA-LIVIA BEJAN
*Seminar Matematic, Universitatea "A. I. Cuza",*
*Iasi 700506, Romania*
*E-mai: bejan@math.tuiasi.ro*

HAJIME URAKAWA
*Division of Mathematics, Graduate School of Information Sciences, Tohoku University,*
*Aoba 09, Sendai, 980-8579, Japan*
*E-mail: urakawa@math.is.tohoku.ac.jp*

In this short note, we give not only a survey of biharmonic submanifolds and biharmonic maps, but also give a Yang-Mills field analogue of biharmonic maps and the Euler-Lagrange equation for the bi-Yang-Mills functional.

*Keywords*: biharmonic submanifolds, biharmonic maps, Yang-Mills fields, bi-Yang-Mills fields

## 1. Introduction

This is a survey and research paper on biharmonic submanifolds, biharmonic maps, and the Yang-Mills field analogue for them. We first give a brief survey on biharmonic submanifolds of a Riemannian manifold, and a biharmonic map between two Riemannian manifolds in Section Two. Then, in Section Three, we give the notion of bi-Yang-Mills functional, and calculate its Euler-Lagrange equation by a similar way to the Yang-Mills equation in [3]. We then give the definition of bi-Yang-Mills field, and show that any Yang-Mills field is also a bi-Yang-Mills field.

This work was done while the second author stayed at Iasi at 2002, and the first author stayed at Sendai at 2003. We thank to the referee who pointed out several msitakes in the first draft.

**MSC 2000.** 53C07, 53C42, 53C43

## 2. Biharmonic submanifolds and maps

We explain a brief history of biharmonic submanifolds and biharmonic maps. We first see biharmonic submanifolds. Let $\Phi : (M, g) \hookrightarrow (\mathbf{R}^n, g_0)$ be an isometric immersion of an $m$-dimensional Riemannian manifold $(M, g)$ into the standard Euclidean space $(\mathbf{R}^n, g_0)$. Let $H$ be the mean curvature vector field of $\Phi$ along $M$. Then, it is well known (cf. [13], p. 340) that

$$H = -\frac{1}{m}\Delta\Phi,$$

where $\Delta$ is the Laplacian of $(M, g)$ and $\Delta\Phi = (\Delta\varphi_1, \ldots, \Delta\varphi_n)$ for $\Phi = (\varphi_1, \ldots, \varphi_n)$. B.Y. Chen introduced (cf. [7]) the following notion.

**Definition 2.1.** The isometric immersion $\Phi : (M, g) \hookrightarrow (\mathbf{R}^n, g_0)$ is *biharmonic* if $\Delta H = 0$, which is equivalent to that $\Delta(\Delta\Phi) = 0$.

By definition, if the immersion $\Phi : (M, g) \hookrightarrow (\mathbf{R}^n, g_0)$ is *harmonic*, by definition, $\Delta\Phi = 0$, then $\Phi$ is biharmonic. He raised the following conjecture:

**B.Y. Chen's Conjecture.** *Any biharmonic submanifold of the Euclidean space is harmonic.*

Recently, Sasahara and Inoguchi (cf. [15], [10]) obtained some progress in this conjecture.

Let us recall the notion of harmonic mappings (cf. [9]). A smooth mapping $\varphi$ of a Riemannian manifolds $(M, g)$ into another Riemannian manifolds $(N, h)$ is called *harmonic* if it is a critical point of the energy functional

$$E(\varphi) = \frac{1}{2}\int_M |d\varphi|^2\, v_g\,,$$

that is, for all smooth one-parameter family of mappings $\varphi^t$ of $M$ into $N$ $(-\epsilon < t < \epsilon)$ with $\varphi^0 = \varphi$, it holds that

$$\left.\frac{d}{dt}\right|_{t=0} E(\varphi^t) = 0,$$

which is equivalent to that $\tau(\varphi) = 0$. Here the tension field $\tau(\varphi)$ is defined by $\tau(\varphi) = \mathrm{trace}(\nabla d\varphi) = \sum_{i=1}^m \{\widetilde{\nabla}_{e_i}(\varphi_* e_i) - \varphi_*(\nabla_{e_i} e_i)\}$, where $\widetilde{\nabla}$ is the induced connection of the induced bundle $\varphi^{-1}TN$ of the tangent bundle $TN$ over $M$ from the Levi-Civita one of $(N, h)$, and $\{e_i\}_{i=1}^m$ is a locally defined orthonormal frame field on $M$ with respect to $g$ $(m = \dim M)$.

**Definition 2.2.** For a smooth mapping $\varphi$ of $(M,g)$ into $(N,h)$, the *bienergy* of $\varphi$ is defined by

$$E_2(\varphi) = \frac{1}{2}\int_M |\tau(\varphi)|^2 v_g.$$

$\varphi$ is said to be *biharmonic* if it is a critical point of $E_2$.

G.Y. Jiang obtained (cf. [11], [12]) the Euler-Lagrange equation for $E_2$ as follows:

**Theorem 2.3.** *Let $\tau_2(\varphi) = J(\tau(\varphi))$, where $J$ is the so called Jacobi operator which is defined by*

$$J(V) = -\sum_{i=1}^{m}(\widetilde{\nabla}_{e_i}\widetilde{\nabla}_{e_i} - \widetilde{\nabla}_{\nabla_{e_i}e_i})V - \sum_{i=1}^{m} R^N(V, \varphi_* e_i)\varphi_* e_i,$$

*for every smooth section $V$ of the induced bundle $\varphi^{-1}TN$ of the tangent bundle $TN$ by $\varphi$, $\widetilde{\nabla}$ is the induced connection on $\varphi^{-1}TN$ and $R^N$ is the curvature tensor of $(N,h)$. Then, we have*

$$\left.\frac{d}{dt}\right|_{t=0} E_2(\varphi) = -\int_M \langle \tau_2(\varphi), V\rangle\, v_g.$$

**Corollary 2.4.** *For an isometric immersion $\Phi : (M,g) \hookrightarrow (\mathbf{R}^n, g_0)$, we have*

$$\tau_2(\Phi) = \Delta\,\tau(\Phi) = \Delta(m\,H).$$

*Thus, $\Phi$ is a biharmonic submanifold if and only if $\Phi$ is a biharmonic map.*

Caddeo, Montaldo and Piu raised (cf. [4]) the conjecture that

**Conjecture.** *The only biharmonic submanifolds of $(N,h)$ whose curvature are nonpositive are the minimal ones.*

Caddeo, Montaldo and Oniciuc showed (cf. [6]) that

**Theorem 2.5.** *Let $M$ be a submanifold of the 3-dimensional space form $N^3(-1)$ of curvaure $-1$. Then, $M$ is biharmonic if and only if it is harmonic.*

Furthermore, they showed (cf. [6]) that

**Theorem 2.6.** *Let $M \subset S^n(a) \times \{b\}$ be a harmonic submanifold with $a^2 + b^2 = 1$ $(0 < a < 1)$. Then, $M \subset S^{n+1}$ is a non-harmonic and biharmonic submanifold if and only if $a = \frac{1}{\sqrt{2}}$ and $b = \pm\frac{1}{\sqrt{2}}$.*

**Corollary 2.7.** *There exist closed orientable embedded non-minimal biharmonic surfaces $M^2$ of arbitrary genus in the standard 4-sphere $S^4$.*

**Corollary 2.8.** *The mapping $\varphi : \mathbf{R}^3 \to \mathbf{R}^6$ defined by*

$$(x, y, z) \mapsto \frac{1}{\sqrt{6}} \left( yz, zx, xy, \frac{1}{2}(x^2 - y^2), \frac{1}{\sqrt{12}}(x^2 + y^2 - 2z^2), \sqrt{3} \right)$$

*yields a non-minimal biharmonic embedding of the 2-dimensional standard projective space $P^2(\mathbf{R})$ into the standard 5-sphere $S^5$.*

In this connection, Oniciuc showed (cf. [14]) that the following theorem:

**Theorem 2.9.** *Let $\varphi : (M, g) \to (N, h)$ be a Riemannian submersion of a compact orientable Riemannian manifold into a Riemannian manifold with nonpositive curvature. Then, $\varphi$ is biharmonic if and only if it is harmonic.*

For the generalized Clifford torus, it is known (cf. [12]) that

**Example.** (Jiang) Let $S^p(\frac{1}{\sqrt{2}}) \times S^q(\frac{1}{\sqrt{2}})$ be the generalized Clifford torus in the standard unit sphere $S^{m+1}$ $(p + q = m)$. Then, it is a non-minimal biharmonic submanifold of $S^{m+1}$ if $p \neq q$.

## 3. Bi-Yang-Mills fields

Now let us begin the Yang-Mills field setting over a compact Riemannian manifold $(M, g)$ following [3], and give also the Yang-Mills field analogue of biharmonic maps. Let $(E, h)$ be a vector bundle over $M$ with an inner product $h$ on $E$. Let $\mathcal{C}(E, h)$ be the set of all connections $\nabla$ of $E$ satisfying

$$Xh(s,t) = h(\nabla_X s, t) + h(s, \nabla_X t), \quad s,\, t \in \Gamma(E),$$

for all smooth vector field $X$ on $M$, where $\Gamma(E)$ be the set of all smooth sections of $E$. For $\nabla \in \mathcal{C}(E, h)$, let $R^{\nabla}$ be its curvature tensor defined by

$$R^{\nabla}(X,Y)s = \nabla_X(\nabla_Y s) - \nabla_Y(\nabla_X s) - \nabla_{[X,Y]} s,$$

for all $s \in \Gamma(E)$ and smooth vector fields $X$ and $Y$ on $M$. Let $F = \mathrm{End}(E, h)$ be the bundle of endomorphisms of $E$ which are skew symmetric with respect to the inner product $h$ on $E$. We define the inner product $\langle\, ,\, \rangle$ on $F$ by

$$\langle \varphi, \psi \rangle = \sum_{i=1}^{r} h(\varphi u_i, \psi u_i), \quad \varphi,\, \psi \in F_x,$$

where $\{u_i\}_{i=1}^r$ is an orthonormal basis of $E_x$ with respect to $h$ ($r = \dim E_x$). Then, let us consider the space of $F$-valued $k$-forms on $M$, denoted by $\Omega^k(F) = \Gamma(\wedge^k T^*M \otimes F)$, with a global inner product $(\,,\,)$ defined by

$$(\alpha, \beta) = \int_M \langle \alpha, \beta \rangle \, v_g,$$

where

$$\langle \alpha, \beta \rangle = \sum_{i_1 < \cdots < i_k} \langle \alpha(e_{i_1}, \ldots, e_{i_k}), \beta(e_{i_1}, \ldots, e_{i_k}) \rangle$$

and $\{e_1, \cdots, e_n\}$ is a locally defined orthonormal frame field on $M$ with respect to $g$.

For every $\nabla \in \mathcal{C}(E, h)$, let $d^\nabla : \Omega^k(F) \to \Omega^{k+1}(F)$ be the exterior differentiation with respect to $\nabla$ (cf. [3]), and $\delta^\nabla : \Omega^k(F) \to \Omega^{k-1}(F)$, the adjoint operator defined by

$$\delta^\nabla \alpha = (-1)^k * d^\nabla * \alpha, \quad \alpha \in \Omega^k(F),$$

where $*$ is the usual star operator $* : \Omega^k(F) \to \Omega^{n-k}(F)$. Then, for $\alpha \in \Omega^{k-1}(F)$ and $\beta \in \Omega^k(F)$,

$$(d^\nabla \alpha, \beta) = (\alpha, \delta^\nabla \beta).$$

**Definition 3.1.** The *bi-Yang-Mills functional* $\mathcal{YM}_2$ is defined by

$$\mathcal{YM}_2(\nabla) = \frac{1}{2} \int_M \|\delta^\nabla R^\nabla\|^2 v_g, \quad \nabla \in \mathcal{C}(E, h),$$

where $\|\delta^\nabla R^\nabla\|$ is the norm of $\delta^\nabla R^\nabla \in \Omega^1(F)$ relative to $\langle\,,\,\rangle$.

**Remark 3.2.** Let us recall the standard definition of the Yang-Mills functional is as follows:

$$\mathcal{YM}(\nabla) = \frac{1}{2} \int_M \|R^\nabla\|^2 v_g, \quad \nabla \in \mathcal{C}(E, h),$$

where $\|R^\nabla\|$ is the norm of $R^\nabla \in \Omega^2(F)$ relative to $\langle\,,\,\rangle$.

**Definition 3.2.** For each $\nabla \in \mathcal{C}(E, h)$, it is a *bi-Yang-Mills field* if it is a critical point of $\mathcal{YM}_2$, i.e., for each smooth one-parameter family $\nabla^t$ $(-\epsilon < t < \epsilon)$ of with $\nabla^0 = \nabla$,

$$\left.\frac{d}{dt}\right|_{t=0} \mathcal{YM}_2(\nabla^t) = 0.$$

We have

**Theorem 3.3.** *Let* $\alpha = \frac{d}{dt}|_{t=0}\nabla^t \in \Omega^1(F)$, *where* $F = End(E, h)$. *Then, we have*

$$\left.\frac{d}{dt}\right|_{t=0} \mathcal{YM}_2(\nabla^t) = \int_M \langle \alpha, (\delta^\nabla d^\nabla + \mathcal{R}^\nabla)(\delta^\nabla R^\nabla)\rangle \, v_g \, .$$

*Here, for* $\beta \in \Omega^1(F)$, $\mathcal{R}^\nabla(\beta) \in \Omega^1(F)$ *is defined by*

$$\mathcal{R}^\nabla(\beta)(X) = \sum_{j=1}^{n} [R^\nabla(e_j, X), \beta(e_j)],$$

*for a smooth vector field* $X$ *on* $M$, *where* $\{e_j\}_{j=1}^n$ *is a locally defined orthonormal frame field on* $M$ *with respect to* $g$. *Thus,* $\nabla$ *is a bi-Yang-Mills field if and only if*

$$(\delta^\nabla d^\nabla + \mathcal{R}^\nabla)(\delta^\nabla R^\nabla) = 0.$$

Let us recall

**Theorem 3.4.** (cf. [3]) *For the Yang-Mills functional, we have*

$$\left.\frac{d}{dt}\right|_{t=0} \mathcal{YM}(\nabla^t) = \int_M \langle \alpha, \delta^\nabla R^\nabla \rangle \, v_g.$$

*Thus,* $\nabla$ *is a Yang-Mills field if and only if* $\delta^\nabla R^\nabla = 0$.

Therefore, we have immediately,

**Corollary 3.5.** *If* $\nabla$ *is a Yang-Mills field, then it is also a bi-Yang-Mills field.*

We raise the following problem:

**Problem.** *Find a bi-Yang-Mills field which is not a Yang-Mills field.*

## 4. Proof of Theorem

We first have

$$\begin{aligned}
&\left.\frac{d}{dt}\right|_{t=0} \mathcal{YM}_2(\nabla^t) \\
&= \int_M \left\langle \frac{d}{dt}\Big|_{t=0} \delta^{\nabla^t} R^{\nabla^t}, \delta^\nabla R^\nabla \right\rangle v_g \\
&= \int_M \left\langle \left(\frac{d}{dt}\Big|_{t=0} \delta^{\nabla^t}\right) R^\nabla + \delta^\nabla \left(\frac{d}{dt}\Big|_{t=0} R^{\nabla^t}\right), \delta^\nabla R^\nabla \right\rangle v_g \qquad (4.1)
\end{aligned}$$

Since we have

$$\frac{d}{dt}\bigg|_{t=0} R^{\nabla^t} = d^{\nabla}\alpha,$$

the second term of the right hand side of (4.1) coincides with

$$\int_M \left\langle \delta^{\nabla}\left(\frac{d}{dt}\bigg|_{t=0} R^{\nabla^t}\right), \delta^{\nabla} R^{\nabla}\right\rangle v_g = \int_M \langle \delta^{\nabla} d^{\nabla}\alpha, \delta^{\nabla} R^{\nabla}\rangle v_g$$
$$= \int_M \langle \alpha, \delta^{\nabla} d^{\nabla}(\delta^{\nabla} R^{\nabla})\rangle v_g. \qquad (4.2)$$

On the other hand, the first term of the right hand side of (4.1) coincides with

$$\int_M \left\langle \left(\frac{d}{dt}\bigg|_{t=0} \delta^{\nabla^t}\right) R^{\nabla}, \delta^{\nabla} R^{\nabla}\right\rangle v_g = \int_M \langle R^{\nabla}, [\alpha \wedge \delta^{\nabla} R^{\nabla}]\rangle v_g$$
$$= \int_M \langle \alpha, \mathcal{R}^{\nabla}(\delta^{\nabla} R^{\nabla})\rangle v_g. \qquad (4.3)$$

Indeed, to see the first equality of (4.3), we only have to show that for all $\varphi \in \Omega^2(F)$ and $\beta \in \Omega^1(F)$,

$$\int_M \left\langle \left(\frac{d}{dt}\bigg|_{t=0} \delta^{\nabla^t}\right)\varphi, \beta\right\rangle v_g = \int_M \left\langle \varphi, \left(\frac{d}{dt}\bigg|_{t=0} d^{\nabla^t}\beta\right)\right\rangle v_g$$

and

$$\left(\frac{d}{dt}\bigg|_{t=0} d^{\nabla^t}\beta\right)(X,Y) = [\alpha \wedge \beta](X,Y), \qquad (4.4)$$

where the right hand side is defined by $[\alpha \wedge \beta](X,Y) = [\alpha(X), \beta(Y)] - [\alpha(Y), \beta(X)]$ (cf. [3]). Then, we have the first equality of (4.3).

Here, we have (4.4) as follows. By definition,

$$(d^{\nabla^t}\beta)(X,Y) = \nabla^t_X(\beta(Y)) - \nabla^t_Y(\beta(X)) - \beta([X,Y]),$$

so that

$$\frac{d}{dt}\bigg|_{t=0}(d^{\nabla^t}\beta)(X,Y) = \frac{d}{dt}\bigg|_{t=0}\nabla^t_X(\beta(Y)) - \frac{d}{dt}\bigg|_{t=0}\nabla^t_Y(\beta(X))$$
$$= [\alpha(X), \beta(Y)] - [\alpha(Y), \beta(X)]$$
$$= [\alpha \wedge \beta](X,Y).$$

Here we want to see the second equality $\frac{d}{dt}|_{t=0}\nabla^t_X(\beta(Y)) = [\alpha(X), \beta(Y)]$ as follows. We put $\nabla^t = \nabla + \alpha^t$ with $\alpha^t \in \Omega^1(F)$ and $\alpha^0 = 0$. Since $\beta(Y) \in \Gamma(F)$ with $F = \mathrm{End}(E,h)$, we have, for all $u \in \Gamma(E)$,

$$(\nabla^t_X(\beta(Y)))(u) = \nabla^t_X(\beta(Y)u) - \beta(Y)(\nabla^t_X u)$$
$$= (\nabla_X + \alpha^t(X))(\beta(Y)u) - \beta(Y)(\nabla_X u + \alpha^t(X)u),$$

which yields that

$$\frac{d}{dt}\bigg|_{t=0}(\nabla^t_X(\beta(Y))(u) = \frac{d}{dt}\bigg|_{t=0}\alpha^t(X)(\beta(Y)u) - \beta(Y)\left(\frac{d}{dt}\bigg|_{t=0}\alpha^t(X)u\right)$$
$$= \alpha(X)(\beta(Y)u) - \beta(Y)(\alpha(X)u)$$
$$= [\alpha(X), \beta(Y)]u,$$

which is the desired.

For the second equality of (4.3), we only have to see that, for all $\beta \in \Omega^1(F)$,

$$\langle R^\nabla, [\alpha \wedge \beta]\rangle = \langle \mathcal{R}^\nabla(\beta), \alpha\rangle,$$

which follows from that the left hand side coincides with

$$\begin{aligned}
&\sum_{i<j}\langle R^\nabla(e_i, e_j), [\alpha \wedge \beta](e_i, e_j)\rangle \\
&= \sum_{i<j}\langle R^\nabla(e_i, e_j), [\alpha(e_i), \beta(e_j)] - [\alpha(e_j), \beta(e_i)]\rangle \\
&= \sum_{i,j=1}^{n}\langle R^\nabla(e_i, e_j), [\alpha(e_i), \beta(e_j)]\rangle \\
&= \sum_{i=1}^{n}\left\langle \sum_{j=1}^{n}[\beta(e_j), R^\nabla(e_i, e_j)], \alpha(e_i)\right\rangle \\
&= \langle \mathcal{R}^\nabla(\beta), \alpha\rangle.
\end{aligned}$$

Thus, we have Theorem 3.3. □

## References


1. P. Baird and D. Kamissoko, *On constructing biharmonic maps and metrics*, Ann. Global Anal. Geom., **23**(2003), 65–75.
2. C.L. Bejan and M. Benyounes, *Kähler manifolds of quasi-constant holomorphic sectional curvature*, a preprint.
3. J.P. Bourguignon and H.B. Lawson, *Stability and isolation phenomena for Yang-Mills fields*, Commun. Math. Phys., **79**(1981), 189–230.
4. R. Caddeo, S. Montald and P. Piu, *On biharmonic maps*, Contemp. Math., **288** (2001), 286–290.
5. R. Caddeo, S. Montald and C. Oniciuc, *Biharmonic submanifolds of $S^3$*, Intern. J. Math., **12** (2001),867–876.
6. ——, *Biharmonic submanifolds in spheres*, Israel J. Math., **130** (2002), 109–123.
7. B.Y. Chen, *Some open problems and conjectures on submanifolds of finite type*, Soochow J. Math., **17** (1991), 169–188.

8. B.Y. Chen and S. Ishikawa, *Biharmonic pseudo-Riemannian submanifolds in pseudo-Euclidean spaces*, Kyushu J. Math., **52** (1998), 167–185.
9. J. Eells and L. Lemaire, *Selected topics in harmonic maps*, Regional Cof. Series in Math., Amer. Math. Soc., 1983.
10. J. Inoguchi, *Submanifolds with harmonic mean curvature in contact 3-manifolds*, a preprint.
11. G.Y. Jiang, *2-harmonic isometric immersions between Riemannian manifolds*, Chinese Ann. Math., **A7** (1986), 130–140.
12. ———, *2-harmonic maps and their first and second variational formula*, Chinese J. Math., **A7** (1986), 389–402.
13. S. Kobayashi and K. Nomizu, *Foundation of Differential Geometry, Vol. II*, Interscience Publ., New York (1969).
14. C. Oniciuc, *Biharmonic maps between Riemannian manifolds*, An. Stiint. Univ. Al. I. Cuza Iasi, **48** (2002), 237–248.
15. T. Sasahara, *Instability of biharmonic Legendre surfaces in Sasakian space forms*, a preprint.
16. H. Urakawa, *Calculus of Variations and Harmonic Maps*, Amer. Math. Soc., Providence, (1993).

# INVARIANT AND ANTI-INVARIANT UNIT VECTOR FIELDS

TRAN QUOC BINH
*Department of Mathematics and Informatics, Debrecen University*
*H-4010 Debrecen, P.O. Box 12, Hungary*
*E-mail: binh@math.klte.hu*

ERIC BOECKX
*Department of Mathematics, Katholieke Universiteit Leuven, Celestijnenlaan 200B, 3001 Leuven, Belgium*
*E-mail: eric.boeckx@wis.kuleuven.ac.be*

LIEVEN VANHECKE
*Department of Mathematics, Katholieke Universiteit Leuven, Celestijnenlaan 200B, 3001 Leuven, Belgium*
*E-mail: lieven.vanhecke@wis.kuleuven.ac.be*

We study unit vector fields on a Riemannian manifold $(M, g)$ which embed $M$ as an invariant or anti-invariant submanifold into its unit tangent bundle with respect to its standard contact metric structure.

*Keywords*: unit vector fields, invariant and anti-invariant submanifolds, minimal submanifolds.

## 1. Introduction

Let $(M, g)$ be a Riemannian manifold and $(T_1M, g_S)$ its unit tangent bundle equipped with the Sasaki metric $g_S$. Every smooth unit vector field $V$ on $M$ determines a mapping from $M$ into $T_1M$, embedding $M$ into its unit tangent bundle. This mapping is an isometry only when $V$ is parallel.

2000 *Mathematics Subject Classification*: 53D10, 53D15, 53C40, 53C42.

Since the unit vector field $V$, the corresponding mapping $V : M \to T_1M$ and the submanifold $V(M) \subset T_1M$ determine each other reciprocally, one naturally comes to the study of 'special' unit vector fields, in the sense that the associated map or the associated submanifold is special in some sense.

In this way, one has defined, e.g., the *energy* and the *volume* of a unit vector field on a compact manifold as the energy of the corresponding mapping and the volume of the corresponding submanifold. Next, *harmonic* and *minimal* unit vector fields appear as the critical points for the associated energy and volume functionals on the space of unit vector fields. The origins of these ideas go back to [8] and have recently bloomed both as concerns theoretical results and as concerns the construction of explicit examples. We refer to [3]–[7] and [9] and the references therein for an (incomplete) survey of the present state of the research in this field. A. Yampolsky is currently making a systematic study of unit vector fields based on the intrinsic and extrinsic properties of the associated submanifolds, focusing mainly on totally geodesic submanifolds and the mean curvature ([11]–[13]).

In this note, we also study unit vector fields starting from the embeddings they generate, but we consider properties related to the standard contact metric structure on the unit tangent bundle $T_1M$. More precisely, we introduce invariant and anti-invariant unit vector fields as those unit vector fields for which the corresponding embedding in $T_1M$ is an invariant or an anti-invariant submanifold with respect to the contact metric structure. We describe these vector fields in a geometric way (Propositions 1 and 5) and present a series of examples.

## 2. Preliminaries

In this section, we give the formulas and definitions needed to work with the standard contact metric structure on the unit tangent bundle. First, we recall the basic concepts in contact geometry. For a more detailed treatment, see [1].

An *almost contact structure* $(\tilde{\xi}, \tilde{\eta}, \tilde{\varphi})$ on a differentiable manifold $M$ consists of a vector field $\tilde{\xi}$, a one-form $\tilde{\eta}$ and a $(1,1)$-tensor field $\tilde{\varphi}$ satisfying

$$\tilde{\eta}(\tilde{\xi}) = 1 \quad \text{and} \quad \tilde{\varphi}^2 = -\mathrm{Id} + \tilde{\eta} \otimes \tilde{\xi}.$$

As a consequence, the dimension of $M$ is odd and $\tilde{\varphi}\tilde{\xi} = 0$, $\tilde{\eta} \circ \tilde{\varphi} = 0$. The structure is *almost contact metric* if, in addition, there is a compatible metric $g$, i.e., a Riemannian metric on $M$ satisfying

$$g(\tilde{\varphi}X, \tilde{\varphi}Y) = g(X,Y) - \tilde{\eta}(X)\tilde{\eta}(Y)$$

for all vector fields $X$ and $Y$ on $M$. Consequently, $\tilde{\eta}(\cdot) = g(\cdot, \tilde{\xi})$ and $\tilde{\varphi}$ is skew-symmetric with respect to $g$. The structure is called *contact metric* if the additional property

$$d\tilde{\eta}(X,Y) = g(X, \tilde{\varphi}Y)$$

holds for all vector fields $X$ and $Y$ on $M$. Then the integral curves of the characteristic vector field $\tilde{\xi}$ are geodesics.

Further, if $\tilde{\xi}$ is a Killing vector field with respect to $g$, then the manifold is *K-contact.* Equivalently, it holds

$$\nabla\tilde{\xi} = -\tilde{\varphi}$$

where $\nabla$ is the Levi Civita connection associated to $g$. Finally, if the Riemann curvature tensor $R$ of $g$ satisfies

$$R(X,Y)\tilde{\xi} = \nabla_X\nabla_Y\tilde{\xi} - \nabla_Y\nabla_X\tilde{\xi} - \nabla_{[X,Y]}\tilde{\xi} = \tilde{\eta}(Y)X - \tilde{\eta}(X)Y$$

for all vector fields $X$ and $Y$, then the contact metric manifold is *Sasakian.* In this case, $\tilde{\xi}$ is necessarily a Killing vector field, hence every Sasakian manifold is $K$-contact.

We will also need the notion of a *D-homothetic transformation* of a contact metric manifold $(M, \tilde{\xi}, \tilde{\eta}, \tilde{\varphi}, g)$. For a (strictly) positive constant $a$, this transformation associates to the contact metric structure $(\tilde{\xi}, \tilde{\eta}, \tilde{\varphi}, g)$ on $M$ another contact metric structure $(\xi', \eta', \varphi', g')$ on the same manifold by setting

$$\eta' = a\tilde{\eta}, \quad \xi' = \frac{1}{a}\tilde{\xi}, \quad \varphi' = \tilde{\varphi}, \quad g' = ag + a(a-1)\tilde{\eta}\otimes\tilde{\eta}.$$

We note that the D-homothetic transformation of a K-contact (respectively, Sasakian) structure is again K-contact (respectively, Sasakian).

Next, we turn our attention to the tangent and the unit tangent bundle of a Riemannian manifold $(M, g)$. We only collect the facts and formulas we need for this note. For a more elaborate exposition, see [2] and the references therein.

Let $(M, g)$ be an $n$-dimensional connected Riemannian manifold with Levi Civita connection $\nabla$. The tangent bundle $TM$ consists of pairs $(x, u)$ where $x$ is a point in $M$ and $u$ a tangent vector to $M$ at $x$. The mapping $\pi \colon TM \to M \colon (x,u) \mapsto x$ is the natural projection from $TM$ to $M$.

It is a well-known fact that the tangent space to $TM$ at $(x, u)$ splits into the direct sum of the vertical subspace $VTM_{(x,u)} = \ker \pi_{*|(x,u)}$ and the horizontal subspace $HTM_{(x,u)}$ with respect to $\nabla$:

$$T_{(x,u)}TM = VTM_{(x,u)} \oplus HTM_{(x,u)}.$$

For $X \in T_xM$, there exists a unique vector $X^h$ at the point $(x,u) \in TM$ such that $X^h \in HTM_{(x,u)}$ and $\pi_*(X^h) = X$. $X^h$ is called the *horizontal lift* of $X$ to $(x,u)$. Similarly, there is a unique vector $X^v$ at the point $(x,u)$ such that $X^v \in VTM_{(x,u)}$ and $X^v(df) = Xf$ for all functions $f$ on $M$. (Here we consider $df$ as a function on $TM$.) $X^v$ is called the *vertical lift* of $X$ to $(x,u)$. The map $X \mapsto X^h$, respectively $X \mapsto X^v$, is an isomorphism between $T_xM$ and $HTM_{(x,u)}$, respectively $T_xM$ and $VTM_{(x,u)}$. In a similar way, one lifts vector fields on $M$ to horizontal or vertical vector fields on $TM$. The expressions in local coordinates for these lifts are given by

$$X^h = \sum (X^i \circ \pi) \frac{\partial}{\partial \bar{x}^i} - \sum u^b((X^a \Gamma^i_{ab}) \circ \pi) \frac{\partial}{\partial u^i}, \tag{1}$$

$$X^v = \sum (X^i \circ \pi) \frac{\partial}{\partial u^i} \tag{2}$$

for a vector field $X = \sum X^i \frac{\partial}{\partial x^i}$ where $\{x^1, \ldots, x^n\}$ are local coordinates around $x$ in $M$ and the local coordinates $\{\bar{x}^1, \ldots, \bar{x}^n; u^1, \ldots, u^n\}$ around $(x,u)$ in $TM$ are defined by

$$\begin{aligned} \bar{x}^i(x,u) &= (x^i \circ \pi)(x,u) = x^i(x), \\ u^i(x,u) &= dx^i(u) = u(x^i). \end{aligned}$$

$\Gamma^i_{ab}$ are the Christoffel symbols for the connection $\nabla$ on $M$.

The tangent bundle $TM$ of a Riemannian manifold $(M,g)$ can be endowed in a natural way with a Riemannian metric $g_S$, the *Sasaki metric,* depending only on the Riemannian structure $g$ of the base manifold $M$. It is uniquely determined by

$$g_S(X^h, Y^h) = g_S(X^v, Y^v) = g(X,Y) \circ \pi, \qquad g_S(X^h, Y^v) = 0$$

for all vector fields $X$ and $Y$ on $M$. There is also an almost complex structure $J$ on $TM$ given by

$$JX^h = X^v, \qquad JX^v = -X^h$$

for all vector fields $X$ on $M$. This makes $(TM, g_S, J)$ into an almost Kähler manifold. It is Kählerian only when $(M,g)$ is flat.

However, our interest lies not with the tangent bundle, but with the unit tangent bundle $T_1M$. The hypersurface $T_1M$ of $TM$ consists of the unit tangent vectors to $(M,g)$ and is given implicitly by the equation $g_x(u,u) = 1$. A unit normal vector $N$ to $T_1M$ at $(x,u) \in T_1M$ is given by the vertical lift of $u$ to $(x,u)$: $N_{|(x,u)} = u^v$.

As the vertical lift of a vector (field) is not tangent to $T_1M$ in general, we define the *tangential lift* of $X \in T_xM$ to $(x,u) \in T_1M$ by

$$X^t_{(x,u)} = (X - g(X,u)\,u)^v.$$

Clearly, the tangent space to $T_1M$ at $(x,u)$ is spanned by vectors of the form $X^h$ and $X^t$ where $X \in T_xM$.

We now describe the natural contact metric structure on $T_1M$. We endow $T_1M$ with the metric induced from the Sasaki metric $g_S$ on $TM$, denoted also by $g_S$. Using the almost complex structure $J$ on $TM$, we define a unit vector field $\xi'$, a one-form $\eta'$ and a $(1,1)$-tensor field $\varphi'$ on $T_1M$ by

$$\xi' = -JN, \qquad JX = \varphi' X + \eta'(X)\,N.$$

Explicitly, this gives

$$\begin{aligned} &\xi' = u^h = u^i \left(\frac{\partial}{\partial x^i}\right)^h \\ &\eta'(X^t) = 0, && \eta'(X^h) = g(X,u), \\ &\varphi'(X^t) = -X^h + g(X,u)\,\xi', && \varphi'(X^h) = X^t. \end{aligned}$$

It is easily checked that $(T_1M, \xi', \eta', \varphi', g_S)$ is an almost contact metric manifold. However, $g_S(X, \varphi' Y) = 2\,d\eta'(X,Y)$, so $(\xi', \eta', \varphi', g_S)$ is not a contact metric structure. This defect can be rectified by taking

$$\xi = 2\xi', \qquad 2\eta = \eta', \qquad \varphi = \varphi', \qquad 4g_{cm} = g_S.$$

Note that the metric $g_{cm}$ is obtained from the one induced from the Sasaki metric on $TM$ by a homothetic change. So it is given explicitly by

$$\begin{aligned} \bar{g}_{cm|(x,u)}(X^t, Y^t) &= \frac{1}{4}\big(g(X,Y) - g(X,u)g(Y,u)\big), \\ \bar{g}_{cm|(x,u)}(X^t, Y^h) &= 0, \\ \bar{g}_{cm|(x,u)}(X^h, Y^h) &= \frac{1}{4} g(X,Y). \end{aligned}$$

We remark that $(T_1M, \xi, \eta, \varphi, g_{cm})$ is K-contact if and only if $(M,g)$ has constant curvature 1, in which case the contact metric structure on $T_1M$ is even Sasakian.

## 3. Submanifolds generated by unit vector fields

Let $V$ be a unit vector field on $(M,g)$. Then $V$ defines an embedding of $(M,g)$ into $(T_1M, g_{cm})$ as $V : M \to T_1M : x \mapsto (x, V_x)$. From the coordinate expressions (1)–(2) for horizontal and vertical lifts, it is easy to see that, given a vector field $X$ on $M$, the tangent vector field $V_*X$ to $V(M)$ can be decomposed as

$$V_*X = X^h + (\nabla_X V)^t = X^h - (A_V X)^t \tag{3}$$

where we have put $A_V = -\nabla V$. It follows from this expression that along the submanifold $V(M)$ the characteristic vector field $\xi_{(x,V_x)} = 2V^h$ is tangent to it if and only if $A_V V = 0$ or, equivalently, $V$ is a geodesic vector field. The normal space contains all vector fields of the form $Y^t + (A_V^t Y)^h$. (Note that $V^t + (A_V^t V)^h = 0$.)

We now study the behaviour of such submanifolds of $T_1M$, defined by a unit vector field on $M$, with respect to the standard contact metric structure of $T_1M$.

**Definition 1 (see [1], e.g.).** A submanifold $N$ of a contact metric space $(\tilde{M}, \tilde{\xi}, \tilde{\eta}, \tilde{\varphi}, \tilde{g})$ is *invariant* if $\tilde{\varphi}(T_pN) \subset T_pN$ for every $p \in N$; it is *anti-invariant* if $\tilde{\varphi}(T_pN) \subset (T_pN)^\perp$ for every $p \in N$.

We note that invariance implies that $\xi$ is tangent to $N$ at each of its points. This need not hold for anti-invariant submanifolds.

With the previous definition in mind, we call a unit vector field $V$ on $M$ *invariant* (respectively, *anti-invariant*) if $V(M)$ is an invariant (respectively, anti-invariant) submanifold of $T_1M$ with its standard contact metric structure.

**Proposition 1.** *A unit vector field $V$ on $(M, g)$ is invariant if and only if $(\bar{\xi} = V, \bar{\eta} = g(\cdot, V), \bar{\varphi} = A_V)$ is an almost contact structure on $M$. In particular, $V$ is a geodesic vector field on $M$ and the dimension of $M$ must be odd. If $V$ is a Killing vector field, then it is invariant if and only if $(M, \bar{\xi} = V, \bar{\eta} = g(\cdot, V), \bar{\varphi} = A_V, g)$ is a K-contact manifold.*

**Proof.** We take arbitrary vector fields $X$ and $Y$ and compute

$$\begin{aligned} &4g_{cm}(\varphi(X^h - (A_V X)^t), Y^t + (A_V^t Y)^h) \\ &= 4g_{cm}(X^t + (A_V X)^h, Y^t + (A_V^t Y)^h) \\ &= g(X, Y) - g(X, V)g(Y, V) + g(A_V{}^2 X, Y) \\ &= g(X - g(X, V)V + A_V{}^2 X, Y). \end{aligned}$$

So, the submanifold $V(M)$ is invariant if and only if $A_V{}^2 = -\mathrm{Id} + g(\cdot, V) \otimes V$ and the structure $(V, g(\cdot, V), A_V)$ is an almost contact structure. Since $\nabla_V V = -A_V V = -\bar{\varphi}\bar{\xi} = 0$, $V$ is a geodesic vector field.

Suppose next that $V$ is also Killing. Then $A_V^t = -A_V$. If we put $\bar{\eta} = g(\cdot\,, V)$ and $\bar{\varphi} = A_V$, we have

$$\begin{aligned} g(\bar{\varphi}X, \bar{\varphi}Y) &= g(A_V X, A_V Y) = g(A_V^t A_V X, Y) = -g(A_V{}^2 X, Y) \\ &= g(X,Y) - g(X,V)g(Y,V) = g(X,Y) - \bar{\eta}(X)\bar{\eta}(Y), \\ 2d\bar{\eta}(X,Y) &= (\nabla_X \bar{\eta})(Y) - (\nabla_Y \bar{\eta})(X) = g(\nabla_X V, Y) - g(\nabla_Y V, X) \\ &= -2g(X, \nabla_Y V) = 2g(X, \bar{\varphi}Y). \end{aligned}$$

So, the structure is a contact metric structure. Since $V$ is Killing, the structure is even K-contact. □

**Remark 1.** When $V$ is not a Killing vector, the metric $g$ is not a compatible metric for the almost contact structure $(\bar{\xi}, \bar{\eta}, \bar{\varphi})$. This situation can occur. Consider, e.g., the three-dimensional unimodular Lie groups $\mathcal{L}$ whose Lie bracket is given by

$$\begin{aligned} [\bar{\xi}, e_1] &= \left(\frac{\alpha}{2} - \frac{1}{2\alpha} - \beta\right) e_2, \\ [\bar{\xi}, e_2] &= \left(\frac{\alpha}{2} - \frac{1}{2\alpha} + \beta\right) e_1, \\ [e_1, e_2] &= \left(\alpha + \frac{1}{\alpha}\right) \bar{\xi} \end{aligned}$$

with respect to a left-invariant orthonormal basis $\{\bar{\xi}, e_1, e_2\}$, where $\alpha \in \mathbb{R}_0$ and $\beta \in \mathbb{R}$. An easy calculation shows

$$\nabla_{\bar{\xi}}\bar{\xi} = 0, \quad \nabla_{e_1}\bar{\xi} = -\alpha\, e_2, \quad \nabla_{e_2}\bar{\xi} = \frac{1}{\alpha}\, e_1.$$

If we define $\bar{\eta} = g(\cdot\,, \bar{\xi})$ and $\bar{\varphi} = -\nabla\bar{\xi}$, then $(\bar{\xi}, \bar{\eta}, \bar{\varphi})$ is an almost contact structure and $\bar{\xi}$ determines an invariant submanifold of $(T_1\mathcal{L}, g_{cm})$. However, for $\alpha \neq \pm 1$, $\bar{\xi}$ is not Killing and $g$ is not a compatible metric.

**Remark 2.** The possible incompatibility of the metric with the operator $A_V$ is not taken into account in the proof of the main theorem in [13]. For this reason, in the result announced "A unit geodesic vector field on the unit sphere $S^{n+1}$ determines a totally geodesic submanifold of its unit tangent bundle if and only if $n$ is even and the vector field is a Hopf vector field." only the sufficiency is proved; the necessity remains open.

**Remark 3.** When we start from a contact metric manifold $(M, \xi, \eta, \varphi, g)$ and we look at the embedding defined by the characteristic vector field $\xi$, then Proposition 1 reduces to Theorem 1 in [10].

Since an invariant submanifold of a contact metric manifold is always minimal ([1, Theorem 8.1]), we have

**Corollary 2 ([9]).** *The characteristic vector field of a K-contact manifold determines a minimal submanifold of its unit tangent bundle.*

Consider now a K-contact structure $(\bar{\xi}, \bar{\eta}, \bar{\varphi}, g)$ on a manifold $M$. By Proposition 1, $\bar{\xi}(M)$ is an invariant submanifold of the contact metric space $(T_1M, \xi, \eta, \varphi, g_{cm})$ and as such inherits a contact metric structure. Since the map $\bar{\xi}$ is a bijection, this raises the question what the precise relation is between the two contact metric structures.

**Proposition 3.** *$(\bar{\xi}(M), \xi, \eta, \varphi, g_{cm})$ is a D-homothetic transform with factor $1/2$ of $(M, \bar{\xi}, \bar{\eta}, \bar{\varphi}, g)$.*

**Proof.** The proof goes by simple computation using the expression (3) and the definition of the standard contact metric structure on $T_1M$. We have

$$\begin{aligned}
2V_*\bar{\xi} &= 2\bar{\xi}^h = \xi, \\
(V^*\eta)(X) &= \eta(V_*X) = \eta(X^h - (\bar{\varphi}X)^t) = \frac{1}{2}\, g(X, \bar{\xi}) = \frac{1}{2}\, \bar{\eta}(X), \\
(V^*\varphi)X &= \varphi(V_*X) = \varphi(X^h - (\bar{\varphi}X)^t) = X^t + (\bar{\varphi}X)^h \\
&= -(\bar{\varphi}(\bar{\varphi}X))^t + (\bar{\varphi}X)^h = V_*(\bar{\varphi}X), \\
(V^*g_{cm})(X,Y) &= g_{cm}(V_*X, V_*Y) = g_{cm}(X^h - (\bar{\varphi}X)^t, Y^h - (\bar{\varphi}Y)^t) \\
&= \frac{1}{4}(g(X,Y) + g(\bar{\varphi}X, \bar{\varphi}Y)) \\
&= \frac{1}{4}(2g(X,Y) - \bar{\eta}(X)\bar{\eta}(Y)) \\
&= \frac{1}{2}\, g(X,Y) + \frac{1}{2}(\frac{1}{2} - 1)\bar{\eta}(X)\bar{\eta}(Y).
\end{aligned}$$

This proves the proposition. □

Since D-homothetic transformations preserve the class of K-contact spaces and of Sasakian spaces, it follows

**Corollary 4.** *If $(M, \bar{\xi}, \bar{\eta}, \bar{\varphi}, g)$ is K-contact (respectively, Sasakian), then the submanifold $\bar{\xi}(M) \subset T_1M$, equipped with the contact metric structure induced by the standard one on $T_1M$, is embedded as a K-contact (respectively, Sasakian) submanifold of the contact metric space $T_1M$.*

Also anti-invariant unit vector fields can be characterized in an easy geometric way.

**Proposition 5.** *A unit vector field $V$ on $(M, g)$ is anti-invariant if and only if $V$ is a geodesic vector field and $V^\perp$ is integrable.*

**Proof.** Take arbitrary vector fields $X$ and $Y$ on $M$. We compute

$$\begin{aligned}&4g_{cm}(\varphi(X^h - (A_V X)^t), Y^h - (A_V Y)^t)\\&\quad = 4g_{cm}(X^t + (A_V X)^h, Y^h - (A_V Y)^t)\\&\quad = g(A_V X, Y) - g(A_V Y, X).\end{aligned}$$

We see that anti-invariance is equivalent to the tensor $A_V$ being symmetric. We now give geometric meaning to this symmetry.

So, suppose that $A_V$ is symmetric. Since $V$ is unit, $A_V X$ is orthogonal to $V$ for all vector fields $X$. But by symmetry, we also have, for $X$ orthogonal to $V$,

$$g(A_V V, X) = g(A_V X, V) = 0$$

and $A_V V = -\nabla_V V = 0$. So, $V$ is indeed a geodesic vector field. (In particular, $\xi$ is tangent to the anti-invariant submanifold $V(M)$ in $T_1 M$.)

Next, take two vector fields $X$ and $Y$ orthogonal to $V$. Then

$$\begin{aligned}g([X, Y], V) &= g(\nabla_X Y, V) - g(\nabla_Y X, V) = -g(Y, \nabla_X V) + g(X, \nabla_Y V)\\&= g(Y, A_V X) - g(X, A_V Y) = 0.\end{aligned}$$

We conclude that $V^\perp$ is integrable.

Conversely, if $V$ is geodesic and $V^\perp$ is integrable, the tensor $A_V$ is basically the second fundamental form for the integral submanifolds of $V^\perp$ and as such is symmetric. Then $V(M)$ is an anti-invariant submanifold of $T_1 M$. □

Using the above proposition, one can now give a lot of examples of anti-invariant submanifolds in contact metric spaces, by considering those generated by appropriate unit vector fields. In contrast to invariant submanifolds, those need not be minimal (but they can be). We mention

- any geodesic unit vector field on a two-dimensional manifold;
- the characteristic vector field $\xi$ of a Kenmotsu manifold: this is an almost contact metric manifold $(M, \xi, \eta, \varphi, g)$ satisfying $(\nabla_X \varphi)Y = g(\varphi X, Y) - \eta(Y)\varphi X$. From this condition, it follows $\nabla_X \xi = X - \eta(X)\xi$ and symmetry of $A_\xi = -\nabla\xi$ follows. Actually, this is (locally) a special case (see [1]) of
- the vector field $\partial/\partial t$ of a warped product manifold $\mathbb{R} \times_f M$, where $t$ is the local coordinate on $\mathbb{R}$;

– radial vector fields on arbitrary Riemannian manifolds: these are the gradient vector fields for the distance function $r(x) = d(p, x)$ with respect to some fixed point $p$ and defined on some pointed neighbourhood around $p$. It was shown in [4] that the corresponding submanifolds in the unit tangent bundle are minimal when the manifold is two-point homogeneous. (For more results in this spirit, see also [5], [9] and the references therein.)

## References

1. D. E. Blair, *Riemannian geometry of contact and symplectic manifolds,* Progress in Mathematics 203, Birkhäuser, 2001.
2. E. Boeckx en L. Vanhecke, Characteristic reflections on unit tangent sphere bundles, *Houston J. Math.* 23 (1997), 427–448.
3. E. Boeckx and L. Vanhecke, Harmonic and minimal vector fields on tangent and unit tangent bundles, *Differential Geom. Appl.* 13 (2000), 77–93.
4. E. Boeckx and L. Vanhecke, Harmonic and minimal radial vector fields, *Acta Math. Hungar.* 90 (2001), 317–331.
5. E. Boeckx and L. Vanhecke, Isoparametric functions and harmonic and minimal unit vector fields, in: Global Differential Geometry: The Mathematical Legacy of Alfred Gray (eds. M. Fernández, J. A. Wolf), *Contemp. Math. 288, Amer. Math. Soc.* (2001), 20–31.
6. O. Gil-Medrano, Relationship between volume and energy of unit vector fields, *Differential Geom. Appl.* 15 (2001), 137–152.
7. O. Gil-Medrano and E. Llinares-Fuster, Minimal unit vector fields, *Tôhoku Math. J.* 54 (2002), 71–84.
8. H. Gluck and W. Ziller, On the volume of a unit vector field on the three-sphere, *Comment. Math. Helvetici* 61 (1986), 177–192.
9. J. C. González-Dávila and L. Vanhecke, Examples of minimal unit vector fields, *Ann. Global Anal. Geom.* 18 (2000), 385–404.
10. Ph. Rukimbira, Criticality of unit contact vector fields, in: *Infinite dimensional Lie groups in geometry and representation theory* (ed. A. Banyaga et al.), *World Scientific* (2002), 105-115.
11. A. Yampolsky, On the mean curvature of a unit vector field, *Publ. Math. Debrecen* 60 (2002), 131–155.
12. A. Yampolsky, On the intrinsic geometry of a unit vector field, *Comment. Math. Univ. Carolinæ* 43 (2002), 299–317.
13. A. Yampolsky, A totally geodesic property of Hopf vector fields, *Acta Math. Hungar.* 101 (2003), 93–112.

# A NEW STRUCTURE ON UNIT TANGENT SPHERE BUNDLES

JONG TAEK CHO

*Department of Mathematics, Chonnam National University,*
*CNU The Institute of Basic Sciences,*
*Kwangju 500-757, Korea*
*E-mail jtcho@chonnam.ac.kr*

SUN HYANG CHUN

*Department of Mathematics,*
*Graduate School Chonnam National University,*
*Kwangju 500-757, Korea*
*E-mail: cshyang@chonnam.ac.kr*

Recently, V. Oproiu [6,7] introduced a new structure on the tangent bundle, which is defined by a function depending on the energy density on $TM$. In this article, we study particularly on the induced structure of the unit tangent sphere bundle $T_1M$ as a hypersurface of $TM$ with the new structure. We first show that $T_1M$ admits a contact metric structure. Moreover, we prove that it is K-contact if and only if $M$ is of constant curvature, further in which case the structure is Sasakian.

*Keywords*: tangent bundles, unit tangent sphere bundles, K-contact structures, Sasakian structures

## 1. Introduction

It is well-known that the tangent bundle $TM$ of a Riemannian manifold $M$ admits a naturally defined Riemannian structure $g_s$, which is called the Sasaki metric, and an almost complex structure $J$ which is compatible with this metric (cf. [8,9]). O. Kowalski [4] proved that the Sasaki metric of a tangent bundle is locally symmetric if and only if the base metric is locally Euclidean. Furthermore, E. Musso and F. Tricerri [5] showed that it

2000 Mathematics Subject Classification: 53C15, 53C25, 53D10

has never a constant scalar curvature unless the base metric is flat. These facts say that the Sasaki metric on a tangent bundle is too rigid. In this context, J. Cheeger and D. Gromoll [2] defined the so-called the Cheeger-Gromoll metric $g_{cg}$. However, for unit tangent sphere bundles, this metric yields nothing new, since $(T_1M, g_{cg})$ is isometric to $(T_{1/\sqrt{2}}M, g_s)$. Recently, V. Oproiu [6,7] introduced a new Riemannian metric on the tangent bundle, which is defined by a function depending on the energy density on $TM$. He gave some Käher Einstein structures on the tangent bundle of a space of constant curvature. In [6] he showed some special cases of the Käher Einstein structures on $TM$ which are in addition locally symmetric or have constant holomorphic sectional curvature.

In this article, we study on the structure of the unit tangent sphere bundle as a hypersurface of the tangent bundle with the Oproiu's structure. We first show that the unit tangent sphere bundle admits a contact Riemannian structure. We prove also that the unit tangent sphere bundle with its induced metric is K-contact if and only if the base manifold has constant curvature, further in which case the structure on $T_1M$ is Sasakian.

All manifolds in the present paper are assumed to be connected and of class $C^\infty$.

## 2. Oproiu's structure on tangent bundles

Let $M = (M, G)$ be a $(n+1)$-dimensional Riemannian manifold, $D$ the associated Levi-Civita connection and $K$ the Riemannian curvature tensor. The tangent bundle of $(M, G)$ is denoted by $TM$ and consists of pairs $(x, u)$ where $x$ is a point in $M$ and $u$ a tangent vector to $M$ at $x$. The mapping $\pi : TM \to M$, $\pi(x, u) = x$ is the natural projection from $TM$ onto $M$. We define $\mathfrak{K} : TTM \to TM$, is called a connection map, by $\mathfrak{K}(X_* u) = D_u X$, where $X : M \to TM$ is a vector field on $M$. Then the tangent space to $TM$ at $(x, u)$ splits into the direct sum of the vertical subspace $VTM_{(x,u)} = \ker \pi_{*|(x,u)}$ and the horizontal subspace $HTM_{(x,u)} = \ker \mathfrak{K}_{|(x,u)}$ :

$$T_{(x,u)}TM = VTM_{(x,u)} \oplus HTM_{(x,u)}.$$

For a vector $X \in T_xM$, there exists a unique vector $X^H \in HTM_{(x,u)}$, which is called the horizontal lift of $X$ to $(x, u)$, such that $\pi_*(X^H) = X$. There is also a unique vector $X^V \in VTM_{(x,u)}$, which is called the vertical lift of $X$ to $(x, u)$, such that $\mathfrak{K}(X^V) = X$. Then every tangent vector $\bar{Z} \in T_{(x,u)}TM$ can be decomposed as $\bar{Z} = X^H + Y^V$ for uniquely determined vectors $X, Y \in T_xM$.

We now take a local coordinate system $(x^1, \cdots, x^{n+1})$ on an open subset $U$ of $M$. On $\pi^{-1}(U)$ we define coordinates $(\bar{x}^1, \ldots, \bar{x}^{n+1}; u^1, \ldots, u^{n+1})$ as follows :

$$\bar{x}^i(x,u) = (x^i \circ \pi)(x,u) = x^i(x),$$
$$u^i(x,u) = dx^i(u) = ux^i$$

for $i = 1, \ldots, n$ and $(x,u) \in \pi^{-1}(U)$. With respect to this coordinate system, the horizontal lift and vertical lift of a vector field $X = \sum X^i \frac{\partial}{\partial x^i}$ on $U$ are given by

$$X^H = \sum (X^i \circ \pi) \frac{\partial}{\partial \bar{x}^i} - \sum u^b((X^a \Gamma^i_{ab}) \circ \pi) \frac{\partial}{\partial u^i}, \tag{2.1}$$
$$X^V = \sum (X^i \circ \pi) \frac{\partial}{\partial u^i},$$

where $\Gamma^i_{jk}$ are the Christoffel symbols associated with the Riemannian metric $G$.

From (2.1), we easily calculate the brackets for horizontal and vertical lifts:

$$[X^H, Y^H] = [X,Y]^H - \{K(X,Y)u\}^V, \tag{2.2}$$
$$[X^H, Y^V] = (D_X Y)^V,$$
$$[X^V, Y^V] = 0$$

for all vector fields $X, Y$ on $M$.

If $f$ is a smooth function on $M$ and $X$ is a vector field on $M$, then

$$X^H(f \circ \pi) = ((\pi_* X^H) f) \circ \pi = (Xf) \circ \pi, \tag{2.3}$$
$$X^V(f \circ \pi) = ((\pi_* X^V) f) \circ \pi = 0.$$

In particular, we write $X = \sum X^i \frac{\partial}{\partial x^i}$ on $U$, and then we have

$$X^H(\bar{x}^i) = (Xx^i) \circ \pi = X^i \circ \pi, \tag{2.4}$$
$$X^V(\bar{x}^i) = 0.$$

Further, from (2.1) we have

$$X^H(u^i) = -\sum u^b (X^a \Gamma^i_{ab}) \circ \pi, \tag{2.5}$$
$$X^V(u^i) = X^i \circ \pi.$$

Finally, for a vector field $Y$ on $M$, we define the function

$$G(Y,u) : TM \to \mathbb{R}, \ (x,u) \mapsto G_x(Y_x, u).$$

In what follows, we will frequently need $X^H(G(Y,u))$ and $X^V(G(Y,u))$. These are given by the following relations:

$$
\begin{aligned}
(2.6)\quad X^H(G(Y,u)) &= \sum X^H(u^i G(Y, \frac{\partial}{\partial x^i}) \circ \pi) \\
&= \sum (X^H u^i) G(Y, \frac{\partial}{\partial x^i}) \circ \pi + \sum u^i (XG(Y, \frac{\partial}{\partial x^i})) \circ \pi \\
&= G(\nabla_X Y, u), \\
X^V(G(Y,u)) &= \sum (X^V u^i) G(Y, \frac{\partial}{\partial x^i}) \circ \pi \\
&= G(Y,X) \circ \pi.
\end{aligned}
$$

It is well-known that the Sasaki metric of tangent bundles is a naturally defined metric. But it is too rigid. For example, O. Kowalski [4] has shown that it is never locally symmetric unless the base metric is locally Euclidean. E. Musso and F. Tricerri [5] have generalized this fact. They have shown that it has a constant scalar curvature if and only if the base metric is flat. In this context, J. Cheeger and D. Gromoll [2] defined the so-called the Cheeger-Gromoll metric $g_{cg}$ on the tangent bundle $TM$ such that

$$
\begin{aligned}
g_{cg\,(x,u)}(X^H, Y^H) &= G_x(X,Y), \\
g_{cg\,(x,u)}(X^V, Y^V) &= \frac{1}{h}\left(G_x(X,Y) + G_x(X,u)G_x(Y,u)\right), \\
g_{cg\,(x,u)}(X^H, Y^V) &= g_{cg\,(x,u)}(X^V, Y^H) = 0
\end{aligned}
$$

for all vector fields $X, Y$ on $M$, where $h(x,u) = 1 + G_x(u,u)$. However, for unit tangent sphere bundles, this metric yields nothing new, since $(T_1M, g_{cg})$ is isometric to $(T_{1/\sqrt{2}}M, g_s)$, where $g_{cg}$ is the Cheeger-Gromoll metric and $g_S$ is the Sasaki metric, respectively on $TM$. This isometry is given by $\psi : T_1M \longrightarrow T_{1/\sqrt{2}}M;\ (x,u) \longmapsto (x, u/\sqrt{2})$.

Now, we give a description of a new Riemannian metric on the tangent bundle $TM$ which is introduced by V. Oproiu [7].

Consider the energy density

$$t(x,u) = \frac{1}{2}G_x(u,u) = \sum \frac{1}{2}G_{ij}(x)u^i u^j$$

defined on $TM$ by the Riemannian metric $G$ on $M$. And let $\alpha : [0,\infty) \to \mathbb{R}$ be positive real-valued smooth functions. We shall find some interesting properties of $TM$ by using a Lagrangian function defined as the integral of a smooth function depending on energy density only, i.e.,

$$L(x,u) = \int \alpha(t)dt.$$

We may consider the symmetric $(0,2)$-tensor field on $TM$, defined by the components,

$$\tilde{G}_{ij} = \frac{\partial^2 L}{\partial u^i \partial u^j} = \alpha(t) G_{ij} + \alpha'(t) \sum G_{hi} u^h G_{kj} u^k. \tag{2.7}$$

We assume that $\alpha'(t) \geq 0$ for $t \in [0, \infty)$. Then we see that the matrix $(\tilde{G}_{ij})$ is symmetric and positive definite, and further we see that it has the inverse with entries

$$H^{kl} = \frac{1}{\alpha} G^{kl} + w u^k u^l, \tag{2.8}$$

where $G^{kl}$ are the components of the inverse of the matrix $(G_{ij})$ and

$$w = w(t) = -\frac{\alpha'}{\alpha(\alpha + 2t\alpha')}.$$

The components $H^{kl}$ define a symmetric $(0,2)$-tensor field on $TM$. Also, we obtain the components $H_{ij}$ of a symmetric $(0,2)$-tensor field from the components $H^{kl}$ as follows:

$$H_{ij} = \sum G_{ik} H^{kl} G_{lj} = \frac{1}{\alpha} G_{ij} + w \sum G_{hi} u^h G_{kj} u^k. \tag{2.9}$$

We define the following Riemannian metric on $TM$ by using two tensor field $\tilde{G}_{ij}$ and $H_{ij}$ on $TM$:

$$\tilde{g} = \sum \tilde{G}_{ij} dx^i dx^j + \sum H_{ij} (du^i + \Gamma^i_{jk} u^k dx^j)(du^j + \Gamma^j_{ik} u^k dx^i). \tag{2.10}$$

Namely, we have the Riemannian metric on $TM$ for vertical and horizontal lifts:

$$\begin{aligned} \tilde{g}_{(x,u)}(X^H, Y^H) &= \alpha G_x(X, Y) + \alpha' G_x(X, u) G_x(Y, u), \\ \tilde{g}_{(x,u)}(X^V, Y^V) &= \frac{1}{\alpha} G_x(X, Y) - \frac{\alpha'}{\alpha(\alpha + 2t\alpha')} G_x(X, u) G_x(Y, u), \\ \tilde{g}_{(x,u)}(X^H, Y^V) &= \tilde{g}_{(x,u)}(X^V, Y^H) = 0. \end{aligned} \tag{2.11}$$

From (2.7)$\sim$(2.9), we obtain the following $(0,2)$-tensor fields on $TM$:

$$\begin{aligned} \tilde{G}^{kl} &= \sum G^{ki} \tilde{G}_{ij} G^{jl} = \alpha G^{kl} + \alpha' u^k u^l, \\ \tilde{G}^i_k &= \sum \tilde{G}^{ih} G_{hk} = \sum \tilde{G}_{kh} G^{hi} = \alpha \delta^i_k + \alpha' \sum u^i G_{hk} u^h, \\ H^i_k &= \sum H^{ih} G_{hk} = \sum H_{kh} G^{hi} = \frac{1}{\alpha} \delta^i_k + w \sum u^i G_{hk} u^h. \end{aligned} \tag{2.12}$$

Using (2.12), an almost complex structure $J$ on $TM$ may be defined by:

$$J\left(\frac{\partial}{\partial x^i}\right)^H = \sum \tilde{G}^k_i \left(\frac{\partial}{\partial x^k}\right)^V, \quad J\left(\frac{\partial}{\partial x^i}\right)^V = \sum -H^k_i \left(\frac{\partial}{\partial x^k}\right)^H. \tag{2.13}$$

That is, we define an almost complex structure $J$ on $TM$ for all horizontal and vertical lifts as the following:

$$(2.14)\qquad \begin{aligned} J_{(x,u)}(X^H) &= \alpha X^V + \alpha' G_x(X,u)u^V, \\ J_{(x,u)}(X^V) &= -\frac{1}{\alpha}X^H + \frac{\alpha'}{\alpha(\alpha+2t\alpha')}G_x(X,u)u^H. \end{aligned}$$

Then we can show that $(TM, \tilde{g}, J)$ is an almost Kählerian manifold. The associated 2-form $\Omega$ is given by

$$\Omega = \sum G_{ij}(du^i + \Gamma^i_{jk}u^k dx^j) \wedge dx^l,$$

that is,

$$\Omega_{(x,u)}(X^H, Y^H) = \Omega_{(x,u)}(X^V, Y^V) = 0, \quad \Omega_{(x,u)}(X^V, Y^H) = G_x(X,Y)$$

for all vector fields $X$ and $Y$ on $M$.

**Theorem 2.1.** ([7]) *$(TM, \tilde{g}, J)$ is a Kählerian manifold if and only if $(M, G)$ has positive constant sectional curvature $c$ and the function $\alpha(t)$ satisfies the ordinary differential equation*

$$2t(\alpha')^2 = c.$$

**Remark.** The tangent bundle $(TM, g_S)$ with Sasaki metric is a Kählerian manifold if and only if $(M, G)$ is locally flat (cf. [3]).

In view of (2.11) and (2.14), we now define a slightly modified(or generalized) Riemannian metric $\tilde{g}$ and an almost complex structure $J$ at $(x, u) \in TM$ as follows :

$$(2.15)\qquad \begin{aligned} \tilde{g}_{(x,u)}(X^H, Y^H) &= \alpha G_x(X,Y) + \beta G_x(X,u)G_x(Y,u), \\ \tilde{g}_{(x,u)}(X^V, Y^V) &= \frac{1}{\alpha}G_x(X,Y) - \frac{\beta}{\alpha(\alpha+2t\beta)}G_x(X,u)G_x(Y,u), \\ \tilde{g}_{(x,u)}(X^H, Y^V) &= \tilde{g}_{(x,u)}(X^V, Y^H) = 0, \end{aligned}$$

$$(2.16)\qquad \begin{aligned} J_{(x,u)}(X^H) &= \alpha X^V + \beta G_x(X,u)u^V, \\ J_{(x,u)}(X^V) &= -\frac{1}{\alpha}X^H + \frac{\beta}{\alpha(\alpha+2t\beta)}G_x(X,u)u^H, \end{aligned}$$

where $\alpha : [0,\infty) \to \mathbb{R}$ be a positive smooth function and $\beta : [0,\infty) \to \mathbb{R}$ be a non-negative smooth function. Then we easily see that the structure defined by (2.15) and (2.16) includes the Sasaki structure as a special case that $\alpha \equiv 1$ and $\beta \equiv 0$. Furthermore, we can also show that $(\tilde{g}, J)$ is an almost Kählerian structure. It was already shown ([6]) that $(TM, \tilde{g}, J)$ is a Kählerian manifold if and only if $(M, G)$ has positive constant sectional curvature $c$ and the function $\beta(t)$ is given by

$$\beta = \frac{c - \alpha\alpha'}{2t\alpha' - \alpha}.$$

## 3. A new structure on unit tangent sphere bundles

First, we give a brief review of several fundamental concepts and formulas on almost contact structure. An odd-dimensional smooth manifold $M^{2n+1}$ has an *almost contact structure* if it admits a vector $\xi$, a 1-form $\eta$ and a $(1,1)$-tensor field $\phi$ satisfying

$$\eta(\xi) = 1 \text{ and } \phi^2 X = -X + \eta(X)\xi.$$

Then there exists a compatible Riemannian metric $g$:

$$g(\phi X, \phi Y) = g(X, Y) - \eta(X)\eta(Y)$$

for all vector fields $X$ and $Y$ on $M$. We call $(\eta, \xi, \phi, g)$ an almost contact metric structure of $M$ and $M = (M; \eta, \xi, \phi, g)$ an almost contact metric manifold. For an almost contact metric manifold $M$ we define its fundamental 2-form $\Phi$ by $\Phi(X, Y) = g(\phi X, Y)$. If

$$\Phi = d\eta,$$

$M$ is called a contact Riemannian manifold. A contact Riemannian manifold for which $\xi$ is a Killing vector field, is called a K-contact manifold. It is easy to see that a contact Riemannian manifold is $K$-contact if and only if $h = 0$. For a contact Riemannian manifold $M$, one may define naturally an almost complex structure $\mathfrak{J}$ on $M \times \mathbb{R}$ by

$$\mathfrak{J}(X, f\frac{d}{dt}) = (\phi X - f\xi, \eta(X)\frac{d}{dt}),$$

where $X$ is a vector field tangent to $M$, $t$ the coordinate of $\mathbb{R}$ and $f$ a function on $M \times \mathbb{R}$. If the almost complex structure $\mathfrak{J}$ is integrable, $M$ is said to be normal or Sasakian. It is known that $M$ is normal if and only if $M$ satisfies

$$[\phi, \phi] + 2d\eta \otimes \xi = 0,$$

where $[\phi, \phi]$ is the Nijenhuis torsion of $\phi$. For more details about almost contact metric structures, we refer to [1].

We consider $T_1M$, the unit tangent sphere bundle, consisting of the unit tangent vectors to $(M, G)$, which is given by the equation $G_x(u, u) = 1$.

Then we see that $t(x,u) = \frac{1}{2}$, and hence $\alpha$ and $\beta$ do not depend on $(x,u)$. A unit normal vector field $N$ on $T_1M$ is the vector field $\sqrt{\alpha+\beta}u^V$. If $X \in T_xM$, we define the tangential lift of $X$ to $(x,u) \in T_1M$ by

$$X^T_{(x,u)} = X^V_{(x,u)} - \frac{1}{\sqrt{\alpha+\beta}} G(X,u) N_{(x,u)}. \tag{3.1}$$

From (2.3), (2.6) and (3.1), we obtain that

$$X^T(f \circ \pi) = 0, \tag{3.2}$$

$$X^T(G(Y,u)) = G(X,Y) - G(X,u)G(Y,u) \tag{3.3}$$

for any smooth function $f$ on $M$ and any vector fields $X, Y$ on $M$. Also from (2.2), (2.6) and (3.1), we obtain the brackets of horizontal and tangential lifts:

$$\begin{aligned} [X^H, Y^H] &= [X,Y]^H - \{K(X,Y)u\}^V, \\ [X^H, Y^T] &= (D_X Y)^T, \\ [X^T, Y^T] &= G(X,u)Y^T - G(Y,u)X^T. \end{aligned} \tag{3.4}$$

Next, from (2.16) and (3.1) we define the Riemannian metric $g'$ on $T_1M$ as follows:

$$\begin{aligned} g'_{(x,u)}(X^H, Y^H) &= \alpha G_x(X,Y) + \beta G_x(X,u) G_x(Y,u), \\ g'_{(x,u)}(X^T, Y^T) &= \frac{1}{\alpha}\left(G_x(X,Y) - G_x(X,u)G_x(Y,u)\right), \\ g'_{(x,u)}(X^T, Y^H) &= g'_{(x,u)}(X^H, Y^T) = 0. \end{aligned} \tag{3.5}$$

Using the almost complex structure $J$ on $TM$, we define a unit vector field $\xi'$, a 1-form $\eta'$ and a $(1,1)$-tensor field $\phi'$ on $T_1M$ by

$$\xi' = -JN, \quad \phi' = J - \eta' \otimes N. \tag{3.6}$$

From this, $\xi'$, $\eta'$ and $\phi'$ are described by

$$\begin{aligned} &\xi' = \frac{1}{\sqrt{\alpha+\beta}} u^H, \\ &\eta'(X^T) = 0, \qquad \eta'(X^H) = \sqrt{\alpha+\beta}\, G(X,u), \\ &\phi'(X^T) = -\frac{1}{\alpha}\left(X^H - \sqrt{\alpha+\beta}\, G(X,u)\xi'\right), \\ &\phi'(X^H) = \alpha X^T. \end{aligned} \tag{3.7}$$

**Proposition 3.1.** *$(T_1M; \xi', \eta', \phi', g')$ is an almost contact Riemannian manifold.*

**Proof.** We shall show that $(\xi', \eta', \phi', \bar{g})$ satisfies

$$\eta'(\xi') = 1, \quad \phi'^2 = -I + \eta' \otimes \xi', \quad g'(\phi'\bar{X}, \phi'\bar{Y}) = g'(\bar{X}, \bar{Y}) - \eta'(\bar{X})\eta'(\bar{Y})$$

for all vector fields $\bar{X}$ and $\bar{Y}$ on $T_1M$. From (3.7), we easily see that

$$\eta'(\xi') = 1, \qquad \phi'(\xi') = 0$$

and

$$\begin{aligned} \phi'^2(X^T) &= -\frac{1}{\alpha}\phi'(X^H) \\ &= -X^T \\ &= -X^T + \eta'(X^T)\xi', \end{aligned} \tag{3.8}$$

$$\begin{aligned} \phi'^2(X^H) &= \alpha\phi'(X^T) \\ &= -X^H + \sqrt{\alpha+\beta}G(X,u)\xi' \\ &= -X^H + \eta'(X^H)\xi'. \end{aligned} \tag{3.9}$$

Together with (3.8) and (3.9), we see that $\phi'^2 = -I + \eta' \otimes \xi'$. From (3.5) and (3.7), it follows that

$$\begin{aligned} & g'(\phi' X^T, \phi' Y^T) \\ &= \frac{1}{\alpha^2} g'(X^H, Y^H) - \frac{\alpha+\beta}{\alpha^2} G(X,u)G(Y,u) \\ &= \frac{1}{\alpha}\left(G(X,Y) - G(X,u)G(Y,u)\right) \\ &= g'(X^T, Y^T) - \eta'(X^T)\eta'(Y^T), \end{aligned} \tag{3.10}$$

$$\begin{aligned} & g'(\phi' X^H, \phi' Y^H) \\ &= \alpha^2 g'(X^T, Y^T) \\ &= \alpha\left(G(X,Y) - G(X,u)G(Y,u)\right) \\ &= \alpha G(X,Y) + \beta G(X,u)G(Y,u) - (\alpha+\beta)G(X,u)G(Y,u) \\ &= g'(X^H, Y^H) - \eta'(X^H)\eta'(Y^H). \end{aligned} \tag{3.11}$$

From (3.10) and (3.11), we see that $g'(\phi'\bar{X}, \phi'\bar{Y}) = g'(\bar{X}, \bar{Y}) - \eta'(\bar{X})\eta'(\bar{Y})$ for all vector fields $\bar{X}$ and $\bar{Y}$ on $T_1M$. Therefore $(T_1M; \xi', \eta', \phi', g')$ is an almost contact Riemannian manifold. □

Also, by using (3.3)∼(3.5) and (3.7) we have

$$\begin{aligned} g'(X^T, \phi X^H) &= \alpha g'(X^T, Y^T) \\ &= G(X,Y) - G(X,u)G(Y,u), \end{aligned} \tag{3.12}$$

$$(3.13)\qquad \begin{aligned} d\eta'(X^T,Y^H) &= \frac{1}{2}\{X^T\eta'(Y^H)-Y^H\eta'(X^T)-\eta'([X^T,Y^H])\}\\ &= \frac{1}{2}\sqrt{\alpha+\beta}X^T(G(X,u))\\ &= \frac{1}{2}\sqrt{\alpha+\beta}(G(X,Y)-G(X,u)G(Y,u)). \end{aligned}$$

Then from (3.12) and (3.13), we find $g'(\bar{X},\phi'\bar{Y})=\frac{2}{\sqrt{\alpha+\beta}}d\eta'(\bar{X},\bar{Y})$ for all vector fields $\bar{X}$ and $\bar{Y}$ on $T_1M$. So, we take $\xi$, $\eta$, $\phi$ and $g$ as follows:

$$\xi=\frac{2}{\sqrt{\alpha+\beta}}\xi',\quad \frac{2}{\sqrt{\alpha+\beta}}\eta=\eta',\quad \phi=\phi',\quad g=\frac{\alpha+\beta}{4}g'.$$

Then we get the contact Riemannian structure $(\phi,\xi,\eta,g)$ on $T_1M$.

Since $\xi=\frac{2}{\alpha+\beta}\,u^H$, by using (3.4) we have

$$(3.14)\qquad \begin{aligned} L_\xi X^H &= \frac{2}{\sqrt{\alpha+\beta}}\Big([u,X]^H+(K_uX)^V+\sum u^bX^a\Gamma^i_{ab}\Big(\frac{\partial}{\partial x^i}\Big)^H\Big),\\ L_\xi X^T &= \frac{2}{\sqrt{\alpha+\beta}}\Big((D_uX)^T-X^H+G(X,u)u^H\Big). \end{aligned}$$

**Theorem 3.2.** *The contact Riemannian structure on $T_1M$ is K-contact if and only if $(M,G)$ has constant curvature $\alpha^2$, in which case the structure on $T_1M$ is Sasakian.*

**Proof.** Suppose that $T_1M$ is K-contact manifold. Then for Riemannian metric $g$ and the characteristic vector field $\xi$, $T_1M$ satisfies $L_\xi g=0$.

For all horizontal and tangential lifts, from (3.14), we obtain

(3.15)

$$\begin{aligned} &(L_\xi g)(X^H,Y^H)\\ &=\xi(g(X^H,Y^H))-g(L_\xi X^H,Y^H)-g(X^H,L_\xi Y^H)\\ &=\frac{1}{2}\Big\{u^H(g'(X^H,Y^H))-g'\Big([u,X]^H+\sum u^bX^a\Gamma^i_{ab}\Big(\frac{\partial}{\partial x^i}\Big)^H,Y^H\Big)\\ &\quad -g'\Big(X^H,[u,Y]^H+\sum u^bY^a\Gamma^i_{ab}\Big(\frac{\partial}{\partial x^i}\Big)^H\Big)\Big\}\\ &=\frac{1}{2}\Big\{\alpha u(G(X,Y))+\beta G(D_uX,u)G(Y,u)+\beta G(X,u)G(D_uY,u)\\ &\quad -(\alpha G([u,X],Y)+\beta G([u,X],u)G(Y,u))\\ &\quad -\Big(\alpha G\Big(\sum u^bX^a\Gamma^i_{ab}\frac{\partial}{\partial x^i},Y\Big)+\beta G\Big(\sum u^bX^a\Gamma^i_{ab}\frac{\partial}{\partial x^i},u\Big)G(Y,u)\Big)\\ &\quad -(\alpha G(X,[u,Y])+\beta G(X,u)G([u,Y],u)) \end{aligned}$$

$$- \Big(\alpha G\Big(X, \sum u^b Y^a \Gamma^i_{ab} \frac{\partial}{\partial x^i}\Big) + \beta G(X,u) G\Big(\sum u^b Y^a \Gamma^i_{ab} \frac{\partial}{\partial x^i}, u\Big)\Big)\Big\}$$
$$= 0,$$

(3.16)

$$\begin{aligned}
&(L_\xi g)(X^H, Y^T) \\
&= \xi(g(X^H, Y^T)) - g(L_\xi X^H, Y^T) - g(X^H, L_\xi Y^T) \\
&= -\frac{1}{2} g'((K_u X)^V, Y^T) + \frac{1}{2} g'\Big(X^H, Y^H - \frac{1}{2t} G(Y,u) u^H\Big) \\
&= -\frac{1}{2}\Big\{\frac{1}{\alpha} G(K_u X, Y) - \alpha G(X,Y) + \alpha G(X,u) G(Y,u)\Big\},
\end{aligned}$$

(3.17)

$$\begin{aligned}
&(L_\xi g)(X^T, Y^T) \\
&= \xi(g(X^T, Y^T)) - g(L_\xi X^T, Y^T) - g(X^T, L_\xi Y^T) \\
&= \frac{1}{2} u^H (g'(X^T, Y^T)) - \frac{1}{2} g'((D_u X)^T, Y^T) - \frac{1}{2} g'(X^T, (D_u Y)^T) \\
&= \frac{1}{2\alpha}\{u(G(X,Y)) - G(D_u X, u) G(Y,u) - G(X,u) G(D_u Y, u) \\
&\quad - G(D_u X, Y) + G(D_u X, u) G(Y,u) \\
&\quad - G(X, D_u Y) + G(X,u) G(D_u Y, u)\} \\
&= 0.
\end{aligned}$$

In (3.16), we take $X$ orthogonal to $u$. Then we have

$$G\Big(\frac{1}{\alpha} K_u X - \alpha X, Y\Big) = 0 \tag{3.18}$$

for all vector field $Y$ on $M$. We suppose that $X = Y$ and $G(X,X) = 1$. Then $G(K_u X, X) = \alpha^2$ for all vector field $X \perp u$ on $M$. Therefore, the base manifold $(M, G)$ is a space of constant curvature $\alpha^2$. Conversely, if $M$ has constant curvature $\alpha^2$, we immediately see that $T_1 M$ has K-contact structure.

On the other hand, let the base manifold $M$ be a space of constant curvature $\alpha^2$. Then the Riemannian curvature tensor $K$ is given by

$$K(X,Y)Z = \alpha^2 (G(Y,Z)X - G(X,Z)Y) \tag{3.19}$$

for all vector fields $X$, $Y$ and $Z$ on $M$. As already mentioned, a Sasakian structure is determined by

$$N_\phi + 2d\eta \otimes \xi = 0, \tag{3.20}$$

where $N_\phi$ denotes the Nijenhuis torsion of $\phi$, that is,

$$N_\phi(X,Y) = \phi^2[X,Y] + [\phi X, \phi Y] - \phi[\phi X, Y] - \phi[X, \phi Y].$$

Then for all horizontal lifts $X^H$ and tangential lifts $X^T$, from (3.19) and (3.20), we obtain

(3.21)

$$\begin{aligned}
&N_\phi(X^H, Y^H) + 2d\eta(X^H, Y^H)\xi \\
&= -[X^H, Y^H] + [\phi X^H, \phi Y^H] - \phi[\phi X^H, Y^H] - \phi[X^H, \phi Y^H] \\
&\quad + X^H(\eta(Y^H))\xi - Y^H(\eta(X^H))\xi \\
&= -[X^H, Y^H] + \alpha^2[X^T, Y^T] - \alpha\phi[X^T, Y^H] - \alpha\phi[X^H, Y^T] \\
&\quad + \frac{\alpha+\beta}{2} X^H(G(Y,u))\xi - \frac{\alpha+\beta}{2} Y^H(G(X,u))\xi \\
&= -[X,Y]^H + \{K(X,Y)u\}^V + \alpha^2(G(X,u)Y^T - G(Y,u)X^T) \\
&\quad + \alpha\phi(D_Y X)^T - \alpha\phi(D_X Y)^T + \frac{\alpha+\beta}{2} G(D_X Y, u)\xi - \frac{\alpha+\beta}{2} G(D_Y X, u)\xi \\
&= \alpha^2(G(X,u)Y^V - G(Y,u)X^V) + \alpha^2(G(X,u)Y^T - G(Y,u)X^T) \\
&= 0,
\end{aligned}$$

(3.22)

$$\begin{aligned}
&N_\phi(X^H, Y^T) + 2d\eta(X^H, Y^T)\xi \\
&= -[X^H, Y^T] + [\phi X^H, \phi Y^T] - \phi[\phi X^H, Y^T] - \phi[X^H, \phi Y^T] - Y^T(\eta(X^H))\xi \\
&= -[X^H, Y^T] - [X^T, Y^H] + X^T(G(Y,u))u^H + G(Y,u)[X^T, Y^H] \\
&\quad - \alpha\phi[X^T, Y^T] + \frac{1}{\alpha}\phi[X^H, Y^H] - \frac{1}{\alpha}G(Y,u)\phi[X^H, u^H] \\
&\quad - \frac{\alpha+\beta}{2} Y^T(G(X,u))\xi - (D_X Y)^T + (D_Y X)^T \\
&\quad + (G(X,Y) - G(X,u)G(Y,u))u^H \\
&\quad + G(Y,u)(-(D_u X)^T + X^H - G(X,u)u^H) - \alpha G(X,u)\phi Y^T \\
&\quad + \alpha G(Y,u)\phi X^T + \frac{1}{\alpha}\phi([X,Y]^H - \{K(X,Y)u\}^V) \\
&\quad + \frac{1}{\alpha}G(Y,u)\phi\Big([u,X]^H + (K_u X)^V + \sum u^b X^a \Gamma^i_{ab}\Big(\frac{\partial}{\partial x^i}\Big)^H\Big) \\
&\quad - \frac{\alpha+\beta}{2}(G(X,Y) - G(X,u)G(Y,u))\xi \\
&= G(X,u)(Y^H - G(Y,u)u^H) + (G(Y,u)X^H - G(X,u)Y^H) \\
&\quad - (X^H - G(X,u)u^H)G(Y,u) \\
&= 0,
\end{aligned}$$

(3.23)

$$
\begin{aligned}
&N_\phi(X^T,Y^T)+2d\eta(X^T,Y^T)\xi\\
&=-[X^T,Y^T]+[\phi X^T,\phi Y^T]-\phi[\phi X^T,Y^T]-\phi[X^T,\phi Y^T]\\
&=-[X^T,Y^T]+\frac{1}{\alpha^2}\Big\{[X^H,Y^H]-X^H(G(Y,u))u^H-G(Y,u)[X^H,u^H]\\
&\quad+Y^H(G(X,u))u^H-G(X,u)[u^H,Y^H]+G(X,u)u^H(G(Y,u))u^H\\
&\quad-G(Y,u)u^H(G(X,u))u^H\Big\}\\
&\quad+\frac{1}{\alpha}\phi([X^H,Y^T]+Y^T(G(X,u))u^H-G(X,u)[u^H,Y^T])\\
&\quad+\frac{1}{\alpha}\phi([X^T,Y^H]-X^T(G(Y,u))u^H-G(Y,u)[X^T,u^H])\\
&=-(G(X,u)Y^T-G(Y,u)X^T)+\frac{1}{\alpha^2}\Big\{[X,Y]^H-\{K(X,Y)u\}^V\\
&\quad-G(D_XY,u)u^H+G(Y,u)\Big([u,X]^H+(K_uX)^V+\sum u^bX^a\Gamma^i_{ab}\Big(\frac{\partial}{\partial x^i}\Big)^H\Big)\\
&\quad+G(D_YX,u)u^H+G(X,u)\Big([u,Y]^H+(K_uY)^V+\sum u^bY^a\Gamma^i_{ab}\Big(\frac{\partial}{\partial x^i}\Big)^H\Big)\\
&\quad+G(X,u)G(D_uY,u)u^H-G(Y,u)G(D_uX,u)u^H\Big\}\\
&\quad+\frac{1}{\alpha}\phi((D_XY)^T-G(X,u)((D_uY)^T-Y^H))\\
&\quad-\frac{1}{\alpha}\phi((D_YX)^T-G(Y,u)((D_uX)^T-X^H))\\
&=-(G(Y,u)X^V-G(X,u)Y^V)+G(Y,u)(X^V-G(X,u)u^v)\\
&\quad-G(X,u)(Y^V-G(Y,u)u^V)\\
&=0.
\end{aligned}
$$

From (3.21) $\sim$ (3.23), we see that $N_\phi(\bar{X},\bar{Y})+2d\eta(\bar{X},\bar{Y})\xi=0$ for all vector fields $\bar{X}$ and $\bar{Y}$ on $T_1M$. Therefore if $M$ is a space of constant curvature $\alpha^2$, $T_1M$ admits a Sasakian structure. □

As already mentioned in section 2, the metric treated includes the standard induced one from Sasaki metric as a special case ($\alpha=1$ and $\beta=0$). So, we get

**Corollary 3.3.** ([11]) *The standard contact Riemannian structure on $T_1M$ which is induced from the tangent bundle with Sasaki metric is K-contact if and only if $(M,G)$ has constant curvature 1, in which case the structure on $T_1M$ is Sasakian.*

## References

1. D.E. Blair, *Contact manifolds in Riemannian geometry*, Lecture Notes in Math. Springer-Verlag, Berlin-Heidelberg-New-York, **509** (1976).
2. J. Cheeger and D. Gromoll, *On the structure of complete manifolds of non-negative curvature*, Ann. of Math. **96** (1972), 413–443.
3. P. Dombroski, *On the geometry of tangent bundle*, J. Reine Angew. Math. **210** (1962), 73–88.
4. O. Kowalski, *Curvature of the induced Riemannian metric of tangent bundle of a Riemannian manifold*, J. Reine Angew. Math. **250** (1971), 124–129.s
5. E. Musso and F. Tricerri, *Riemannian metric on tangent bundles*, Ann. Math. Pura Appl. **150** (1998), 1–19.
6. V. Oproiu, *Some Kähler structrues on the tangent bundle of a space form*, Proceedings of the Third International Workshop on Differential Geometry and its Applications and the First German-Romanian Seminar on Geometry (Sibiu, 1997). Gen. Math. **5** (1997), 297–312.
7. V. Oproiu and N. Papaghiuc, *A Kaehler structrue on the nonzero tangent bundle of a space form*, Differential Geom. Appl. **11** (1999), 1–12.
8. S. Sasaki, *On the differential geometry of tangent bundles of Reimannian manifolds*, Tohoku Math. J. **10** (1958), 338–354.
9. S. Sasaki, *On the differential geometry of tangent bundles of Reimannian manifolds II*, Tohoku Math. J. **14** (1962), 146–155.
10. M. Sekizawa, *Curvatures of tangent bundles with Cheeger-Gromoll metric*, Tokyo J. Math. **14** (1991), 407–417.
11. Y. Tashiro, *On contact structures of tangent sphere bundles*, Tôhoku Math. J. **21** (1969) 117–143.

# APPLICATION OF BOCHNER-WEIZENBOCK FORMULAS TO SYMPLECTIC AND COMPLEX PAIRS TO BE KÄHLER PAIRS IN DIMENSION FOUR

IAGO GARCíA-RAMíREZ, EDUARDO GARCíA-RíO*
*Department of Geometry and Topology, Faculty of Mathematics,*
*University of Santiago de Compostela,*
*15782 Santiago de Compostela, Spain.*
*E-mails: xtiagogr@usc.es, xtedugr@usc.es*

YASUO MATSUSHITA
*Section of Mathematics, School of Engineering,*
*University of Shiga Prefecture,*
*Hikone 522-8533, Japan.*
*E-mail: yasuo@mech.usp.ac.jp*

Bochner's techniques are used to obtain curvature conditions which ensure that a four-dimensional symplectic or complex pair is a Kähler one in terms of nonnegativity or nonpositivity of sectional curvature.

*Keywords*: Almost Hermitian structures, complex, symplectic and Kähler pairs

## 1. Introduction

We call a 4-dimensional Riemannian manifold $(M, g)$ *double almost Hermitian* (DAH) if $M$ admits a pair of an almost complex structure $J_1$ and an opposite almost complex structure $J_2$, which commute with each other: $J_1 J_2 = J_2 J_1$, and define linear isometries at the tangent space level, i.e., $g(J_1 X, J_1 Y) = g(X, Y)$, $g(J_2 X, J_2 Y) = g(X, Y)$ for all vector fields $X, Y \in \mathfrak{X}(M)$. Note that the orientation of $J_1$ (resp. $J_2$) coincides with the normal orientation (resp. opposite orientation) of the manifold. The

*Supported by project BFM 2003-02949 (Spain)
2000 *Mathematics Subject Classification.* 53C55

condition for a 4-manifold to be DAH is that the 4-manifold admits a nonsingular field of oriented tangent 2-planes [13]. (It is known that this condition is equivalent to the existence of an indefinite metric of neutral signature $(++--)$, with the structure group $SO_0(2,2)$ (cf. [12])) For a DAH 4-manifold $M = (M, g, J_1, J_2)$, we can define two kinds of Kähler forms $\Omega_1$ and $\Omega_2$ as follows: for arbitrary vector fields $X$, $Y$ on $M$

$$\Omega_1(X,Y) = g(J_1X,Y), \qquad \Omega_2(X,Y) = g(J_2X,Y). \tag{1}$$

Our objects in the present paper are some specific Riemannian 4-manifolds in the family of DAH 4-manifolds.

A *symplectic pair* on a DAH 4-manifold $M$ is a pair of symplectic Kähler forms $\Omega_1$ and $\Omega_2$, with the properties

$$\Omega_1 \wedge \Omega_2 = 0, \quad \Omega_1 \wedge \Omega_1 = -\Omega_2 \wedge \Omega_2 = 2\,\omega \quad (\omega\text{: volume 4-form of } M). \tag{2}$$

A DAH 4-manifold is called *double almost Kähler* (DAK) if it admits a symplectic pair. Such a symplectic pair is defined, apart from any Riemannian metric, in a recent paper by Bande and Kotschick [1]. A *complex pair* on a DAH 4-manifold $M$ is a pair $(J_1, J_2)$ both of which are integrable. A DAH 4-manifold is called *double Hermitian* (DH) if it admits a complex pair. A *Kähler pair* on a DAH 4-manifold $M$ is a pair of two kinds of triples $(g, J_1, \Omega_1)$ and $(g, J_2, \Omega_2)$, both of which are Kähler, i.e., $(\Omega_1, \Omega_2)$ is a symplectic pair, and also $(J_1, J_2)$ a complex pair. A DAH 4-manifold is called *double Kähler* (DK) if it admits a Kähler pair.

The purpose of the present paper is to show some curvature conditions forcing such a symplectic pair and a complex pair to be a Kähler pair, which are obtained by applying the Bochner-Weizenbock formulas.

## 2. Symplectic pairs, complex pairs and Kähler pairs

The condition for a compact oriented 4-manifold to admits a pair $(J_1, J_2)$ of an almost complex structure $J_1$ and an opposite almost complex structure $J_2$ (equivalently to admit a field of 2-planes) is expressed in terms of the Euler characteristic $\chi$ and the Hirzebruch index $\tau$ of the manifold as follows.

**Theorem 1.** [13, Theorem 2] (A) *Let $M$ be a 4-manifold whose intersection form $\mu$ is indefinite. Then $M$ admits a pair $(J_1, J_2)$ if and only if $\tau[M]$ and $\chi[M]$ satisfy a pair of conditions*

$$\tau[M] + \chi[M] \equiv 0 \bmod 4, \qquad \tau[M] - \chi[M] \equiv 0 \bmod 4. \tag{3}$$

(B) *Let $M$ be a 4-manifold whose intersection form $\mu$ is definite, then $M$ admits a pair $(J_1, J_2)$ if and only if $\tau[M]$ and $\chi[M]$ satisfy the pair of conditions (3), and*

$$|\tau[M]| + \chi[M] \geq 0. \tag{4}$$

Therefore, every 4-manifold which satisfies the above conditions is a DAH 4-manifold. For a DAH 4-manifold $M = (M, g, J_1, J_2)$, we can define two kinds of Chern numbers $(c_1^2(J_1), c_2(J_1))$ for $J_1$ and $(c_1^2(J_2), c_2(J_2))$ for $J_2$ (cf. [5] and [13, §5] for details). These Chern numbers can be expressed in terms of $\tau[M]$ and $\chi[M]$ as follows:

$$\begin{cases} c_1^2(J_1) = 3\tau[M] + 2\chi[M] \\ c_2(J_1) = \chi[M], \end{cases} \qquad \begin{cases} c_1^2(J_2) = -3\tau[M] + 2\chi[M] \\ c_2(J_2) = \chi[M]. \end{cases} \tag{5}$$

From these relations, the opposite Chern numbers can be expressed in terms of $(c_1^2(J_1), c_2(J_1))$ as follows:

$$c_1^2(J_2) = -c_1^2(J_1) + 4c_2(J_1), \qquad c_2(J_2) = c_2(J_1). \tag{6}$$

### 2.1. *Symplectic pairs*

The present subsection is a brief review of Kotschick's symplectic pairs [1] and Geiges' symplectic couples [7]. Their definitions, apart from any Riemannian metrics, consist of two symplectic structures $\omega_1$ and $\omega_2$ on a 4-manifold as follows:

| symplectic pair | (conformal) symplectic couple | |
|---|---|---|
| $d\omega_1 = d\omega_2 = 0$ | $d\omega_1 = d\omega_2 = 0$ | (7) |
| $\omega_1 \wedge \omega_2 = 0$ | $\omega_1 \wedge \omega_2 = 0$ | (8) |
| $\omega_1 \wedge \omega_1 = -\omega_2 \wedge \omega_2$ | $\omega_1 \wedge \omega_1 = \omega_2 \wedge \omega_2$ | (9) |

If we consider these pair and couple, together with a suitable Riemannian metric as shown in §1, we can easily make clear the difference between them.

For a symplectic couple $(\Omega_1, \Omega_2)$, Geiges showed that there exists an integrable almost complex structure, denoted by $J_0$, and therefore, the bundle $\Lambda^2(M)$ of 2-forms splits into three subbundles as follows:

$$\Lambda^2(M) = \mathbb{R}\Omega_0 \oplus L_0 \oplus \Lambda^2_-, \tag{10}$$

where

$$\mathbb{R}\Omega_0 = \{ \alpha \in \Lambda^2(M) \mid *\alpha = \alpha,\ J_0\alpha = \alpha \}, \quad \dim \mathbb{R}\Omega_0 = 1 \tag{11}$$
$$L_0 = \{ \alpha \in \Lambda^2(M) \mid *\alpha = \alpha,\ J_0\alpha = -\alpha \}, \quad \dim L_0 = 2 \tag{12}$$
$$\Lambda^2_- = \{ \alpha \in \Lambda^2(M) \mid *\alpha = -\alpha \}, \quad \dim \Lambda_- = 3. \tag{13}$$

Then we see that $\Omega_1,\ \Omega_2 \in L_0$. We can choose an orthonormal frame $\{e^i\}$, $(i = 1, \ldots, 4)$, of 1-forms so that $\Omega_1$, $\Omega_2$ are written as follows: $\Omega_1 = e^1 \wedge e^3 + e^4 \wedge e^2$ and $\Omega_2 = e^1 \wedge e^4 + e^2 \wedge e^3$.

For a DAH 4-manifold $M = (M, g, J_1, J_2)$, the bundle $\Lambda^2(M)$ of 2-forms splits into four subbundles as follows:

$$\Lambda^2(M) = \mathbb{R}\Omega_1 \oplus L_1 \oplus \mathbb{R}\Omega_2 \oplus L_2, \tag{14}$$

where for $i = 1, 2$,

$$\mathbb{R}\Omega_i = \{ \alpha \in \Lambda^2(M) \mid *\alpha = (-1)^{i-1}\alpha,\ J_i\alpha = \alpha \}, \quad \dim \mathbb{R}\Omega_i = 1 \tag{15}$$
$$L_i = \{ \alpha \in \Lambda^2(M) \mid *\alpha = (-1)^{i-1}\alpha,\ J_i\alpha = -\alpha \}, \quad \dim L_i = 2. \tag{16}$$

Then we see that $\Omega_1 \in \mathbb{R}\Omega_1$ and $\Omega_2 \in \mathbb{R}\Omega_2$. We can choose an orthonormal frame $\{e^i\}$, $(i = 1, \ldots, 4)$, of 1-forms so that $\Omega_1$, $\Omega_2$ are written as follows: $\Omega_1 = e^1 \wedge e^2 + e^3 \wedge e^4$ and $\Omega_2 = e^1 \wedge e^2 - e^3 \wedge e^4$.

Finally, recall that a DAK manifold is foliated by two complementary minimal foliations citeBKo and the converse is true in dimension four

**Theorem 2.** [3, Theorem 3.6] *Let $(M, g)$ be a four-dimensional DAH manifold, then it is DAK if and only if there exist two orthogonal $J_i$-invariant minimal foliations.*

## 2.2. *Complex pairs*

A compact complex surface (simply a surface) has an integrable almost complex structure $J_1$. A surface is a DAH 4-manifold if it admit also an opposite almost complex structure $J_2$. We already know that a surface is also a DAH 4-manifold if and only if $c_2(J_1) = \chi$ is even [4, Proposition 5]. The following fact is due to Beauville [2] (cf. also [14]).

**Theorem 3.** (A) *If a compact DAH 4-manifold $M$, with $\chi[M] \geq 0$, admits a complex pair $(J_1, J_2)$, then the Chern numbers $(c_1^2(J_1), c_2(J_1))$ must satisfy*

$$0 \leq c_2(J_1) \leq c_1^2(J_2) \leq 3c_2(J_1). \tag{17}$$

(B) *If a compact DAH 4-manifold $M$, with $\chi[M] < 0$, admits a complex pair $(J_1, J_2)$, then $M$ is a ruled surface of genus $g \geq 2$.*

From (A), we immediately see that a K3 surface cannot admit a complex pair, since $c_1^2(J_1) = 0$, $c_2(J_1) = 24$, which do not satisfy the condition (17) (cf. [4]).

**Remark 1.** For any given Riemannian manifold $(M, g)$, its tangent bundle $TM$ has a natural metric which makes the projection $\pi : TM \to M$ a Riemannian submersion; the Sasaki metric (see [9] and the references therein). Moreover, if the base is equipped with an orthogonal almost complex structure $J$, one may lift it to obtain an almost complex pair on the tangent bundle, just by considering

$$J_i(X^V) = (JX)^V, \qquad J_i(X^H) = (-1)^{i-1}(JX)^H, \qquad i = 1, 2.$$

Now, it follows that $J_1$ and $J_2$ are integrable almost complex structures on $TM$ provided that $M$ is a two-dimensional Kähler manifold (i.e., an oriented surface). However none of $J_1$, $J_2$ is Kähler unless $M^2$ is flat.

### 2.3. *Kähler pairs*

A DAH structure is said to be DK if both almost Hermitian structures $(g, J_1)$ and $(g, J_2)$ are Kählerian. Here note that, as a special case of Theorem 2, we have

**Theorem 4.** [3] *A DAH manifold is DK if and only if it is locally isometric to a product of two Kähler manifolds.*

Therefore, in dimension four such condition is equivalent to the existence of two parallel orthogonal $J_i$-invariant foliations. This lead to the following

**Theorem 5.** *A 4-dimensional Lie group $G$ admits a left-invariant Kähler pair if and only if its Lie algebra $\mathfrak{g}$ corresponds to one of the following four-dimensional Lie algebras:*

*Type I*: $\quad [X,Y] = \alpha X + \beta Y, \qquad [Z,T] = \gamma Z + \delta T$

*Type II*: $\quad [X,T] = \alpha Y, \qquad [Y,T] = -\alpha X, \qquad [Z,T] = \beta Z,$

*Type III*: $$\begin{cases} [X,Z] = \alpha Y, \qquad [X,T] = \beta Y, \qquad [Y,Z] = -\alpha X \\ [Y,T] = -\beta X, \qquad [Z,T] = -\dfrac{\gamma\beta}{\alpha} Z + \gamma T. \end{cases}$$

**Proof.** First of all, recall that $(M^4, g, J_1, J_2)$ is a symplectic pair if and only if $\mathfrak{X}^+(M)$ and $\mathfrak{X}^-(M)$ define minimal complementary foliations on

$M$, where $\mathfrak{X}^{\pm}(M) = \{ X \in \mathfrak{X}(M) \mid i_X \Omega_{(\pm)} = 0 \}$ and $\Omega_{(\pm)} = \frac{1}{2}(\Omega_1 \mp \Omega_2)$. Therefore, if $G$ admits two complementary left-invariant minimal foliations, its Lie algebra admits two complementary subalgebras, $\mathfrak{g} = \mathfrak{g}_1 \oplus \mathfrak{g}_2$. Put

$$\mathfrak{g}_1 = \{X, Y\} \qquad \mathfrak{g}_2 = \{Z, T\}$$

with $[X,Y] = aX + bY$ and $[Z,T] = cZ + dT$. Moreover, since each subalgebra is minimal if and only if the corresponding mean curvatures vanish, we have that $\mathfrak{g}_1$ is minimal if and only if

$$\langle \nabla_X X, Z\rangle + \langle \nabla_Y Y, Z\rangle = 0, \qquad \langle \nabla_X X, T\rangle + \langle \nabla_Y Y, T\rangle = 0,$$

and $\mathfrak{g}_2$ is minimal if and only if

$$\langle \nabla_Z Z, X\rangle + \langle \nabla_T T, X\rangle = 0, \qquad \langle \nabla_Z Z, Y\rangle + \langle \nabla_T T, Y\rangle = 0.$$

Hence

$$\begin{aligned} \langle [Z,X], X\rangle &= -\langle [Z,Y], Y\rangle, & \langle [X,Z], Z\rangle &= -\langle [X,T], T\rangle, \\ \langle [T,X], X\rangle &= -\langle [T,Y], Y\rangle, & \langle [Y,Z], Z\rangle &= -\langle [Y,T], T\rangle, \end{aligned}$$

which shows that the Lie algebra $\mathfrak{g}$ satisfies

$$\begin{aligned} [X,Y] &= aX + bY, & [X,Z] &= eX + fY + gZ + hT, \\ [X,T] &= iX + jY + kZ - gT, & [Y,Z] &= mX - eY + nZ + oT, \\ [Y,T] &= qX - iY + pZ - nT, & [Z,T] &= cZ + dT. \end{aligned}$$

Then, the Jacobi identity ($[A,[B,C]] + [C,[A,B]] + [B,[C,A]] = 0$) implies the following constraints:

$$\begin{aligned} & fn + jo + be + eg + hi - af = 0, && ch + eg + fn + hi + jo - dg = 0, \\ & ae + bm + gm + hq - en - io = 0, && in + ai + bq + km - ep - gq = 0, \\ & dk + gi + jn + cg - ek - fp = 0, && dp + ep + gq + cn - km - in = 0, \\ & fp + bi + ek - jn - aj - gi = 0, && co + gm + hq - en - dn - io = 0, \\ & ko - ag - bn - hp = 0, && fq - ce - di - jm = 0, \\ & 2go + bo + ah - 2hn = 0, && 2kn + bp + ak - 2gp = 0, \\ & 2fi + cf + dj - 2ej = 0, && 2eq + cm + dq - 2im = 0. \end{aligned} \tag{18}$$

Now, let $J_1$ and $J_2$ be almost complex structures associated to the splitting $\mathfrak{g} = \mathfrak{g}_1 \oplus \mathfrak{g}_2$ defined by

$$J_i(X) = Y, \qquad J_i(Z) = (-1)^{i-1}T, \qquad i = 1, 2.$$

Next, since $G$ is equipped with a left-invariant Kähler pair, both almost complex structures $J_1$ and $J_2$ must be integrable, and therefore the Nijenhuis tensors $N_{J_1}$ and $N_{J_2}$ must vanish. For the integrability condition of

$J_1$, we have

$$N_{J_1}(X,Z) = (q+j-2e)X - (m+f+2i)Y \\ + (p+o-2g)Z - (k+h+2n)T = 0,$$

and using that $N_{J_1}(J_1\tilde{X},\tilde{Y}) = -J_1 N_{J_1}(\tilde{X},\tilde{Y})$ for arbitrary $\tilde{X}$, $\tilde{Y}$, one obtains that $J_1$ is integrable if and only if

$$q+j-2e=0,\ m+f+2i=0,\ p+o-2g=0,\ k+h+2n=0. \qquad (19)$$

Similarly, from $N_{J_2}=0$, we see that $J_2$ is integrable if and only if

$$q+j+2e=0,\ p+o+2g=0,\ 2i-f-m=0,\ 2n-h-k=0. \qquad (20)$$

These conditions (19) and (20), together, reduce to

$$e=g=n=i=0,\ m=-f,\ q=-j,\ k=-h,\ o=-p. \qquad (21)$$

Then, the constraints in (18) further reduce to

$$pj+af=0,\ pj+ch=0,\ bf+hj=0,\ bj-hf=0,\ dh+fp=0, \\ dp-hf=0,\ fp-aj=0,\ cp+hj=0,\ bp-ah=0,\ cf+dj=0.$$

We have finally five solutions to (18), which give the following five possibilities (trivial Lie brackets are omitted):

i) $f=h=j=p=0$

$\rightarrow\quad [X,Y]=aX+bY,\quad [Z,T]=cZ+dT$

ii) $a=b=d=f=h=p=0$

$\rightarrow\quad [Y,T]=-jX,\quad [X,T]=jY,\quad [Z,T]=cZ$

iii) $b=c=d=f=h=j=0$

$\rightarrow\quad [X,Y]=aX,\quad [Y,Z]=-pT,\quad [Y,T]=pZ$

iv) $a=b=h=p=0,\ c=-\dfrac{dj}{f}$

$\rightarrow\quad [Y,Z]=-fX,\quad [X,Z]=fY,\quad [Y,T]=-jX,$

$[X,T]=jY,\quad [Z,T]=-\dfrac{dj}{f}Z+dT$

v) $c=d=f=j=0,\ a=\dfrac{bp}{h}$

$\rightarrow\quad [X,Y]=\dfrac{bp}{h}X+bY,\quad [Y,Z]=-pT,\quad [X,Z]=hT,$

$[Y,T]=pZ,\quad [X,T]=-hZ.$

We can observe that solutions ii) and iii) are isomorphic, as well as solutions iv) and v), both of which are given by a mapping $\alpha : \mathfrak{g} \to \mathfrak{g}$ as follows:

$$\psi(X) = Z, \quad \psi(Y) = T, \quad \psi(Z) = X, \quad \psi(T) = Y. \tag{22}$$

We thus arrive at three types of the essentially different Lie algebras Types I, II, and III, as desired. □

**Remark 2.** These three Lie algebras admitting a Kählerian pair are solvable, but not nilpotent (unless the abelian algebra).

## 3. Curvature conditions for a DAH 4-manifold to be DK

For a given Riemannian metric $g$, we use the convention for the curvature tensor $R_{XY} = [\nabla_X, \nabla_Y] - \nabla_{[X,Y]}$, where $\nabla$ stands for the Levi Civita connection of the metric and $K(X \wedge Y) = g(R_{X,Y}Y, X)$ denotes the sectional curvature of the two-plane spanned by the orthonormal vectors $X$, $Y$.

### 3.1. *Symplectic pairs and sectional curvatures*

First of all, observe that since $\Omega_1$ and $\Omega_2$ induce opposite orientations on $M$, both are invariant and skew-invariant, respectively, by the Hodge star operator

$$*\Omega_1 = \Omega_1, \qquad *\Omega_2 = -\Omega_2.$$

Hence it follows from the expression of the Laplacian ($\Delta = d\delta + \delta d$) that the 2-forms $\Omega_1$ and $\Omega_2$ are harmonic if and only if they are closed. Then, we have

**Theorem 6.** *Let $(M, g, \Omega_1, \Omega_2)$ be a 4-dimensional symplectic pair of non-negative sectional curvature. Then $(M, g, \Omega_1, \Omega_2)$ is a Kählerian pair.*

**Proof.** For any $p$-form $\phi$ on a Riemannian manifold, its Laplacian satisfies the Weitzenbock formula [16]

$$\Delta\phi = -\nabla^2\phi + \widetilde{R}\phi, \tag{23}$$

where $\nabla^2\phi = \sum_i \nabla^2_{e_i,e_i}\phi$ with respect to a local orthonormal frame $\{e_i\}$ and $\widetilde{R}$ is the curvature operator $\widetilde{R}\phi = \sum_{i<j} R_{ij} \cdot [(\theta_i \wedge \theta_j) \cdot \phi]$, where $\{\theta_i\}$ is a local frame of 1-forms. Moreover

$$-\frac{1}{2}\Delta\|\phi\|^2 = p!\,\|\nabla\phi\|^2 + \langle\nabla^2\phi, \phi\rangle \tag{24}$$

holds for any $p$-form.

Hence for any 4-dimensional symplectic pair we have

$$\nabla^2\Omega_i = \widetilde{R}\Omega_i, \qquad i = 1, 2.$$

Also, since $\|\Omega_i\|^2 = \langle \Omega_i, \Omega_i \rangle$ is constant, one has that $\Delta\|\Omega_i\|^2 = 0$, and thus (24) shows that

$$\|\nabla\Omega_i\|^2 = -\frac{1}{2}\langle \nabla^2\Omega_i, \Omega_i\rangle = -\frac{1}{2}\langle \widetilde{R}\Omega_i, \Omega_i\rangle, \qquad i = 1, 2.$$

Moreover, it follows from [6] that $\langle \widetilde{R}\Omega_i, \Omega_i\rangle = 4\mathrm{tr}\,(R|_{(\mathfrak{u}^i)^{\perp}_{\frac{}{2}}})$.

In order to analyze $\mathrm{tr}(R_{(\mathfrak{u}^{(i)})^{\perp}_{2}})$, we follow [6] and [15], and use the identification $(\mathfrak{u}^{(i)})^{\perp}_{2} \equiv [[\bigwedge^{(2,0)}_{(i)} \oplus \bigwedge^{(0,2)}_{(i)}]]$, where $[[\cdot]]$ represents the real part and $\bigwedge^{r,s}_{(i)}$ denotes the $J_i$-complex forms of degree $(r, s)$. We now choose $\{e_1, J_1e_1, e_2, J_1e_2\}$ $(= \{e_1, J_2e_1, e_2, -J_2e_2\})$ as an adapted local orthonormal frame, and put

$$\alpha_1 = \frac{1}{\sqrt{2}}(e_1 \wedge e_2 - J_1e_1 \wedge J_1e_2), \quad \alpha_2 = \frac{1}{\sqrt{2}}(e_1 \wedge J_1e_2 + J_1e_1 \wedge e_2),$$

to obtain $\mathrm{tr}\,(R_{(\mathfrak{u}^{(i)})^{\perp}_{2}}) = R(\alpha_1)\alpha_1 + R(\alpha_2)\alpha_2$, where

$$\begin{aligned} R(\alpha_1)\alpha_1 &= -\frac{1}{2}\Big\{R(e_1, e_2, e_1, e_2) - R(e_1, e_2, J_1e_1, J_1e_2) \\ &\qquad - R(J_1e_1, J_1e_2, e_1, e_2) + R(J_1e_1, J_1e_2, J_1e_1, J_1e_2)\Big\} \\ &= \frac{1}{2}\Big\{K(e_1, e_2) + K(J_1e_1, J_1e_2) + 2\,R(e_1, e_2, J_1e_1, J_1e_2)\Big\} \end{aligned}$$

$$\begin{aligned} R(\alpha_2)\alpha_2 &= -\frac{1}{2}\Big\{R(e_1, J_1e_2, e_1, J_1e_2) + R(e_1, J_1e_2, J_1e_1, e_2) \\ &\qquad + R(J_1e_1, e_2, e_1, J_1e_2) + R(J_1e_1, e_2, J_1e_1, e_2)\Big\} \\ &= \frac{1}{2}\Big\{K(e_1, J_1e_2) + K(J_1e_1, e_2) - 2\,R(e_1, J_1e_2, J_1e_1, e_2)\Big\}. \end{aligned}$$

Hence

$$\begin{aligned} \mathrm{tr}\,(R|_{(\mathfrak{u}^1)^{\perp}_{2}}) = \frac{1}{2}\{&K(e_1, e_2) + K(e_1, J_1e_2) + K(J_1e_1, e_2) + K(J_1e_1, J_1e_2) \\ &- 2R(e_1, J_1e_2, J_1e_1, e_2) + 2R(e_1, e_2, J_1e_1, J_1e_2)\}. \end{aligned}$$

Proceeding in the same way with $J_2$, we obtain that

$$\begin{aligned} &\|\nabla\Omega_1\|^2 + \|\nabla\Omega_2\|^2 \\ &= -2\,[K(e_1, e_2) + K(e_1, J_1e_2) + K(J_1e_1, e_2) + K(J_1e_1, J_1e_2)]\,. \end{aligned}$$

Now, it is clear that nonnegative sectional curvature implies $\|\nabla\Omega_i\|^2 = 0$ for $i = 1, 2$, and thus that $(M, g, \Omega_1, \Omega_2)$ is a Kählerian pair. □

**Remark 3.** It is not difficult to show examples of non-Kählerian symplectic pairs of nonpositive sectional curvature. Consider the 4-dimensional Lie algebra defined by the Lie brackets

$$[X,Y] = Y \qquad [Y,Z] = -Y \qquad [X,Z] = X - Z$$
$$[Y,T] = 0 \qquad [X,T] = T \qquad [Z,T] = T,$$

and let $\omega_X$, $\omega_Y$, $\omega_Z$, $\omega_T$ the dual one-forms. Then

$$\Omega_1 = \omega_X \wedge \omega_Y + \omega_Z \wedge \omega_T, \qquad \Omega_2 = \omega_X \wedge \omega_Y - \omega_Z \wedge \omega_T$$

is a symplectic pair which is not Kählerian since the associated almost complex structures $J_i(X) = Y$, $J_i(Z) = (-1)^{i-1}T$ are not integrable. Moreover, the sectional curvature of the adapted metric is nonpositive (indeed one has after some calculations that $K(X,Z)+K(X,T)+K(Y,Z)+K(Y,T) = -4$).

### 3.2. *Complex pairs of nonpositive sectional curvature*

By integrating the Weitzenbock formula (23) associated with the Kähler form of an almost Hermitian structure, one obtains the corresponding Bochner formula as (see [6], [10])

$$\int_M \langle \widetilde{R}\Omega, \Omega \rangle = \int_M \{\|d\Omega\|^2 + \|\delta\Omega\|^2 - 2\|\nabla\Omega\|^2\}. \tag{25}$$

Recall that $\nabla\Omega$ is a 3-tensor with the symmetries $\nabla\Omega(X,Y,Z) = -\nabla\Omega(X,Z,Y) = -\nabla\Omega(X,JY,JZ)$. Putting $\mathfrak{W}$ for the space of 3-tensors with such symmetries, the group $U(n)$ acts on $\mathfrak{W}$ and it decomposes into four irreducible $U(n)$-modules (exactly two irreducible modules in dimension four), $\mathfrak{W} = \mathfrak{W}_1 + \mathfrak{W}_2 + \mathfrak{W}_3 + \mathfrak{W}_4$ (cf. [8]). Put $\nabla\Omega = (\nabla\Omega)_1 + (\nabla\Omega)_2 + (\nabla\Omega)_3 + (\nabla\Omega)_4$ to get

$$\|d\Omega\|^2 = 6\|(\nabla\Omega)_1\|^2 + 2\|(\nabla\Omega)_3\|^2 + 2\|(\nabla\Omega)_4\|^2, \quad \|\delta\Omega\|^2 = 2(n-1)\|(\nabla\Omega)_4\|^2.$$

Finally, the Bochner formula for almost Hermitian manifolds is as follows (cf. [6])

$$2E_1 - E_2 + (n-1)E_4 = \frac{1}{2}\int_M (\tau - \tau^*), \quad \left( E_j = \int_M \|(\nabla\Omega)_j\|^2 \right). \tag{26}$$

**Theorem 7.** *Let $(M, g, J_1, J_2)$ be a compact 4-dimensional complex pair of non positive sectional curvature. Then $(M, g, J_1, J_2)$ is a Kählerian pair.*

**Proof.** Note that an almost Hermitian structure is integrable if and only if $\|(\nabla\Omega)_2\|^2 = 0$ and moreover $\|(\nabla\Omega)_1\|^2 = \|(\nabla\Omega)_3\|^2 = 0$ in dimension four (cf. [8]). Hence, for a compact complex pair, the Bochner formula becomes

$$\int_M \|(\nabla\Omega_1)_4\|^2 + \int_M \|(\nabla\Omega_2)_4\|^2 = 2\int_M \{\mathrm{tr}\,(R|_{(u^1)_2^\perp}) + \mathrm{tr}\,(R|_{(u^1)_2^\perp})\},$$

and therefore, just proceeding as in the proof of Theorem 6, it follows that

$$\int_M \|(\nabla\Omega_1)_4\|^2 + \int_M \|(\nabla\Omega_2)_4\|^2$$
$$= 2\int_M \{K(e_1,e_2) + K(e_1,J_1e_2) + K(J_1e_1,e_2) + K(J_1e_1,J_1,e_2)\}.$$

Thus, one sees that nonpositivity on the sectional curvature forces the complex pair to be Kähler. □

**Remark 4.** Examples of complex pairs which are not Kählerian can be constructed as follows. The manifold of Kodaira-Thurston $M$ is obtained as the quotient of the nilpotent Lie group $G$ with Lie algebra $\mathfrak{g} = \{X,Y,Z,T\}$ satisfying $[X,Y] = Z$ by a discrete subgroup $M = \Gamma\backslash G$. A complex pair on $M$ is obtained by projecting the complex structures $J_1$ and $J_2$ defined on the Lie algebra by

$$J_i(X) = Y, \qquad J_i(Z) = (-1)^{i-1}T, \qquad i = 1,2.$$

Moreover, note that the complex pair above is not Kählerian and that the sectional curvature is nonnegative. Indeed, if $g$ is an inner product on $\mathfrak{g}$ making $\{X,Y,Z,T\}$ orthonormal, it projects to a DAH structure on $M$ which is DH and the sectional curvatures are not nonpositive.

### 3.3. *Mixed pairs*

Recall that the bisectional curvature of a plane $\pi$ with respect to an almost Hermitian structure $(g,J)$ is given by $B^J(X,Y) = R(X,JX,JY,Y)$, where $\{X,Y\}$ is an orthonormal base of $\pi$. If the almost Hermitian structure $(g,J)$ is Kähler, the bisectional curvature reduces to $B^J(X,Y) = K(X,JY) + K(X,Y)$, and proceeding as in Theorem 6, we have

**Theorem 8.** *Let $(M,g,J_1,J_2)$ be a 4-dimensional Kähler and symplectic pair. If the bisectional curvature $B^{J_1} \geq 0$, then $(M,g,J_1,J_2)$ is a Kähler pair.*

**Remark 5.** Let $G$ be the Lie group associated to the Lie algebra $\mathfrak{g} = \{X, Y, Z, T\}$ defined by

$$[X,T] = X, \quad [Y,Z] = -2X, \quad [Y,T] = -Y, \quad [Z,T] = 2Z.$$

equipped with the left invariant almost Hermitian pair $(g, J_1, J_2)$ given by

$$J_i(X) = Y, \quad J_i(Z) = (-1)^{i-1}T,$$
$$\Omega_i = \omega_X \wedge \omega_Y + (-1)^{i-1}\omega_Z \wedge \omega_T, \qquad i = 1, 2.$$

Further note that $(g, J_1, J_2)$ is Kähler and opposite almost Kähler, but not opposite complex, since $J_2$ is not integrable. Moreover, note that the bisectional curvature $B^{J_1}(X, Z) = -2 < 0$.

Finally, proceeding as in previous theorems, one also has the following

**Theorem 9.**
(i) *Let $(M, g, J_1, J_2)$ be a compact 4-dimensional Kähler and complex pair. If the bisectional curvature $B^{J_1} \leq 0$, then $(M, g, J_1, J_2)$ is a Kähler pair.*
(ii) *Let $(M, g, J_1, J_2)$ be a compact 4-dimensional symplectic and complex pair. If the bisectional curvature $B^{J_1} \leq 0$, then $(M, g, J_1, J_2)$ is a Kähler pair.*

## References

1. G. Bande, D. Kotschick; The geometry of symplectic pairs, *Trans. Amer. Math. Soc.*, to appear.
2. A. Beauville; Surfaces complexes et orientation, *Asterisque* **126** (1985), 41–43.
3. A. Bonome, R. Castro, E. García-Río, L. Hervella, Y. Matsushita; Almost complex manifolds with holomorphic distributions *Rend. Mat. Appl.* **14** (1994), 506–513.
4. A. Bonome, R. Castro, E. García-Río, L. Hervella, Y. Matsushita; The Kähler-Einstein metrics on a K3 surface cannot be almost Kähler with respect to an opposite almost complex structure, *Kodai Math. J.* **18** (1995), 567–589.
5. A. Bonome, R. Castro, E. García-Río, L. Hervella, Y. Matsushita; Pseudo-Chern classes and opposite Chern classes of indefinite almost Hermitian manifolds, *Acta Math. Hungar.* **75** (1997), 299–316.
6. G. Bor, L. Hernández Lamoneda; Bochner formulae for orthogonal $G$-structures on compact manifolds, *Differential Geom. Appl.* **15** (2001), 265–286.
7. H. Geiges; Symplectic couples on 4-manifolds, *Duke Math. J.* **85** (1996), 701–711.

8. A. Gray, L. Hervella; The sixteen classes of almost Hermitian manifolds and their linear invariants, *Ann. Mat. Pura Appl.* **(4) 123** (1980), 35–58.
9. S. Gudmunsdsson, E. Kappos; On the geometry of tangent bundles, *Exp. Math.* **20** (2002), 1–41.
10. L. Hernández Lamoneda; Curvature vs. almost Hermitian structures, *Geom. Ded.* **79** (2000), 205–218.
11. D. Kotschick; Orientations and geometrisations of compact complex surfaces, *Bull. London Math. Soc.* **29** (1997), 145–149.
12. Y. Matsushita, P. Law; Hitchin-Thorpe Type Inequalities for Pseudo-Riemannian 4-Manifolds of Metric Signature (++−−), *Geom. Ded.* **87** (2001), 65–89.
13. Y. Matsushita; Fields of oriented 2-plane and two kinds of almost complex structures on compact 4-manifolds, *Math. Z.* **207** (1991), 281–291.
14. Y. Miyaoka; On the Chern numbers of surfaces of general type, *Invent. Math.* **42** (1977), 225–237.
15. S. Salamon; *Riemannian geometry and holonomy groups*, Pitman Research Notes Math. **201**, Longman, New York, 1989.
16. H. Wu; The Bochner Technique in Differential Geometry, *Math Rep.* 3, Harwood Academic, London, (1988), 289–253.
17. W.T. Wu; Sur les classes caractéristiques des structures fibrées sphériques, *Actual. Sci. Industr.* **1183** (1952)

# COMPLETE $K$-CURVATURE HOMOGENEOUS PSEUDO-RIEMANNIAN MANIFOLDS 0-MODELED ON AN INDECOMPOSIBLE SYMMETRIC SPACE

PETER B GILKEY

*Mathematics Department, University of Oregon,*
*Eugene Or 97403 USA.*
*E-mail: gilkey@darkwing.uoregon.edu*

STANA NIKČEVIĆ

*Mathematical Institute, SANU, Knez Mihailova 35, p.p. 367,*
*11001 Belgrade, Serbia and Montenegro.*
*E-mail: stanan@mi.sanu.ac.yu*

For $k \geq 2$, we exhibit complete $k$-curvature homogeneous neutral signature pseudo-Riemannian manifolds which are not locally affine homogeneous (and hence not locally homogeneous). The curvature tensor of these manifolds is modeled on that of an indecomposible symmetric space. All the local scalar Weyl curvature invariants of these manifolds vanish.

*Keywords*: Affine $k$-curvature homogeneous, $k$-curvature homogeneous, homogeneous space, symmetric space, Weyl invariants, vanishing scalar invariants

## 1. Introduction

### 1.1. *Affine manifolds*

Let $\mathcal{A} := (M, \nabla)$ be an *affine manifold* where $\nabla$ is a torsion free connection on a smooth manifold $M$. Let $\mathcal{R}_A$ be the associated curvature operator:

$$\mathcal{R}_{\mathcal{A}}(\xi_1, \xi_2)\xi_3 := (\nabla_{\xi_1}\nabla_{\xi_2} - \nabla_{\xi_2}\nabla_{\xi_1} - \nabla_{[\xi_1,\xi_2]})\xi_3.$$

Let $\nabla^i\mathcal{R}_{\mathcal{A}}$ be the $i^{\text{th}}$ covariant derivative of the curvature operator. If $P \in M$, let $\nabla^i\mathcal{R}_{\mathcal{A},P}$ be the restriction of $\nabla^i\mathcal{R}_{\mathcal{A}}$ to $T_PM$. Consider the

2000 *Mathematics Subject Classification.* 53B20.

following algebraic structure which encodes the covariant derivatives of the curvature operator up to order $k$:

$$\mathfrak{A}^k(\mathcal{A}, P) := (T_P M, \mathcal{R}_{\mathcal{A},P}, \ldots, \nabla^k \mathcal{R}_{\mathcal{A},P}).$$

We say that $\phi : \mathfrak{A}^k(\mathcal{A}_1, P_1) \to \mathfrak{A}^k(\mathcal{A}_2, P_2)$ is an *affine isomorphism* if $\phi$ is a linear map from $T_{P_1}M_1$ to $T_{P_2}M_2$ satisfing

$$\phi^*\{\nabla_2^i \mathcal{R}_{\mathcal{A}_2,P_2}\} = \nabla_1^i \mathcal{R}_{\mathcal{A}_1,P_1} \quad \text{for} \quad 0 \le i \le k.$$

### 1.2. *Pseudo-Riemannian manifolds*

If $\mathcal{M} := (M, g)$ is a pseudo-Riemannian manifold of signature $(p, q)$ and of dimension $m = p + q$, let $\nabla$ be the Levi-Civita connection, let $\mathcal{A}(\mathcal{M}) := (M, \nabla)$ be the underlying affine structure, and let

$$R_{\mathcal{M}}(\xi_1, \xi_2, \xi_3, \xi_4) := g(\mathcal{R}_{\mathcal{A}}(\xi_1, \xi_2)\xi_3, \xi_4)$$

be the curvature tensor; $R_{\mathcal{M}} \in \otimes^4 T^*M$. Similarly, let $\nabla^i R_{\mathcal{M}}$ be the $i^{\text{th}}$ covariant derivative of the curvature tensor. Let

$$\mathfrak{M}^k(\mathcal{M}, P) := (T_P M, g_P, R_{\mathcal{M},P}, \ldots, \nabla^k R_{\mathcal{M},P}).$$

One says that $\phi : \mathfrak{M}^k(\mathcal{M}_1, P_1) \to \mathfrak{M}^k(\mathcal{M}_2, P_2)$ is an isomorphism if $\phi$ is a linear isomorphism from $T_{P_1}M_1$ to $T_{P_2}M_2$ so that

$$\phi^*\{g_{2,P_2}\} = g_{1,P_1} \quad \text{and} \quad \phi^*\{\nabla_2^i R_{\mathcal{M}_2,P_2}\} = \nabla_1^i R_{\mathcal{M}_1,P_1} \quad \text{for} \quad 0 \le i \le k.$$

In this situation, the metric permits one to raise indices and conclude as well that

$$\phi^*\{\nabla_2^i \mathcal{R}_{\mathcal{M}_2,P_2}\} = \nabla_1^i \mathcal{R}_{\mathcal{M}_1,P_1} \quad \text{for} \quad 0 \le i \le k.$$

Thus $\phi$ is also an isomorphism from $\mathfrak{A}^k(\mathcal{A}(\mathcal{M}_1), P_1)$ to $\mathfrak{A}^k(\mathcal{A}(\mathcal{M}_2), P_2)$ of the underlying affine structure.

We shall frequently simplify the notation by setting $\mathcal{R} = \mathcal{R}_{\mathcal{A}}$ or $R = R_{\mathcal{M}}$ when no confusion is likely to result.

### 1.3. *Various notions of homogeneity*

One is often interested in manifolds with a great deal of geometric symmetry. Sometimes this symmetry arises from a transitive group action; such manifolds are called *homogeneous*.

**Definition 1.1.**

(1) An affine manifold $\mathcal{A} = (M, \nabla)$ is said to be *locally affine homogeneous* if given $P, Q \in M$, there is a diffeomorphism $\Phi_{P,Q}$ from a neighborhood of $P$ to a neighborhood of $Q$ so $\Phi_{P,Q}^* \nabla = \nabla$ and so $\Phi(P) = Q$.

(2) A pseudo-Riemannian manifold $\mathcal{M} = (M, g)$ is said to be *locally homogeneous* if given $P, Q \in M$, there is a diffeomorphism $\Phi_{P,Q}$ from a neighborhood of $P$ to a neighborhood of $Q$ so $\Phi_{P,Q}^* g = g$ and so $\Phi(P) = Q$.

There are, however, other less restrictive notions of symmetry arising from the curvature operator and curvature tensor:

**Definition 1.2.**

(1) One says that an affine manifold $\mathcal{A}$ is *affine k-curvature homogeneous* if $\mathfrak{A}^k(\mathcal{A}, P)$ and $\mathfrak{A}^k(\mathcal{A}, Q)$ are isomorphic for any $P, Q \in M$.

(2) One says that a pseudo-Riemannian manifold $\mathcal{M}$ is *k-curvature homogeneous* if $\mathfrak{M}^k(\mathcal{M}, P)$ and $\mathfrak{M}^k(\mathcal{M}, Q)$ are isomorphic for any $P, Q \in M$.

One is interested finding manifolds which are affine $k$-curvature homogeneous but not locally affine homogeneous or which are $k$-curvature homogeneous but not locally homogeneous.

### 1.4. *Previous results*

There are 2-curvature homogeneous affine manifolds which are not locally affine homogeneous [9,14,15,16,21]. In the Riemannian setting ($p = 0$), Takagi [29] constructed 0-curvature homogeneous complete non-compact manifolds which are not locally homogeneous; compact examples were exhibited subsequently by Ferus, Karcher, and Münzer [8]. Many other examples are known [7,17,18,19,30,31,33,34]. There are no known Riemannian manifolds which are 1-curvature homogeneous but not locally homogeneous. In the Lorentzian setting ($p = 1$) 0-curvature homogeneous manifolds which are not locally homogeneous were constructed by Cahen et al. [6]; 1-curvature homogeneous manifolds which are not locally homogeneous were constructed by Bueken and Djorić [4] and by Bueken and Vanhecke [5].

### 1.5. *Curvature homogeneity and homogeneity*

It is clear that local homogeneity implies $k$-curvature homogeneity for any $k$. The following result, due to Singer [26] in the Riemannian setting and to F. Podesta and A. Spiro [23] in the general context, provides a partial converse:

**Theorem 1.3** (Singer, Podesta-Spiro). *There exists an integer $k_{p,q}$ so that if $\mathcal{M}$ is a complete simply connected pseudo-Riemannian manifold of signature $(p,q)$ which is $k_{p,q}$-curvature homogeneous, then $(M,g)$ is homogeneous.*

These constants were first studied in the Riemannian setting. Singer [26] showed $k_{0,m} < \frac{1}{2}m(m-1)$; subsequently Yamato [35] and Gromov [13] established the bounds $3m-5$ and $\frac{3}{2}m-1$ for $k_{0,m}$, respectively. Sekigawa, Suga, and Vanhecke [27,28] showed any 1-curvature homogeneous complete simply connected Riemannian manifold of dimension $m < 5$ is homogeneous; thus $k_{0,2} = k_{0,3} = k_{0,4} = 1$. We refer to the discussion in Boeckx, Vanhecke, and Kowalski [2] for further details concerning $k$-curvature homogeneous manifolds in the Riemannian setting; Opozda [22] has established an analogue of Theorem 1.3 in the affine setting. Observe that our definition of $k_{p,q}$ differs slightly from that given elsewhere by certain authors.

We constructed [11] complete metrics of neutral signature $(p+3, p+3)$ on $\mathbb{R}^{2p+6}$ for any $p \geq 0$ which are $p+2$-curvature homogeneous but not affine $p+3$-curvature homogeneous [11]. The discussion there shows $k_{p,q} \geq \min\{p,q\}$.

### 1.6. *Scalar invariants*

One can use the metric to contract indices in pairs and form *scalar Weyl invariants*. Adopt the Einstein convention and sum over repeated indices. Let $R_{i_1i_2i_3i_4}$ denote the components of the curvature tensor. The scalar curvature $\tau$ and the norm of the Ricci tensor $|\rho|^2$ are given respectively by:

$$\tau = g^{i_1i_2}g^{j_1j_2}R_{i_1j_1j_2i_2} \quad \text{and} \quad |\rho|^2 = g^{i_1j_1}g^{i_2j_2}g^{i_3j_3}g^{i_4j_4}R_{i_1i_2i_3j_1}R_{i_4j_2j_3j_4}.$$

There is a related result concerning scalar invariants:

**Theorem 1.4** (Prüfer, Tricerri, and Vanhecke [24]). *If all local scalar Weyl invariants up to order $\frac{1}{2}m(m-1)$ are constant on a Riemannian manifold $\mathcal{M}$, then $\mathcal{M}$ is locally homogeneous and $\mathcal{M}$ is determined up to local isometry by these invariants.*

This result fails in the pseudo-Riemannian setting; Koutras and McIntosh [20] gave examples of non-flat manifolds all of whose scalar Weyl invariants vanish; see also related examples by Pravda, Pravdová, Coley, and Milson [25].

## 1.7. *Riemannian manifolds modeled on homogeneous spaces*

One says that $\mathcal{M}$ is *k-modeled* on a homogeneous pseudo-Riemannian manifold $\mathcal{N}$ if $\mathfrak{M}^k(\mathcal{M},P)$ and $\mathfrak{M}^k(\mathcal{N},Q)$ are isomorphic for any $P \in M$ and $Q \in N$; the precise $Q \in N$ being irrelevant as $\mathcal{N}$ is homogeneous. One has the following results in the Riemannian and Lorentzian settings:

**Theorem 1.5.**

(1) (Tricerri and Vanhecke [32]) *If a Riemannian manifold $\mathcal{M}$ is 0-modeled on an irreducible symmetric space $\mathcal{S}$, then $\mathcal{M}$ is locally isometric to $\mathcal{S}$.*
(2) (Cahen et al. [6]) *If a Lorentzian manifold $\mathcal{M}$ is 0-modelled on an irreducible symmetric space, then $\mathcal{M}$ has constant sectional curvature.*

There is a bit of technical fuss here. Recall that a pseudo-Riemannian manifold $\mathcal{M}$ is said to be *irreducible* if the holonomy representation is irreducible, i.e. if $T_PM$ does not have any proper non-trivial subspace which is invariant under the holonomy representation for any (and hence for all) $P \in M$; $\mathcal{M}$ is said to be *indecomposible* if there does not exist a non-trivial decomposition of $T_PM$ which is invariant under the holonomy representation.

These two notions are equivalent in the Riemannian setting but are not equivalent in the higher signature setting. It is known that there are 1-curvature homogeneous 3-dimensional Lorentzian manifolds which are modeled on an indecomposible symmetric space (which is not irreducible) but which are not locally homogeneous; see [3,4,5,6] for further details.

In this paper, we turn to the question of constructing pseudo-Riemannian manifolds which are 0-curvature modeled on an indecomposible symmetric space and which are $k$-curvature homogeneous for arbitrarily large $k$; our construction is motivated by the examples described in [11]. We shall be defining several tensors. To simplify the discussion, we only give the non-zero entries in these tensors up to the usual symmetries.

## 1.8. *The pseudo-Riemannian manifolds $\mathcal{M}_{6+4p,f}$*

For $p \geq 1$, let

$$(x, z_0, \ldots, z_p, \tilde{z}_0, \ldots, \tilde{z}_p, x^*, z_0^*, \ldots, z_p^*, \tilde{z}_0^*, \ldots, \tilde{z}_p^*)$$

be coordinates on $\mathbb{R}^{6+4p}$. If $f$ is a smooth function on $\mathbb{R}^{p+1}$, a *generalized plane wave* manifold $\mathcal{M}_{6+4p,f} := (\mathbb{R}^{6+4p}, g_{6+4p,f})$ of neutral signature $(3+$

$2p, 3+2p)$ may be defined by setting:

$$g_{6+4p,f}(\partial_x, \partial_x) = -2\{f(z_0, \dots, z_p) + z_0\tilde{z}_0 + \dots + z_p\tilde{z}_p\}, \quad \text{and}$$
$$g_{6+4p,f}(\partial_x, \partial_{x^*}) = g_{6+4p,f}(\partial_{z_i}, \partial_{z_i^*}) = g_{6+4p,f}(\partial_{\tilde{z}_i}, \partial_{\tilde{z}_i^*}) = 1.$$

A word on notation. The *dual variables* $\{x^*, z_i^*, \tilde{z}_i^*\}$ enter only rather trivially; Theorem 1.6 below will imply that $\nabla^i R(\cdot)$ vanishes if any entry belongs to the span of $\{\partial_{x^*}, \partial_{z_i^*}, \partial_{\tilde{z}_i^*}\}$. Thus $\mathcal{M}_{6+4p,f}$ has a parallel totally isotropic distribution of maximal dimension. The dependence of the metric on the variables $\{\tilde{z}_0, \dots, \tilde{z}_p\}$ is fixed and ensures that the 0-model space is an indecomposible symmetric space. The crucial variables are $\{x, z_0, \dots, z_p\}$.

### 1.9. *The geometry of the manifolds $\mathcal{M}_{6+4p,f}$*

**Theorem 1.6.**

(1) *All geodesics in $\mathcal{M}_{6+4p,f}$ extend for infinite time.*

(2) $\exp_{P,\mathcal{M}_{6+4p,f}} : T_P\mathbb{R}^{6+4p} \to \mathbb{R}^{6+4p}$ *is a diffeomorphism for all* $P \in \mathbb{R}^{6+4p}$.

(3) *The non-zero components of $\nabla^k R$ are:*

$$\nabla^k R(\partial_x, \partial_{\xi_1}, \partial_{\xi_2}, \partial_x; \partial_{\xi_3}, \dots, \partial_{\xi_{k+2}}) = -\tfrac{1}{2}(\partial_{\xi_1} \cdots \partial_{\xi_{k+2}}) g_{6+4p,f}(\partial_x, \partial_x)$$

*for* $\xi_i \in \{z_0, \dots, z_p, \tilde{z}_0, \dots, \tilde{z}_p\}$.

(4) *All scalar Weyl invariants of $\mathcal{M}_{6+4p,f}$ vanish.*

(5) *$\mathcal{M}_{6+4p,f}$ is a symmetric space if and only if $f$ is at most quadratic.*

### 1.10. *The symmetric space $\mathcal{S}_{6+4p}$*

**Theorem 1.7.** *Let $\mathcal{S}_{6+4p} := \mathcal{M}_{6+4p,0}$ be defined by $f = 0$. Then:*

(1) *$\mathcal{S}_{6+4p}$ is an indecomposible symmetric space.*

(2) *$\mathcal{M}_{6+4p,f}$ is 0-modeled on $\mathcal{S}_{6+4p}$ for any $f = f(z_0, ..., z_p)$.*

### 1.11. *The homogeneous spaces $\mathcal{H}_{6+4p,k}$*

Theorems 1.6 and 1.7 show that Theorems 1.4 and 1.5 fail in the higher signature context. There are other interesting properties that this family of manifolds has. Construct a sequence of pseudo-Riemannian manifolds $\mathcal{H}_{6+4p,k} := \mathcal{M}_{6+4p,f_k}$ by defining:

$$f_k(z_0, \dots, z_p) := z_1 z_0^2 + \dots + z_k z_0^{k+1} \quad \text{if} \quad 1 \le k \le p,$$

and as exceptional cases

$$f_{p+1}(z_0,\ldots,z_p) = z_1z_0^2 + \cdots + z_pz_0^{p+1} + z_0^{p+3}, \quad \text{and}$$
$$f_{p+2}(z_0,\ldots,z_p) = z_1z_0^2 + \cdots + z_pz_0^{p+1} + e^{z_0}.$$

The following result shows that the local isometry type of a homogeneous space need not be determined by the first few covariant derivatives of the curvature tensor:

**Theorem 1.8.** *Let $1 \le k \le p+2$. Then:*

(1) *$\mathcal{H}_{6+4p,k}$ is 0-modeled on the indecomposible symmetric space $\mathcal{S}_{6+4p}$.*

(2) *If $j < k$, then*
   (a) *$\mathcal{H}_{6+4p,k}$ is $j$-modeled on $\mathcal{H}_{6+4p,j}$.*
   (b) *$\mathcal{H}_{6+4p,j}$ is not $k$-modeled on $\mathcal{H}_{6+4p,k}$.*

(3) *$\mathcal{H}_{6+4p,k}$ is a homogeneous space which is not symmetric.*

## 1.12. *The manifolds $\mathcal{N}_{6+4p,\psi}$*

Let $\psi \in C^\infty(\mathbb{R})$ satisfy

$$\psi^{(p+3)}(z_0) > 0 \quad \text{and} \quad \psi^{(p+4)}(z_0) > 0 \quad \text{for all} \quad z_0 \in \mathbb{R}.$$

Let $\mathcal{N}_{6+4p,\psi} := \mathcal{M}_{6+4p,f_\psi}$ where

$$f_\psi := z_1z_0^2 + \cdots + z_pz_0^{p+1} + \psi(z_0).$$

The following Theorem shows that

$$\alpha^k_{6+4p,\psi}(P) := \psi^{(k+p+3)}\{\psi^{(p+3)}\}^{k-1}\{\psi^{(p+4)}\}^{-k}(P) \quad \text{for} \quad k \ge 2$$

forms a collection of affine invariants which determines the isometry types of these manifolds; these invariants are not of Weyl type. Again, this does not happen in the Riemannian setting.

**Theorem 1.9.** *Suppose that $\psi_i$ are real analytic for $i = 1,2$ and that $\psi_i^{(p+3)}$ and $\psi_i^{(p+4)}$ are positive. The following assertions are equivalent:*

(1) *There exists a local diffeomorphism $\phi$ from $\mathcal{N}_{6+4p,\psi_1}$ to $\mathcal{N}_{6+4p,\psi_2}$ with $\phi(P_1) = P_2$ and $\phi^*\nabla_{\mathcal{N}_{6+4p,\psi_2}} = \nabla_{\mathcal{N}_{6+4p,\psi_1}}$.*

(2) *We have $\alpha^k_{6+4p,\psi_1}(P_1) = \alpha^k_{6+4p,\psi_2}(P_2)$ for $k \ge 2$.*

(3) *There exists an isometry $\phi : \mathcal{N}_{6+4p,\psi_1} \to \mathcal{N}_{6+4p,\psi_2}$ with $\phi(P_1) = P_2$.*

### 1.13. *Curvature and affine homogeneity*

One has the following Theorem:

**Theorem 1.10.** *Assume that* $\psi^{(p+3)}$ *and* $\psi^{(p+4)}$ *are positive. Then:*

(1) $\mathcal{N}_{6+4p,\psi}$ *is* 0*-modeled on the indecomposible symmetric space* $\mathcal{S}_{6+4p}$.

(2) $\mathcal{N}_{6+4p,\psi}$ *is* $j$*-modeled on the homogeneous space* $\mathcal{H}_{6+4p,j}$ *for* $1 \leq j \leq p+2$.

(3) $\mathcal{N}_{6+4p,\psi}$ *is* $(p+2)$*-curvature homogeneous.*

(4) *The following conditions are equivalent:*
  (a) $\mathcal{N}_{6+4p,\psi}$ *is homogeneous.*
  (b) $\mathcal{N}_{6+4p,\psi}$ *is affine* $(p+3)$*-curvature homogeneous.*
  (c) $\alpha^2_{6+4p,\psi}$ *is constant.*
  (d) $\psi^{(p+3)} = ae^{bz_0}$ *for some* $a, b \neq 0$.

Taking $\psi = e^{z_0} + e^{2z_0}$ constructs a manifold which is $(p+2)$-modeled on the homogeneous space $\mathcal{N}_{6+4p,e^{z_0}}$, which is curvature 0-modeled on the indecomposible symmetric space $\mathcal{S}_{6+4p}$, and which is not affine $(p+3)$-curvature homogeneous and hence not affine homogeneous.

## 2. Completeness

**Proof of Theorem 1.6.** To simplify the notation a bit, we introduce the variables

$$s = (s_1, \ldots, s_{2+2p}) := (z_0, \ldots, z_p, \tilde{z}_0, \ldots, \tilde{z}_p), \quad \text{and}$$
$$s^* = (s_1^*, \ldots, s_{2+2p}^*) := (z_0^*, \ldots, z_p^*, \tilde{z}_0^*, \ldots, \tilde{z}_p^*).$$

Let $1 \leq i \leq 2+2p$. The metric then takes the form

$$g_{6+4p,f}(\partial_x, \partial_x) = -2F(s) \quad \text{and} \quad g_{6+4p,f}(\partial_x, \partial_{x^*}) = g_{6+4p,f}(\partial_{s_i}, \partial_{s_i^*}) = 1.$$

for $F := f(z_0, \ldots, z_p) + z_0\tilde{z}_0 + z_1\tilde{z}_1 + \cdots + z_p\tilde{z}_p$. We compute the non-zero Christoffel symbols of the first and second kinds:

$$g_{6+4p,f}(\nabla_{\partial_x}\partial_x, \partial_{s_i}) = \partial_{s_i}F,$$
$$g_{6+4p,f}(\nabla_{\partial_x}\partial_{s_i}, \partial_x) = g_{6+4p,f}(\nabla_{\partial_{s_i}}\partial_x, \partial_x) = -\partial_{s_i}F,$$
$$\nabla_{\partial_x}\partial_x = \textstyle\sum_i \partial_{s_i}F \cdot \partial_{s_i^*}, \quad \text{and}$$
$$\nabla_{\partial_x}\partial_{s_i} = \nabla_{\partial_{s_i}}\partial_x = -\partial_{s_i}F \cdot \partial_{x^*}.$$

The curve $\gamma(t) = (x(t), s(t), x^*(t), s^*(t))$ is a geodesic if and only if

$$0 = \ddot{x}, \quad 0 = \ddot{s}_i, \quad 0 = \ddot{x}^* - 2\dot{x}\textstyle\sum_i \dot{s}_i \partial_{s_i} F, \quad \text{and} \quad 0 = \ddot{s}_i^* + \dot{x}\dot{x}\partial_{s_i} F.$$

We solve the geodesic equation with initial conditions $\gamma(0) = (\alpha, \xi, \alpha^*, \xi^*)$ and $\dot{\gamma}(0) = (\beta, \eta, \beta^*, \eta^*)$ by setting:

$$\begin{aligned}
&x(t) = \alpha + \beta t, \qquad s_i(t) = \xi_i + t\eta_i, \\
&x^*(t) = \alpha^* + \beta^* t + 2\beta \textstyle\int_0^t \int_0^\tau \{\sum_i \eta_i \partial_{s_i} F(\xi + t\eta)\} d\sigma d\tau, \\
&s_i^*(t) = \xi_i^* + t\eta_i^* - \beta^2 \textstyle\int_0^t \int_0^\tau \partial_{s_i} F(\xi + t\eta) d\sigma d\tau.
\end{aligned}$$

The solution exists for all time. Furthermore, there exists a unique geodesic with $\gamma(0) = P$ and $\gamma(1) = Q$; this establishes Assertions (1) and (2).

Since $\nabla\partial_{x^*} = \nabla\partial_{s_i^*} = 0$, Assertion (3) follows as the quadratic terms in the Christoffel symbols play no role in the covariant derivatives. Let

$$\begin{aligned}
\mathcal{V}_1 &:= \operatorname{Span}\{\partial_x - \tfrac{1}{2} g_{6+4p,f}(\partial_x, \partial_x)\partial_{x^*}, \partial_{s_1}, \dots, \partial_{s_p}\}, \\
\mathcal{V}_2 &:= \operatorname{Span}\{\partial_{x^*}, \partial_{s_1^*}, \dots, \partial_{s_p^*}\}.
\end{aligned}$$

This decomposes $\mathbb{R}^{6+4p} = \mathcal{V}_1 \oplus \mathcal{V}_2$ as the direct sum of two totally isotropic subspaces. Since $\nabla^i R$ vanishes if any entry belongs to $\mathcal{V}_2$, $\nabla^i R$ is supported on $\mathcal{V}_1$. As $\mathcal{V}_1$ is totally isotropic, Assertion (4) follows. Assertion (5) is immediate from Assertion (3). □

## 3. A 0-model for $\mathcal{M}_{6+4p,f}$

It is convenient to work in the purely algebraic setting. Let $V$ be an $m$ dimensional vector space. Let

$$\mathfrak{M}^k := (V, \langle\cdot,\cdot\rangle, A^0, \dots, A^k)$$

where $\langle\cdot,\cdot\rangle$ is a non-degenerate inner product on $V$ and where $A^i \in \otimes^{4+i} V^*$ satisfies the appropriate symmetries of the covariant derivatives of the curvature tensor; if $k = \infty$, then the sequence is infinite. We say that $\mathfrak{M}$ is a $k$-model for $\mathcal{M} = (M, g)$ if for each point $P \in M$, there is an isomorphism $\phi : T_P M \to V$ so that

$$\phi^*\langle\cdot,\cdot\rangle = g_P \quad \text{and} \quad \phi^* A^i = \nabla^i R_P \quad \text{for} \quad 0 \le i \le k.$$

Clearly $\mathcal{M}$ is $k$-curvature homogeneous if and only if it admits a $k$-model as one could take $\mathfrak{M}^k := \mathfrak{M}^k(\mathcal{M}, P)$ for any $P \in M$.

### 3.1. *Models for the manifolds $\mathcal{M}_{6+4p,f}$*

Let

$$\{X, Z_0, \ldots, Z_p, \tilde{Z}_0, \ldots, \tilde{Z}_p, X^*, Z_0^*, \ldots, Z_p^*, \tilde{Z}_0^*, \ldots, \tilde{Z}_p^*\}$$

be a basis for $\mathbb{R}^{6+4p}$. Define a hyperbolic inner product on $\mathbb{R}^{6+4p}$ by pairing ordinary variables with the corresponding dual variables:

$$\langle X, X^* \rangle = \langle Z_i, Z_i^* \rangle = \langle \tilde{Z}_i, \tilde{Z}_i^* \rangle = 1 \quad \text{for} \quad 0 \leq i \leq p.$$

Define an algebraic curvature tensor $A^0$ supported on $\text{Span}\{X, Z_i, \tilde{Z}_i\}$ by:

$$A^0(X, Z_i, \tilde{Z}_i, X) = 1 \quad \text{for} \quad 0 \leq i \leq p.$$

Define higher order covariant derivative curvature tensors $A^i$ for $1 \leq i \leq p$ by:

$$\begin{aligned} &A^i(X, Z_0, Z_i, X; Z_0, \ldots, Z_0) = 1, \\ &A^i(X, Z_0, Z_0, X; Z_i, Z_0, \ldots, Z_0) = 1, \ldots, \\ &A^i(X, Z_0, Z_0, X; Z_0, \ldots, Z_0, Z_i) = 1. \end{aligned}$$

The vectors $\{Z_i, \tilde{Z}_i\}$ for $0 \leq i \leq p$ are linked by $A^0$; the vectors $Z_0$ and $Z_i$ are linked by $A^i$ for $1 \leq i \leq p$. Set

$$\begin{aligned} &A^{p+1}(X, Z_0, Z_0, X; Z_0, \ldots, Z_0) = 1, \quad \text{and} \\ &A^{p+2}(X, Z_0, Z_0, X; Z_0, \ldots, Z_0) = 1. \end{aligned}$$

For $0 \leq k \leq p+2$, we define models:

$$\mathfrak{M}_{6+4p}^k := (\mathbb{R}^{6+4p}, \langle \cdot, \cdot \rangle, A^0, \ldots, A^k).$$

**Proof of Theorem 1.7.** Let $0 \leq i, j \leq p$. By Theorem 1.6,

$$R(\partial_x, \partial_{z_i}, \partial_{\tilde{z}_i}, \partial_x) = 1 \quad \text{and} \quad R(\partial_x, \partial_{z_i}, \partial_{z_j}, \partial_x) = \partial_{z_i}\partial_{z_j} F$$

where $F = f(z_0, \ldots, z_p) + z_0\tilde{z}_0 + \cdots + z_p\tilde{z}_p$. We set

$$\begin{aligned} &X := \partial_x + F\partial_{x^*}, && X^* := \partial_{x^*}, \\ &Z_i := \partial_{z_i} - \frac{1}{2}\textstyle\sum_j \partial_{z_i}\partial_{z_j} f \cdot \partial_{\tilde{z}_j}, && Z_i^* := \partial_{z_i^*}, \\ &\tilde{Z}_i := \partial_{\tilde{z}_i}, && \tilde{Z}_i^* := \partial_{\tilde{z}_i^*} + \tfrac{1}{2}\textstyle\sum_j \partial_{z_i}\partial_{z_j} f \cdot \partial_{z_j^*}. \end{aligned} \tag{3.a}$$

We show that $\mathfrak{M}_{6+4p}^0$ is a 0-model for $\mathcal{M}_{6+4p,f}$ by noting that the non-zero components of $g_{6+4p,f}$ and $R$ are then given by

$$\begin{aligned} &g_{6+4p,f}(X, X^*) = g_{6+4p,f}(Z_i, Z_i^*) = g_{6+4p,f}(\tilde{Z}_i, \tilde{Z}_i^*) = 1, \quad \text{and} \\ &R(X, Z_i, \tilde{Z}_i, X) = 1 \quad \text{for} \quad 0 \leq i \leq p. \end{aligned} \tag{3.b}$$

By Theorem 1.6, $\mathcal{S}_{6+4p}$ is a symmetric space. As $\mathfrak{M}^0_{6+4p}$ is a 0-model for $\mathcal{S}_{6+4p}$ and $\mathfrak{M}^0_{6+4p}$ is a 0-model for $\mathcal{M}_{6+4p,f}$, $\mathcal{S}_{6+4p}$ is a 0-model for $\mathcal{M}_{6+4p,f}$. To show complete the proof, we must only show $\mathfrak{M}^0_{6+4p}$ is indecomposible.

Suppose we have a non-trivial decomposition $\mathbb{R}^{6+4p} = V_1 \oplus V_2$ such that

$$A^0 = A^0_1 \oplus A^0_2 \quad \text{and} \quad \langle \cdot, \cdot \rangle = \langle \cdot, \cdot \rangle_1 \oplus \langle \cdot, \cdot \rangle_2.$$

We argue for a contradiction. Denote the natural projections induced by this decomposition by $\pi_i : \mathbb{R}^{6+4p} \to V_i$. Since

$$1 = \langle X, X^* \rangle = \langle \pi_1 X, X^* \rangle + \langle \pi_2 X, X^* \rangle$$

we may assume without loss of generality $\langle \pi_1 X, X^* \rangle \neq 0$. Set $\alpha := \pi_1(X)$. Let $\beta \in (X^*)^\perp \cap V_2$. Then $A^0(\alpha, \cdot, \beta, \alpha) = 0$ as $\alpha \in V_1$ and $\beta \in V_2$. Since $\beta$ doesn't involve $X$,

$$\begin{aligned} 0 &= A^0(\alpha, Z_i, \beta, \alpha) = \langle \alpha, X^* \rangle^2 \langle \beta, \tilde{Z}^*_i \rangle, \quad \text{and} \\ 0 &= A^0(\alpha, \tilde{Z}_i, \beta, \alpha) = \langle \alpha, X^* \rangle^2 \langle \beta, Z^*_i \rangle. \end{aligned}$$

Consequently $\langle \beta, X^* \rangle = 0$, $\langle \beta, Z^*_i \rangle = 0$, and $\langle \beta, \tilde{Z}^*_i \rangle = 0$. Thus

$$\beta \in \operatorname{Span}\{X^*, Z^*_0, \dots, Z^*_p, \tilde{Z}^*_0, \dots, \tilde{Z}^*_p\}$$

so $(X^*)^\perp \cap V_2$ is totally isotropic. Since the restriction of $\langle \cdot, \cdot \rangle$ to $V_2$ is non-degenerate and since

$$\dim\{(X^*)^\perp \cap V_2\} \geq \dim\{V_2\} - 1,$$

we conclude that $\dim\{V_2\} = 2$. Furthermore there must exist an element of $V_2$ not in $(X^*)^\perp$. We can therefore interchange the roles of $V_1$ and $V_2$ to see that $\dim\{V_1\} = 2$. Consequently $6 + 4p = \dim\{V_1\} + \dim\{V_2\} = 4$ which provides the desired contradiction. □

Theorem 1.8 (1) and Theorem 1.10 (1) are specials cases of Theorem 1.7 (2). Theorem 1.8 (2b) follows since $\nabla^j R_{\mathcal{H}_{6+4p,k}} = 0$ if $j > k$ whereas $\nabla^j R_{\mathcal{H}_{6+4p,j}} \neq 0$. Theorem 1.8 (2a) and Theorem 1.10 (2, 3) will follow from the following result.

**Lemma 3.1.**

(1) *If $f = \psi(z_0) + z_1 z_0^2 + \cdots + z_k z_0^{k+1}$ for $1 \leq k \leq p$, then $\mathfrak{M}^k_{6+4p}$ is a $k$-model for $\mathcal{M}_{6+4p,f}$.*

(2) *If $f = \psi(z_0) + z_1 z_0^2 + \cdots + z_p z_0^{p+1}$ and if $\psi^{(p+3)}$ is positive on $\mathbb{R}$, then $\mathfrak{M}^{p+1}_{6+4p}$ is a $(p+1)$-model for $\mathcal{M}_{6+4p,f}$.*

(3) *If $f = \psi(z_0) + z_1 z_0^2 + \cdots + z_p z_0^{p+1}$ and if $\psi^{(p+3)}$ and $\psi^{(p+4)}$ are positive on $\mathbb{R}$, then $\mathfrak{M}^{p+2}_{6+4p}$ is a $(p+2)$-model for $\mathcal{M}_{6+4p,f}$.*

**Proof.** We adopt the notation of Equation (3.a). The normalizations of Equation (3.b) are then satisfied. Suppose $f = \psi(z_0) + z_1 z_0^2 + \cdots + z_k z_0^{k+1}$. If $1 \le i \le k$ and $1 \le j \le p$,

$$\begin{aligned}
&\nabla^i R(X, Z_0, Z_0, X; Z_0, \dots, Z_0) = \varepsilon_i,\\
&\nabla^i R(X, Z_0, Z_j, X; Z_0, \dots, Z_0) = \varepsilon_{j,i},\\
&\nabla^i R(X, Z_0, Z_0, X; Z_0, \dots, Z_j, \dots, Z_0) = \varepsilon_{j,i}
\end{aligned}$$

where $\varepsilon_i = (\partial_{z_0})^{i+2} f$ and $\varepsilon_{j,i} = (\partial_{z_0})^{i+1}\partial_{z_j} f$. Note that

$$\varepsilon_{i,i} \ne 0 \quad \text{for} \quad 1 \le i \le k \quad \text{and} \quad \varepsilon_{j,i} = 0 \quad \text{for} \quad 1 \le j < i \le k.$$

To prove Assertion (1), we must define a new frame $\{{}^1X, {}^1X^*, {}^1Z_i, {}^1\tilde{Z}_i, {}^1Z_i^*, {}^1\tilde{Z}_i^*\}$ so that in addition to the relations of Equation (3.b), the only non-zero components of $\nabla^i R$ are given by

$$\begin{aligned}
&\nabla^i R({}^1X, {}^1Z_0, {}^1Z_i, {}^1X; {}^1Z_0, \dots, {}^1Z_0) = \cdots\\
&= \nabla^i R({}^1X, {}^1Z_0, {}^1Z_0, {}^1X; {}^1Z_0, \dots, {}^1Z_0, {}^1Z_i) = 1.
\end{aligned} \tag{3.c}$$

Set

$${}^1X := X \quad \text{and} \quad {}^1Z_0 := Z_0 + a_1 Z_1 + \cdots + a_k Z_k.$$

To ensure $\nabla^\ell R({}^1X, {}^1Z_0, {}^1Z_0, {}^1X; {}^1Z_0, \dots, {}^1Z_0) = 0$ for $1 \le \ell \le k$, we must have:

$$\begin{aligned}
&0 = \varepsilon_k + (k+2)\varepsilon_{k,k} a_k,\\
&0 = \varepsilon_{k-1} + (k+1)\{\varepsilon_{k,k-1} a_k + \varepsilon_{k-1,k-1} a_{k-1}\}, \dots\\
&0 = \varepsilon_1 + 3\{\varepsilon_{k,1} a_k + \cdots + \varepsilon_{1,1} a_1\}.
\end{aligned}$$

Because $\varepsilon_{i,i} \ne 0$ for $1 \le i \le k$, this upper triangular system of equations is recursively solvable for $a_k, \dots, a_1$.

To ensure that

$$\begin{aligned}
&\nabla^i R({}^1X, {}^1Z_0, {}^1Z_i, {}^1X; {}^1Z_0, \dots, {}^1Z_0) = \cdots = 1, \quad \text{and}\\
&\nabla^i R({}^1X, {}^1Z_0, {}^1Z_j, {}^1X; {}^1Z_0, \dots, {}^1Z_0) = \cdots = 0 \quad \text{for} \quad i \ne j,
\end{aligned}$$

we set ${}^1Z_i = Z_i$ for $k < i \le p$, while for $1 \le i \le k$, we set

$${}^1Z_1 = a_{1,1} Z_1,\ {}^1Z_2 = a_{2,1} Z_1 + a_{2,2} Z_2,\ \dots\ {}^1Z_k = a_{k,1} Z_1 + \cdots + a_{k,k} Z_k.$$

To ensure that ${}^1Z_k$ is properly normalized, the following relations must hold:

$$\begin{aligned}
&1 = a_{k,k}\varepsilon_{k,k},\\
&0 = a_{k,k-1}\varepsilon_{k-1,k-1} + a_{k,k}\varepsilon_{k,k-1}, \dots\\
&0 = a_{k,1}\varepsilon_{1,1} + \cdots + a_{k,k}\varepsilon_{k,1}.
\end{aligned}$$

This determines ${}^1Z_k$. We continue in this fashion to determine the remaining coefficients. This ensures the proper normalizations for $\nabla^i R$ for $1 \le i \le k$.

We now return to the relations of Equation (3.b) for $g$ and $R$. We regard $R(X,\cdot,\cdot,X)$ as defining a neutral signature inner product on

$$\mathrm{Span}\{\partial_{z_0},\dots,\partial_{z_p},\partial_{\tilde z_0},\dots,\partial_{\tilde z_p}\}.$$

Since ${}^1X = \partial_x + F\partial_{x^*}$ and since $\{{}^1Z_0,\dots,{}^1Z_p\} \subset \mathrm{Span}\{Z_0,\dots,Z_p\}$ we may choose

$$\{{}^1\tilde Z_0,\dots,{}^1\tilde Z_p\} \subset \mathrm{Span}\{\tilde Z_0,\dots,\tilde Z_p\}$$

so the only non-zero components of $R$ are $R({}^1X,{}^1Z_i,{}^1\tilde Z_i,X) = 1$. Finally, we choose a dual basis

$$\{{}^1X^*,{}^1Z_0^*,\dots,{}^1Z_p^*,{}^1\tilde Z_0^*,\dots,{}^1\tilde Z_p^*\} \subset \mathrm{Span}\{X^*,Z_0^*,\dots,Z_p^*,\tilde Z_0^*,\dots,\tilde Z_p^*\}$$

so the non-zero components of the metric $g$ are

$$g({}^1X,{}^1X^*) = g({}^1Z_i,{}^1Z_i^*) = g({}^1\tilde Z_i,{}^1\tilde Z_i^*) = 1.$$

Assertion (1) of the Lemma now follows.

There is a final bit of flexibility that we use in proving Assertions (2) and (3) of the Lemma. The relations of Equation (3.b) continue to hold. We rescale the basis we have constructed by setting:

$$\begin{aligned} &{}^2X = \varepsilon\cdot{}^1X, && {}^2X^* = \varepsilon^{-1}\cdot{}^1X^*, && {}^2Z_i = \varepsilon_i\cdot{}^1Z_i,\\ &{}^2Z_i^* = \varepsilon_i^{-1}\cdot{}^1Z_i^*, && {}^2\tilde Z_i = \varepsilon^{-2}\varepsilon_i^{-1}\cdot{}^1\tilde Z_i, && {}^2\tilde Z_i^* = \varepsilon^2\varepsilon_i\cdot{}^1\tilde Z_i^*. \end{aligned}$$

The non-zero components of $g$ and of $R$ are

$$\begin{aligned} &g({}^2X,{}^2X^*) = g({}^2Z_i,{}^2Z_i^*) = g({}^2\tilde Z_i,{}^2\tilde Z_i^*) = 1,\\ &R({}^2X,{}^2Z_i,{}^2\tilde Z_i,{}^2X) = 1. \end{aligned}$$

for $0 \le i \le p$. The non-zero components of $\nabla^i R$ for $1 \le i \le p$ are

$$\begin{aligned} &\nabla^i R({}^2X,{}^2Z_0,{}^2Z_i,{}^2X;{}^2Z_0,\dots,{}^2Z_0) = \cdots\\ &\quad = \nabla^i R({}^2X,{}^2Z_0,{}^2Z_0,{}^2X;{}^2Z_0,\dots,{}^2Z_i) = \varepsilon^2\varepsilon_i\varepsilon_0^{i+1}. \end{aligned}$$

The non-zero components of $\nabla^{p+1}R$ and $\nabla^{p+2}R$ are:

$$\begin{aligned} &\nabla^{p+1}R(({}^2X,{}^2Z_0,{}^2Z_0,{}^2X;{}^2Z_0,\dots,{}^2Z_0) = \varepsilon^2\varepsilon_0^{p+3}\psi^{(p+3)},\\ &\nabla^{p+2}R(({}^2X,{}^2Z_0,{}^2Z_0,{}^2X;{}^2Z_0,\dots,{}^2Z_0) = \varepsilon^2\varepsilon_0^{p+4}\psi^{(p+4)}. \end{aligned}$$

We set $\varepsilon_i := \varepsilon^{-2}\varepsilon_0^{-i-1}$ for $1 \le i \le p$ to ensure $\nabla^i R$ has the proper normalization for $1 \le i \le p$. Suppose that $\psi^{(p+3)}$ is positive on $\mathbb{R}$. We normalize $\nabla^{p+1}R$ and prove Assertion (2) of the Lemma by setting:

$$\varepsilon_0 = 1 \quad \text{and} \quad \varepsilon = \{\psi^{(p+3)}\}^{-1/2}.$$

If additionally $\psi^{(p+4)}$ is positive on $\mathbb{R}$, we may set

$$\varepsilon_0 := \psi^{(p+3)}\{\psi^{(p+4)}\}^{-1} \quad \text{and} \quad \varepsilon = \{\varepsilon_0^{p+3}\psi^{(p+3)}\}^{-1/2}$$

to ensure that both $\nabla^{p+1}$ and $\nabla^{p+2}$ are normalized appropriately. This establishes Assertion (3) of the Lemma. □

## 4. Isometries

Let $\mathfrak{M}^\infty(\mathcal{M}, P) = (T_P M, g_{\mathcal{M}}, R_{\mathcal{M},P}, \ldots, \nabla^i R_{\mathcal{M},P}, \ldots)$ be the full model at a point $P$ of a pseudo-Riemannian manifold $\mathcal{M}$. This encodes complete information about the isometry type of the manifold under certain circumstances:

**Lemma 4.1.** *Let $\mathcal{M}_i := (M_i, g_i)$ be real analytic pseudo-Riemannian manifolds for $i = 1, 2$. Assume there exist points $P_i \in M_i$ so $\exp_{P_i,\mathcal{M}_i} : T_{P_i}M_i \to M_i$ is a diffeomorphism and so there exists an isomorphism $\Phi$ between $\mathfrak{M}^\infty(\mathcal{M}_1, P_1)$ and $\mathfrak{M}^\infty(\mathcal{M}_2, P_2)$. Then $\phi := \exp_{P_2,\mathcal{M}_2} \circ \Phi \circ \exp_{P_1,\mathcal{M}_1}^{-1}$ is an isometry from $\mathcal{M}_1$ to $\mathcal{M}_2$.*

**Proof.** Belger and Kowalski [1] note about analytic pseudo-Riemannian metrics that the "metric $g$ is uniquely determined, up to local isometry, by the tensors $R$, $\nabla R$, $\ldots$, $\nabla^k R$, $\ldots$ at one point."; see also Gray [12] for related work. The desired result now follows. □

**Proof of Theorem 1.9.** Let $f = \psi(z_0) + z_1 z_0^2 + \cdots + z_p z_0^{p+1}$. We assume $\psi^{(p+3)}$ and $\psi^{(p+4)}$ are positive. If $k \ge p+1$, then the non-zero components of the curvature operator $\nabla^k\mathcal{R}$ are given by

$$(\nabla_{\partial_{z_0}})^k\mathcal{R}(\partial_x, \partial_{z_0})\partial_{z_0} = -(\nabla_{\partial_{z_0}})^k\mathcal{R}(\partial_{z_0}, \partial_x)\partial_{z_0} = \psi^{(k+2)}\partial_{x^*}, \quad \text{and}$$
$$(\nabla_{\partial_{z_0}})^k\mathcal{R}(\partial_x, \partial_{z_0})\partial_x = -(\nabla_{\partial_{z_0}})^k\mathcal{R}(\partial_{z_0}, \partial_x)\partial_x = -\psi^{(k+2)}\partial_{z_0^*}.$$

Choose $X, Z_0 \in T_P\mathbb{R}^{6+4p}$ and $\Theta \in T_P^*(\mathbb{R}^{6+4p})$ so:

$$\Theta\{(\nabla_{Z_0})^{p+1}\mathcal{R}(X, Z_0)X\} \neq 0. \tag{4.a}$$

For example one could take $\Theta = dz_0^*$, $X = \partial_x$ and $Z_0 = \partial_{z_0}$. Equation (4.a) is an invariant of the affine $p+1$-model as it does not depend on the metric and is preserved by local affine isomorphisms. Expand

$$X = a\partial_x + a^*\partial_{x^*} + \textstyle\sum_i\{a_i\partial_{z_i} + \tilde{a}_i\partial_{\tilde{z}_i} + a_i^*\partial_{z_i^*} + \tilde{a}_i^*\partial_{\tilde{z}_i^*}\},$$
$$Z_0 = b\partial_x + b^*\partial_{x^*} + \textstyle\sum_i\{b_i\partial_{z_i} + \tilde{b}_i\partial_{\tilde{z}_i} + b_i^*\partial_{z_i^*} + \tilde{b}_i^*\partial_{\tilde{z}_i^*}\}.$$

If $k \geq p+1$,

$$\Theta\{(\nabla_{Z_0})^k\mathcal{R}(X,Z_0)X\} = (ab_0 - ba_0)b_0^k\Theta\{(\nabla_{\partial_{z_0}})^k\mathcal{R}(\partial_x,\partial_{z_0})(a\partial_x + a_0\partial_{z_0}\}$$
$$= (ba_0 - ab_0)b_0^k\psi^{(k+2)}\Theta(a\partial_{z_0^*} - a_0\partial_{x^*}).$$

By hypothesis this is non-zero when $k = p+1$. Thus

$$a \neq 0, \quad b_0 \neq 0, \quad ba_0 - ab_0 \neq 0, \quad \text{and} \quad \Theta(a\partial_{z_0^*} - a_0\partial_{x^*}) \neq 0.$$

Set $\gamma := \Theta(a\partial_{z_0^*} - a_0\partial_{x^*})$. We may now compute:

$$\frac{\Theta\{(\nabla_{Z_0})^{k+p+1}\mathcal{R}(X,Z_0)Z_0\}\{\Theta\{(\nabla_{Z_0})^{p+1}\mathcal{R}(X,Z_0)Z_0\}\}^{k-1}}{\{\Theta\{(\nabla_{Z_0})^{p+2}\mathcal{R}(X,Z_0)Z_0\}\}^k}$$
$$= \frac{(ba_0 - ab_0)b_0^{k+p+1}\psi^{(k+p+3)}\gamma)\cdot\{(ba_0 - ab_0)b_0^{p+1}\psi^{(p+3)}\gamma\}^{k-1}}{\{(ba_0 - ab_0)b_0^{p+2}\psi^{(p+4)}\gamma\}^k}$$
$$= \psi^{(k+p+3)}\{\psi^{(p+3)}\}^{k-1}\{\psi^{(p+4)}\}^{-k} = \alpha_{6+4p,\psi}^k.$$

This shows that $\alpha_{6+4p,\psi}^k$ is an affine invariant. Consequently Assertion (1) implies Assertion (2) in Theorem 1.9.

We now show Assertion (2) implies Assertion (3) in Theorem 1.9; this will complete the proof as it is immediate that Assertion (3) implies Assertion (1). By Lemma 3.1 (3), we can choose a basis $\{X, X^*, Z_i, \tilde{Z}_i, Z_i^*, \tilde{Z}_i^*\}$ which normalizes $g_{6+4p,f}$ and $\nabla^i R$ appropriately for $0 \leq i \leq p+2$. Since

$$\nabla^i R(X, Z_0, Z_0, X; Z_0, \ldots, Z_0) = 1 \quad \text{for} \quad i = p+1, p+2,$$

we have

$$\nabla^{k+p+1}R(X, Z_0, Z_0, X; Z_0, \ldots, Z_0) = \alpha_{6+4p,\psi}^k \quad \text{for} \quad k \geq 2.$$

This shows that the higher covariant derivatives are controlled by $\alpha_{6+4p,\psi}^k$. Consequently if $\alpha_{6+4p,\psi_1}^k(P_1) = \alpha_{6+4p,\psi_2}^k(P_2)$ for $k \geq 2$, there is an isomorphism between $\mathfrak{M}^\infty(\mathcal{N}_{6+4p,\psi_1}, P_1)$ and $\mathfrak{M}^\infty(\mathcal{N}_{6+4p,\psi_2}, P_2)$ and hence by Lemma 4.1 an isometry between $(\mathcal{N}_{6+4p,\psi_1}, P_1)$ and $(\mathcal{N}_{6+4p,\psi_2}, P_2)$ as desired. □

**Proof of Theorem 1.8 (4).** Let $f_k := z_1 z_0^2 + \cdots + z_k z_0^{k+1}$. Let $P_i \subset \mathbb{R}^{6+4p}$. By Lemma 4.1,

$$\mathfrak{M}^i(\mathcal{M}_{6+4p,f_k}, P_1) \approx \mathfrak{M}^i(\mathcal{M}_{6+4p,f_k}, P_2) \tag{4.b}$$

for $i = k$. Since $\nabla^j R = 0$ for $j > k$, we may take $i = \infty$ in Equation (4.b). Thus by Lemma 4.1, there is an isometry of $\mathcal{M}_{6+4p,f_k}$ taking $P_1$ to $P_2$. This shows $\mathcal{M}_{6+4p,f}$ is a homogeneous space. The argument is the same if $f = z_1 z_0^2 + \cdots + z_p z_0^{p+1} + z_0^{p+3}$ where we start with $i = p+1$ in Equation (4.b).

If $f = z_1 z_0^2 + \cdots + z_p z_0^{p+1} + ae^{bz_0}$, then

$$\alpha^k_{6+4p,\psi} = b^{k+p+3} b^{(p+3)(k-1)} b^{(p+4)(-k)}$$

is independent of the point in question. We use Theorem 1.9 (2) to see $\mathcal{M}_{6+4p,f}$ is a homogeneous space. □

**Proof of Theorem 1.10 (4).** By Theorem 1.9, (4*a*) ⇒ (4*b*) ⇒ (4*c*). Set $h = \psi^{(p+3)}$. If (4c) holds, then

$$k = \alpha^2_{6+4p,\psi} = h^{(2)} h \{h^{(1)}\}^{-2}.$$

We integrate the relation $h^{(2)} h = k h^{(1)} h^{(1)}$ to see there exist $(a, b)$ so

$$h(z_0) = \begin{cases} ae^{bz_0} & \text{if} \quad k = 1, \\ a(z_0 + b)^{1/(1-k)} & \text{if} \quad k \neq 1. \end{cases}$$

Since $a(z_0 + b)^{1/(1-k)}$ vanishes when $z_0 = -b$, these solutions are ruled out by the assumption $h$ is always positive and smooth. Consequently $h(z_0) = ae^{bz_0}$ and (4d) holds. By Theorem 1.9, (4*d*) ⇒ (4*a*). □

## Acknowledgments

Research of P. Gilkey partially supported by the Max Planck Institute in the Mathematical Sciences (Leipzig). Research of S. Nikčević partially supported by MM 1646 (Srbija) and by the DAAD (Germany). Both authors wish to express their thanks to the Technical University of Berlin where parts of research reported here were conducted.

## References

1. M. Belger and O. Kowalski, Riemannian metrics with the prescribed curvature tensor and all its covariant derivatives at one point, *Math. Nachr.* **168** (1994), 209–225.
2. E. Boeckx, O. Kowalski and L. Vanhecke, *Riemannian manifolds of conullity two*, World Scientific (1996).

3. P. Bueken, On curvature homogeneous three-dimensional Lorentzian manifolds, *J. Geom. Phys.* **22** (1997), 349–362.
4. P. Bueken and M. Djorić, Three-dimensional Lorentz metrics and curvature homogeneity of order one, *Ann. Global Anal. Geom.* **18** (2000), 85–103.
5. P. Bueken and L. Vanhecke, Examples of curvature homogeneous Lorentz metrics, *Classical Quantum Gravity* **14** (1997), L93–L96.
6. M. Cahen, J. Leroy, M. Parker, F. Tricerri and L. Vanhecke, Lorentz manifolds modeled on a Lorentz symmetric space, *J. Geom. Phys.* **7** (1990), 571–581.
7. G. Calvaruso, R.A. Marinosci and D. Perrone, Three-dimensional curvature homogeneou hypersurfaces, *Arch. Math. (Brno)* **36** (2000), 269–278.
8. D. Ferus, H. Karcher and H. Münzner, Cliffordalgebren und neue isoparametrische Hyperflächen, *Math. Z.* **177** (1981), 479–502.
9. E. García-Río, D. Kupeli, M.E. Vázquez-Abal and R. Vázquez-Lorenzo, Affine Osserman connections and their Riemann extensions. *Differential Geom. Appl.* **11** (1999), 145–153.
10. P. Gilkey and S. Nikčević, Curvature homogeneous spacelike Jordan Osserman pseudo-Riemannian manifolds, *Classical Quantum Gravity* **21** (2004), 497–507.
11. ______, Complete $k$-curvature homogeneous pseudo-Riemannian manifolds; math. DG/0405024.
12. A. Gray, The volume of a small geodesic ball of a Riemannian manifold, *Mich. Math. J.* **20** (1973), 329–344.
13. M. Gromov, *Partial differential relations*, Ergeb. Math. Grenzgeb 3. Folge, Band 9, Springer-Verlag (1986).
14. O. Kowalski, B. Opozda and Z. Vlášek, Curvature homogeneity of affine connections on two-dimensional manifolds, *Colloq. Math.* **81** (1999), 123–139.
15. ______, A classification of locally homogeneous affine connections with skew-symmetric Ricci tensor on 2 dimensional manifolds, *Monatsh. Math.* **130** (2000), 109–125.
16. ______, A classification of locally homogeneous connections on 2-dimensional manifolds via group-theoretical approach, *Central European Journal of Mathematics* **2** (2004), 87–102.
17. O. Kowalski and F. Prüfer, Curvature tensors in dimension four which do not belong to any curvature homogeneous space, *Arch. Math. (Brno)* **30** (1994), 45–57.
18. O. Kowalski, F. Tricerri and L. Vanhecke, Curvature homogeneous Riemannian manifolds *J. Math. Pures Appl.* **71** (1992), 471–501.
19. ______, Curvature homogeneous spaces with a solvable Lie group as homogeneous model, *J. Math. Soc. Japan* **44** (1992), 461–484.
20. A. Koutras and C. McIntosh, A metric with no symmetries or invariants, *Classical Quantum Gravity* **13** (1996), L47-L49.
21. B. Opozda, On curvature homogeneous and locally homogeneous affine connections, *Proc. Amer. Math. Soc.* **124** (1996), 1889–1893.
22. ______, Affine versions of Singer's theorem on locally homogeneous spaces,

*Ann. Global Anal. Geom.* **15** (1997), 187–199.

23. F. Podesta and A. Spiro, Introduzione ai Gruppi di Trasformazioni, *Volume of the Preprint Series of the Mathematics Department* "V. Volterra" of the University of Ancona, Via delle Brecce Bianche, Ancona, ITALY (1996).
24. F. Prüfer, F. Tricerri and L. Vanhecke, Curvature invariants, differential operators and local homogeneity, *Trans. Am. Math. Soc.* **348** (1996), 4643–4652.
25. V. Pravda, A. Pravdová, A. Coley and R. Milson, All spacetimes with vanishing curvature invariants, *Classical Quantum Gravity* **19** (2002), 6213–6236.
26. I.M. Singer, Infinitesimally homogeneous spaces, *Commun. Pure Appl. Math.* **13** (1960), 685–697.
27. K. Sekigawa, H. Suga and L. Vanhecke, Four-dimensional curvature homogeneous spaces, *Commentat. Math. Univ. Carol.* **33** (1992), 261–268.
28. ———, Curvature homogeneity for four-dimensional manifolds, *J. Korean Math. Soc.* **32** (1995), 93–101.
29. H. Takagi, On curvature homogeneity of Riemannian manifolds, *Tôhoku Math. J.* **26** (1974), 581–585.
30. A. Tomassini, Curvature homogeneous metrics on principal fibre bundles, *Ann. Mat. Pura Appl.* **172** (1997), 287–295.
31. F. Tricerri, Riemannian manifolds with the same curvature as a homogeneous space, and a conjecture of Gromov, *Geometry Conference* (Parma, 1988). Riv. Mat. Univ. Parma (4) **14** (1988), 91–104.
32. F. Tricerri and L. Vanhecke, Variétés riemanniennes dont le tenseur de courbure est celui d'un espace symétrique riemannien irréductible, *C. R. Acad. Sci.*, Paris, Sér. I **302** (1986), 233–235.
33. K. Tsukada, Curvature homogeneous hypersurfaces immersed in a real space form, *Tohoku Math. J.* **40** (1988), 221–244.
34. L. Vanhecke, Curvature homogeneity and related problems, *Proceedings of the Workshop on Recent Topics in Differential Geometry* (Puerto de la Cruz, 1990), 103–122, Informes, 32, Univ. La Laguna, La Laguna, 1991.
35. K. Yamato, Algebraic Riemann manifolds, *Nagoya Math. J.* **115** (1989), 87–104.

# MIYAOKA-YAU INEQUALITY AND COMPLEX HYPERBOLICITY

MITSUHIRO ITOH
*Institute of Mathematics, University of Tsukuba, 305-8571, JAPAN*
*E-mail: itohm@sakura.cc.tsukuba.ac.jp*

## 1. Main Theorems

In [13] we raised the following problem.

**Problem.** Whether a compact, connected, oriented Einstein self-dual 4-manifold of negative Ricci curvature is either a real hyperbolic or complex hyperbolic?

Here an oriented Riemannian 4-manifold is self-dual when its anti-self-dual Weyl curvature $W^-$ vanishes. As is known ([2]), the self-duality implies the integrability of the canonically defined almost complex structure on the twistor space.

If the 4-manifold $(M, g)$ has further $W^+ = 0$, then it must be real hyperbolic, since an Einstein, conformally flat manifold is a real space form.

Examples are compact quotients $\tilde{M}/\Gamma$ where $\tilde{M}$ is a real hyperbolic 4-manifold $\mathbf{RH}^4$ or a complex hyperbolic space $\mathbf{CH}^2$. So, the problem asks whether these examples exhaust such 4-manifolds or not.

Further there are the evidences for the problem in terms of rigidity. In fact, we have

**Theorem.** (Besson, Courtois and Gallot [3], LeBrun [11]) *Let $(M, g_0)$ be a compact real (or complex) hyperbolic 4-manifold. Then $g_0$ is a unique Einstein metric on $M$ up to homothety.*

The real hypebolic case was proved by Besson, Courtois and Gallot by using the volume entropy and the complex case by C. LeBrun in terms of the Seiberg-Witten theory.

Before investigating the above problem, we should refer to the Ricci positive case and the Ricci flat case.

Ricci positive Einstein, self-dual 4-manifolds are, up to homothety, only the standard 4-sphere $S^4$ and the complex projective plane $\mathbf{CP}^2$ with the Fubini-Study metric, as was shown by N. Hitchin ([6]).

On the other hand, Ricci flat, self-dual 4-manifolds admit in brief a "hyperkähler" structure, with respect to the reversed orientation so that a compact, connected, oriented Ricci flat anti-self-dual (i.e., $W^+ = 0$) 4-manifold is either a flat torus or a finite quotient of a Ricci flat K3 Kähler surface ([5]).

Together with ellipticity, flatness of spaces, hyperbolicity of space should complete the trichotomy of geometry, which is lucid in surfaces.

One of our main results is:

**Theorem 1.** *Let $(M, g)$ be a compact, connected oriented Einstein, self-dual (i.e., $W^- = 0$) 4-manifold with negative scalar curvature $s_g$.*

*Assume that $M$ satisfies $\chi(M) - 3\tau(M) \geq 0$ (Miyaoka-Yau inequality).*

*If, moreover, the modified Yamabe constant $\tilde{Y}([g])$ is non-negative for the modified scalar curvature $s_g - 6w^-(W_g^+)$, then $(M, g)$ is a compact quotient of the complex hyperbolic 2-space $\mathbf{CH}^2$. Here $\tilde{Y}([g])$ is the infimum over $g' \in [g]$ of the modified Yamabe functional.*

Here, the modified scalar curvature and the modified Yamabe functional are modified notion of the scalar curvature and the Yamabe functional which are conformally defined and will be explained in the next section. This theorem will be seen also in [9].

Note $\chi(M) - 3\tau(M) = (3c_2 - c_1^2)(M) = 0$ holds for a compact 4-manifold $(M, g)$ which is a quotient of complex hyperbolic space.

**Remark.** As will be seen in proving Theorem 1 in §§ 3, 4, we can weaken the Einstein hypothesis of Theorem 1 as simply the scalar curvature of $g$ being negative constant.

Another result which we present is the following related to the Goldberg conjecture.

**Theorem 2.** *Let $(M, g, J)$ be a compact, Einstein almost Kähler 4-manifold with negative scalar curvature $s_g$. If $M$ satisfies the Miyaoka-Yau equality $\chi(M) - 3\tau(M) = 0$, then $(M, g, J)$ is complex hyperbolic.*

The Miyaoka-Yau inequality holds for every compact, Einstein almost Kähler 4-manifold (see also [1]), because of the $L^2$-inequality of the scalar curvature and the self-dual Weyl curvature.

## 2. The modified scalar curvature and the $L^2$-curvature inequality

Recall the Yamabe functional associated with the scalar curvature for a given smooth conformal class $[g]$ as

$$Q : [g] \ni \tilde{g} \mapsto \int_M s_{\tilde{g}} dv_{\tilde{g}} / (\int_M dv_{\tilde{g}})^{2/p}, \quad p = 2n/(n-2).$$

A metric which attains the infimum of the $Q$ is called a Yamabe metric. Since the Yamabe problem is completely solved, there exists for each conformal class $[g]$ a Yamabe metric having constant scalar curvature.

Replace the scalar curvature term in $Q$ by a modified scalar curvature, i.e., the scalar curvature term together with a constant multiple of the largest (least) eigenvalue $w^{\pm}(W_g)$ of the Weyl curvature $W_g : \Lambda^2 \longrightarrow \Lambda^2$. Then we obtain a modified functional in terms of the modified scalar curvature;

$$\tilde{Q}(\tilde{g}) = \int_M \sigma_{\tilde{g}} dv_{\tilde{g}} / (\int_M dv_{\tilde{g}})^{2/p}, \quad \sigma_{\tilde{g}} = s_{\tilde{g}} + t w^{\pm}(W_{\tilde{g}}).$$

The infimum of $\{\, \tilde{Q}(g') \mid g' \in [g] \,\}$ is called the modified Yamabe constant, and a metric which attains the infimum is called a modified Yamabe metric. In general, a modified Yamabe metric is not necessarily smooth. It is in class $C^{2,\alpha}$. There exists a sufficient condition for the existence of modified Yamabe metric given by the best Sobolev constant of the round sphere. Refer to [9] for the details. See also [4] and [12] for its references.

When $\dim M = 4$ we decompose $W$ as $W = W^+ + W^-$ into the (anti-) self-dual Weyl curvature $W^{\pm} : \Lambda^2_{\pm} \longrightarrow \Lambda^2_{\pm}$ to consider the 4-dimensional modified scalar curvature in terms of $W_g^+$ as $\sigma_g^{\pm,t} = s_g + t w^{\pm}(W_g^+)$.

Assume that $(M, g)$ is a Kähler 4-manifold. Then $W^+$ is completely determined and thus for $t = -6$ $\sigma_g^{-,t} = s_g - 6w^-(W_g^+)$ is identically zero, when $s_g \leq 0$. Otherwise, when $s_g \geq 0$, $\sigma_g^{+,t} = s_g - 6w^+(W_g^+) = 0$.

Remark that the $*$-scalar curvature of an almost Kähler 4-manifold $(M, g, J)$ is also considered as a modified scalar curvature associated to the conformal class $[g]$, since $s_g^* = \frac{1}{3}\{s_g + 12\langle W_g^+(\theta), \theta\rangle / |\theta|_g^2\}$ (see [12] for this formula and note that $\theta$ is the Kähler form of $(g, J)$) obeys the conformal covariance rule (see [9] for the precise notion of modified scalar curvature). For the definition of the star scalar curvature the reader is referred to [14] and [1].

The following is an easy exercise.

**Lemma 1.** *It holds*

$$(w^{\pm}(W_g^+))^2 \leq \frac{2}{3}|W_g^+|^2$$

*and equality is achieved if and only if in an appropriate orthonormal base* $W_g^+$ *has the diagonal form* $W_g^+ = a\ \mathrm{diag}(1, -\frac{1}{2}, -\frac{1}{2})$, *where* $a = w^{\pm}(W_g^+)$. *Here the square-norm of* $W_g^+$ *is given by* $\sum_{i<j,k<\ell}(W^+(e_i, e_j, e_k, e_\ell))^2$

**Lemma 2.** *Let* $(M, g)$ *be a compact, connected oriented Riemannian 4-manifold of non-positive Yamabe constant. If the modified Yamabe constant* $\tilde{Y}([g]) = \inf_{\tilde{g}\in[g]} \tilde{Q}(\tilde{g})$ *is non-negative for the modified scalar curvature* $\sigma_g^- = s_g - 6w^-(W_g^+)$, *then for a metric* $g'$ *of constant scalar curvature* $s_{g'}$ *in* $[g]$ *it holds*

$$\int_M s_{g'}^2 dv_{g'} \leq 24 \int_M |W_{g'}^+|^2 dv_{g'}.$$

*Moreover, equality holds in this inequality if and only if* $W_{g'}^+$ *is diagonalizable everywhere as*

$$W_{g'}^+ = \frac{s_{g'}}{6}\ \mathrm{diag}\,(1, -\frac{1}{2}, -\frac{1}{2})$$

*so that* $w^-(W_{g'}^+) = \frac{s_{g'}}{6}$.

The proof will be given as follows. Since, from Lemma 1

$$s_{g'}^2 = s_{g'}(\sigma_{g'}^- + 6w^-(W_{g'}^+)) \leq s_{g'}\sigma_{g'}^- + 2\sqrt{6}(-s_{g'})|W_{g'}^+|,$$

we have the integration over $M$ as

$$\int_M s_{g'}^2 dv_{g'} \leq 2\sqrt{6}\int_M (-s_{g'})|W_{g'}^+|dv_{g'} \leq 2\sqrt{6}(\int_M s^2 dv)^{1/2}\,(\int_M |W^+|^2 dv)^{1/2}$$

to obtain the $L^2$-inequality in Lemma 2. If equality holds, then $|W_{g'}^+|$ is a constant multiple of $s_{g'}$, which is also constant and further the equality in Lemma 1 holds for $w^-(W^+)$ so that the self-dual Weyl curvature $W_{g'}^+$ must have the above diagonal form.

## 3. The self-dual Weyl curvature and the Kähler-Einstein condition

The following is a key in proving Theorem 1.

**Theorem 3.** *Let* $M$ *be a compact, connected oriented 4-manifold with a metric* $g$ *of negative Yamabe constant. If the modified Yamabe constant* $\tilde{Y}([g])$ *is non-negative for the modified scalar curvature* $\sigma_g^- = s_g -$

*$6w^-(W_g^+)$, then the $L^2$-square norm satisfies*

$$\int_M |W_g^+|^2 dv_g \geq \frac{4}{3}\, \pi^2(2\chi + 3\tau)(M)$$

*Furthermore, equality attains if and only if $M$ admits an Einstein metric $g'$ conformal to $g$ such that $(M, g')$ is either Kähler-Einstein or a quotient of a Kähler-Einstein 4-manifold by an anti-holomorphic involution.*

This theorem is a negative curvature version of M. Gursky's theorem([4]).

**Proof.** Choose a metric $g'$ conformal to $g$ having constant scalar curvature. Then from Lemma 2 $\int s_{g'}^2 dv \leq 24 \int |W^+|^2 dv$. The $L^2$-inequality, a conformal invariant, is then obtained by using the Chern-Weil characteristic formula

$$(2\chi + 3\tau)(M) = \frac{1}{4\pi^2} \int_M \{2|W^+|^2 + \frac{1}{24} s_{g'}^2 - \frac{1}{2}|\mathrm{Ric}^0|^2\} dv_{g'}.$$

Consider the equality situation. Then the equality implies vanishing of the trace-free Ricci tensor, namely the metric $g'$ is Einstein and also from Lemma 2 that the modified scalar curvature $s_{g'} - 6w^-(W_{g'}^+)$ vanishes and further $W_{g'}^+$ is diagonalizable as $\frac{s_{g'}}{6}(1, -1/2, -1/2)$ and hence the least eigenvalue $w^-$ of $W_{g'}^+$ is $s_{g'}/6$ with multiplicity one so that its eigenvectors yield over $M$ a real line bundle, a subbundle of the bundle $\Lambda_+^2(M)$. Then $M$ or its double covering, if necessary, admits a globally defined self-dual 2-form $\theta$ of norm $\sqrt{2}$ satisfying

$$W_{g'}^+(\theta) = \frac{s_g'}{6}\,\theta.$$

It is not difficult to show that $\theta$ is covariant constant with respect to the Levi-Civita connection of $g'$.

In fact, we take the rough Laplacian $\nabla^*\nabla$ of $W^+(\theta)$ to have

$$\nabla^*\nabla(W^+(\theta)) = (\nabla^*\nabla W^+)(\theta) + 2(\nabla W^+)(\nabla^*\theta) + W^+(\nabla^*\nabla\theta)$$

whose LHS is equal to $1/6\, s_{g'}(\nabla^*\nabla\theta)$. The first term of the RHS reduces to

$$(\nabla^*\nabla W^+)(\theta) = -\frac{1}{12} s_{g'}^2\, \theta - \{W^+, W^+\}\, \theta,$$

since $W_{g'}^+$ satisfies the divergence equation: $\delta W^+ = 0$ (this is because $g'$ is Einstein) so that one has the Weitzenböck formula (Proposition 3.6, [7])

$$0 = 2d^+\delta W^+ = \nabla^*\nabla W^+ + \frac{1}{2} s_{g'} W^+ + \{W^+, W^+\}.$$

Here we have $\{W^+, W^+\}\,\theta = -\frac{1}{12}s_{g'}^2\,\theta$. So, $(\nabla^*\nabla W^+)(\theta) = 0$.

Take next the inner product of $\nabla^*\nabla(W^+(\theta))$ with $\theta$ and integrate over $M$. So we have

$$\frac{1}{6}\int_M s_{g'}\langle\nabla^*\nabla\theta, \theta\rangle dv = 2\int_M \langle\nabla W^+(\nabla^*\theta), \theta\rangle dv + \int_M \langle W^+(\nabla^*\nabla\theta), \theta\rangle dv.$$

The second term of the RHS is $1/6\int_M s_{g'}\langle\nabla^*\nabla\theta, \theta\rangle dv$, since $W^+ : \Lambda_+^2 \longrightarrow \Lambda_+^2$ is self-adjoint. So

$$\int_M \langle\nabla W^+(\nabla^*\theta), \theta\rangle dv = 0.$$

By Stokes' theorem the LHS reduces to

$$\begin{aligned}&\int_M \langle W^+(\nabla\theta), \nabla\theta\rangle dv - \int_M \langle W^+(\nabla^*\nabla\theta), \theta\rangle dv\\ &= \int_M \langle W^+(\nabla\theta), \nabla\theta\rangle dv - \frac{1}{6}\int_M s_{g'}|\nabla\theta|^2 dv.\end{aligned}$$

For the first term of the RHS notice that $|\theta|^2 = 2$ everywhere so $\nabla_X\theta \in \Lambda_+^2$ is an eigenvector of $W^+$ of eigenvalue $-1/12\, s_{g'}$ and then

$$W^+(\nabla_X\theta) = -\frac{1}{12}s_{g'}\nabla_X\theta.$$

Therefore we obtain

$$-\frac{1}{4}\int_M s_{g'}|\nabla\theta|^2 dv_{g'} = 0,$$

that is, $\theta$ is covariant constant so that the almost complex structure $J$ given by $\theta(X, Y) = g(JX, Y)$ satisfies $\nabla J = 0$. Thus we obtain that $(g', J)$ is Kähler-Einstein. Conversely, it is not hard to see that for a Kähler-Einstein structure $(g', J)$ it holds $\int |W^+|^2 dv = 4/3\,\pi^2(2\chi + 3\tau)(M)$. □

## 4. Proof of Theorem 1

Our metric $g$ on the 4-manifold $M$ is Einstein and $s_g < 0$ and has $\tilde{Y}([g]) \geq 0$. So, we can apply Theorem 3 to $g$ and have the $L^2$-inequality

$$\int_M |W_g^+|^2 dv_g \geq \frac{4}{3}\,\pi^2(2\chi + 3\tau)(M).$$

Therefore from the inequality $(\chi - 3\tau)(M) \geq 0$ we have

$$\int_M |W_g^+|^2 dv_g \geq 12\pi^2\tau(M).$$

The Hirzebruch signature theorem asserts on the other hand

$$\int_M (|W_g^+|^2 - |W_g^-|^2)dv_g = \int_M |W_g^+|^2 dv_g = 12\pi^2\tau(M),$$

because $W_g^- = 0$. So we get $\chi = 3\tau$ and hence

$$\int_M |W_g^+|^2 dv_g = \frac{4}{3}\pi^2(2\chi + 3\tau)(M)$$

to conclude from Theorem 3 that $(M, g)$ or its double covering is Kähler-Einstein.

In a Kähler manifold of real dimension four the anti-self-dual Weyl curvature $W^-$ coincides with the Bochner curvature $B$. So one has $B = 0$ and hence from the uniformization theorem given by Kamishima ([10]) $(M, g)$ or its double covering is complex hyperbolic, since $s_g$ is negative.

## 5. Proof of Theorem 2

**Lemma 3.** *Let $(M, g)$ be a compact, connected oriented Einstein 4-manifold. If*

$$\int_M s_g^2 dv_g \geq 24 \int_M |W_g^+|^2 dv_g, \qquad (*)$$

*then*

$$\chi(M) \geq 3\tau(M).$$

*Furthermore, equality $\chi(M) = 3\tau(M)$ holds if and only if $\int s_g^2 dv_g = 24 \int |W_g^+|^2 dv_g$ and $W_g^- = 0$.*

This is from the Chern-Weil characteristic formula

$$(\chi - 3\tau)(M) = \frac{1}{4 \times 48\pi^2} \int_M (s_g^2 - 24|W_g^+|^2)dv_g + \frac{3}{8\pi^2} \int_M |W_g^-|^2 dv_g$$

Let $(M, g, J)$ be now a compact, connected, Einstein almost Kähler 4-manifold with $s_g < 0$. By making use of the argument of the Seiberg-Witten theory we have the Miyaoka-Yau inequality:

$$\chi(M) \geq 3\tau(M).$$

It is proven in [8] and [11] that by applying the Lichnerowicz formula to the twisted Dirac equation $D_A\psi = 0$ one obtains for any metric on $M$

$$\int_M s_g^2 dv_g \geq 32\pi^2(2\chi + 3\tau)(M). \qquad (**)$$

From the Chern-Weil characteristic formula applied to the $g$ which is almost Kähler, Einstein the RHS is written as

$$8\int_M \{2|W_g^+|^2 + \frac{1}{24}s_g^2 - \frac{1}{2}|\mathrm{Ric}^0|^2\} dv_g$$

so that (∗∗) is equivalent to the $L^2$-inequality (∗). Therefore one has from Lemma 3 that $\chi(M) \geq 3\tau(M)$.

However, $\chi(M) = 3\tau(M)$ is assumed in Theorem 2. So from the equality statement of Lemma 3 one has $\int s_g^2 dv_g = 24\int |W_g^+|^2 dv_g = 32\pi^2(2\chi + 3\tau)(M)$ and $W_g^- = 0$. Hence it is concluded that $(g, J)$ is Kähler from Theorem 1 in [8] and having vanishing Bochner curvature. So Theorem 2 is verified.

## References

1. J. Armstrong, On four-dimensional almost Kähler manifolds, Quart. J. Math. Oxford, 48(1997), 405–415.
2. M.F. Atiyah, N.J. Hitchin and I.M. Singer, Self-duality in four-dimensional Riemannian geometry, Proc. R. S. London, 362(1978), 425–461.
3. G. Besson, G. Courtois and S. Gallot, Entropies et rigidités des espaces localement symétriques de courbure strictement négative, Geom. Func. Anal., 5(1995), 731–799.
4. M. Gursky, Four-manifolds with $\delta W^+ = 0$ and Einstein constants of the sphere, Math. Ann., 318(2000), 417–431.
5. N. Hitchin, On compact four-dimensional Einstein manifolds, J. D. Geom., 9(1974), 435–442.
6. N. Hitchin, Kähler twistor spaces, Proc. London Math. Soc., 43(1981), 133–150.
7. M. Itoh, The Weitzenböck formula for the Bach operator, Nagoya Math. J., 137(1995), 149–181.
8. M. Itoh, Almost Kähler 4-manifolds, $L^2$-scalar Curvature Functional and Seiberg-Witten Equations, International Journ. Math., 15(2004), 573–580.
9. M. Itoh, The Modified Yamabe Problem and Geometry of Modified Scalar Curvatures, to appear in Journ. Geometric Analysis.
10. Y. Kamishima, Uniformization of Kähler manifolds with vanishing Bochner tensor, Acta Math., 172(1994), 299–308.
11. C. LeBrun, Einstein metrics and Mostow rigidity, Math. Res. Lett., 2(1995), 1–8.
12. C. LeBrun, Ricci curvature, minimal volumes, and Seiberg-Witten theory, Invent. math., 145(2001), 279–316.
13. R. Miyaoka and M. Kotani, *Mathematics in the 21st century unscaled peaks of geometry*, (in Japanese) Nihon-Hyoronsha, Tokyo, 2004.
14. K. Sekigawa, On some compact Einstein almost Kähler manifolds, J. Math. Soc. Japan, 39(1987), 677–684.

# NEUTRAL BI-HERMITIAN GRAY SURFACES

WŁODZIMIERZ JELONEK
*Institute of Mathematics*
*Technical University of Cracow*
*Warszawska 24*
*31-155 Kraków, POLAND.*
*E-mail: wjelon@usk.pk.edu.pl*

The aim of this paper is to give examples of compact neutral 4-manifolds $(M, g)$ whose Ricci tensor $\rho$ satisfies the relation $\nabla_X \rho(X, X) = \frac{1}{3} X\tau g(X, X)$. We present also a family of new Einstein bi-Hermitian neutral metrics on ruled surfaces of genus $g > 1$.

## 0. Introduction

In the present paper we are concerned with the class of neutral semi-Riemannian 4-manifolds $(M, g)$ whose Ricci tensor $\rho$ satisfies the condition

$$(*) \qquad \nabla_X \rho(X, X) = \frac{1}{3} X\tau g(X, X)$$

where $\tau$ is the scalar curvature of $(M, g)$. These class of manifolds was introduced by A. Gray (see [G], [Be]). For general facts and some results concerning neutral 4-manifolds we refer to [K], [Pe], [A], [D], [M-L]. In [M] the condition for the existence of two opposite almost complex structures on 4-manifolds are studied. Note that in [A] (p. 187) it is stated that there are no known neutral Hermitian-Einstein metrics on ruled surfaces except trivial ones. In this note we give among others many new, non-trivial examples of such manifolds. The main subject of the paper is the construction of new Einstein and Gray neutral metrics on ruled surfaces. The methods however are very similar to those used by me in [J-1]-[J-4].

In our papers [J-2]-[J-4] we have described Riemannian metrics $g$ on compact complex surfaces $(M, J)$ such that $(M, g)$ satisfies the condition

---

MS classification:53C05,53C20,53C25. The paper was partially supported by KBN grant 2P0 3A 02324.

(*) and has $J$-invariant Ricci tensor. In particular we have given a complete classification of bi-Hermitian Gray surfaces on ruled surfaces of genus $g \geq 1$ and surfaces which are of co-homogeneity 1 on ruled surfaces of genus $g = 0$. We denote by $g$ also the Riemannian metric but it should not cause any misunderstandings. The technics used in [J-3] can be applied to describe a large class of neutral bi-Hermitian Gray surfaces. The equations, describing such surfaces whose Ricci tensor has exactly two real eigenvalues such that the corresponding two dimensional eigendistributions are space-like and time-like, are described by the same equations as in [J-3], [J-4]. In that way we present a large class of explicit examples of neutral bi-Hermitian Gray surfaces. In particular we get a large class of Einstein neutral bi-Hermitian surfaces on ruled surfaces of genus $g > 1$.

## 1. Neutral $\mathcal{AC}^{\perp}$-surfaces

By an $\mathcal{AC}^{\perp}$- manifold (see [Be], [G]) we mean a semi-Riemannian manifold $(M, g)$ satisfying the condition

$$(*) \qquad \mathfrak{C}_{XYZ}\nabla_X\rho(Y,Z) = \frac{2}{(\dim M + 2)}\mathfrak{C}_{XYZ}X\tau g(Y,Z),$$

where $\rho$ and $\tau$ are the Ricci tensor and the scalar curvature of $(M, g)$ respectively and $\mathfrak{C}$ means the cyclic sum. In this paper, an $\mathcal{AC}^{\perp}$-manifold with neutral metric is also called a neutral Gray manifold. A Riemannian manifold $(M, g)$ is an $\mathcal{AC}^{\perp}$ manifold if and only if the Ricci endomorphism Ric of $(M, g)$ is of the form $\text{Ric} = S + \frac{2}{n+2}\tau Id$ where $S$ is a Killing tensor and $n = \dim M$. Let us recall that a symmetric (1,1) tensor $S$ on a semi-Riemannian manifold $(M, g)$ is called a Killing tensor if $g(\nabla S(X, X), X) = 0$ for all $X \in TM$ and that a semi-Riemannian manifold whose Ricci tensor is a Killing tensor is called an $\mathcal{A}$-manifold.

On a semi-Riemannian manifold $(M, g)$ a distribution $\mathcal{D} \subset TM$ is called umbilical (see [J-3]) if $\mathcal{D}$ is non-degenerate (i.e. a metric $g$ is non-degenerate on $\mathcal{D}$) and $\nabla_X X_{|\mathcal{D}^{\perp}} = g(X, X)\xi$ for every $X \in \Gamma(\mathcal{D})$, where $X_{|\mathcal{D}^{\perp}}$ is the $\mathcal{D}^{\perp}$ component of $X$ with respect to the orthogonal decomposition $TM = \mathcal{D} \oplus \mathcal{D}^{\perp}$. The vector field $\xi$ is called the mean curvature normal of $\mathcal{D}$. The foliation tangent to involutive distribution $\mathcal{D}$ is called totally geodesic if its every leaf is a totally geodesic (i.e. $\nabla_X X \in \mathcal{D}$ for any $X \in \mathcal{D}$) non-degenerate submanifold of $(M, g)$. In the sequel we shall not distinguish between $\mathcal{D}$ and a foliation tangent to $\mathcal{D}$ and we shall also say that $\mathcal{D}$ is totally geodesic in such a case.

It is not difficult to prove exactly as in the Riemannian case (see [J-3]) the following lemma:

**Lemma 1.** *Let $S \in End(TM)$ be a (1,1) tensor on a neutral semi-Riemannian 4-manifold $(M, g)$. Let us assume that $S$ has exactly two everywhere different eigenvalues $\lambda$, $\mu$ of the same multiplicity 2, i.e.* $\dim \mathcal{D}_\lambda = \dim \mathcal{D}_\mu = 2$, *where $\mathcal{D}_\lambda$, $\mathcal{D}_\mu$ are non-degenerate eigendistributions of $S$ corresponding to $\lambda$, $\mu$ respectively. Then $S$ is a Killing tensor if and only if both distributions $\mathcal{D}_\lambda$ and $\mathcal{D}_\mu$ are umbilical with mean curvature normal equal respectively*

$$\xi_\mu = \frac{\nabla \mu}{2(\lambda - \mu)}, \quad \xi_\lambda = \frac{\nabla \lambda}{2(\mu - \lambda)}.$$

We shall call a bi-Hermitian surface with bi-Hermitian Ricci tensor simply as bi-Hermitian surface. If $(M, g)$ is also an $\mathcal{AC}^\perp$ manifold, then we call it a neutral bi-Hermitian Gray surface.

**Proposition 1.** *Let us assume that $(M, g)$ is a simply connected neutral Gray 4-manifold and the Ricci tensor $S$ ($\rho(X, Y) = g(SX, Y)$) has exactly two real eigenvalues $\lambda$, $\mu$ and no null eigenvectors. Then there exist two Hermitian complex structures $J$, $\bar{J}$ commuting with $S$ and $(M, g)$ is a bi-Hermitian neutral Gray surface.*

**Proof.** Analogous to [J-1]. □

We shall prove

**Proposition 2.** *Let $(M, g)$ be a 4-dimensional neutral semi-Riemannian manifold. Let $\mathcal{D}$ be a two dimensional totally geodesic bundle-like and space-like foliation on $M$. Then $M$ admits (up to 4-fold covering) two opposite Hermitian structures $J, \bar{J}$ such that $J|\mathcal{D} = -\bar{J}|\mathcal{D}$, $J|\mathcal{D}^\perp = \bar{J}|\mathcal{D}^\perp$. The nullity distributions of both $J$, $\bar{J}$ contain $\mathcal{D}$.*

**Proof.** To prove the first part of the Proposition it is enough to show that on $M$ there exists a Killing tensor with eigendistributions $\mathcal{D}$, $\mathcal{D}^\perp$. Since the foliation $\mathcal{D}$ is totally geodesic and bundle-like it follows that $\nabla_X X \in \Gamma(\mathcal{D})$ (resp.) $\nabla_X X \in \Gamma(\mathcal{D}^\perp)$ if $X \in \Gamma(\mathcal{D})$ (resp.) $X \in \Gamma(\mathcal{D}^\perp)$. Consequently a tensor $S$ defined by

$$SX = \lambda X \quad \text{if} \quad X \in \mathcal{D}$$
$$SX = \mu X \quad \text{if} \quad X \in \mathcal{D}^\perp$$

where $\lambda \neq \mu$ are two different real numbers is a smooth Killing tensor. We can assume (up to 4-fold covering) that the distributions $\mathcal{D}$, $\mathcal{D}^\perp$ are orientable. Let us denote by $J$ the only almost Hermitian structure which preserves $\mathcal{D}$, $\mathcal{D}^\perp$ and agrees with their orientations. We define $\bar{J}$ by $J|\mathcal{D} = -\bar{J}|\mathcal{D}$, $J|\mathcal{D}^\perp = \bar{J}|\mathcal{D}^\perp$. From Proposition 1 it follows that both $J$, $\bar{J}$ are Hermitian. Now let $\{E_1, E_2\}$ be a local orthonormal frame on $\mathcal{D}$. Then

$$\nabla J(E_1, E_1) + J(\nabla_{E_1} E_1) = \nabla_{E_1} E_2.$$

It follows that $\nabla J(E_1, E_1) \in \mathcal{D}$ and consequently $\nabla_{E_1} J = 0$. Analogously $\nabla_{E_2} J = 0$. □

Now we give a theorem, whose proof is analogous as in the Riemannian case (see [M-S], [B], [S]).

**Theorem 1.** *Let us consider the manifold $U = (a, b) \times P$, where $(P, g_0)$ is a 3-dimensional semi-Riemannian $\mathcal{A}$-manifold (a principle $S^1$ bundle $p : P \to \Sigma$) over a Riemannian surface $(\Sigma, g_{can})$ of constant sectional curvature $K$, with an $\mathcal{AC}^\perp$-metric*

$$g = dt^2 + f(t)^2\theta^2 - g(t)^2 p^* g_{can}, \tag{1.5}$$

*where $g_0 = \theta^2 - p^* g_{can}$, $f, g \in C^\infty(a, b)$ and $\theta$ is the connection form of $P$. Then the metric $g$ on $U$ extends to a smooth $\mathcal{AC}^\perp$ metric on the ruled surface $M$ which is a $\mathbb{C}P^1$-bundle over $\Sigma$ and such that $U$ is an open and dense subset of $M$ if and only if the functions $f, g \in C^\infty(a, b)$ satisfy the conditions:*

(a) $f(a) = f(b) = 0$, $f'(a) = 1$, $f'(b) = -1$,

(b) $g(a) \neq 0 \neq g(b)$, $g'(a) = g'(b) = 0$.

**Remark.** Let us note that the metric (1.5) induces a semi-Riemannian metric on $M$ if the functions $f$, $g$ satisfy:

(i) $f$ is positive on $(a, b)$, and $f$ is odd at $a$ and $b$, i.e. $f$ is the restriction of a function $f$ on $\mathbb{R}$ satisfying $f(a+t) = -f(a-t)$, $f(b+t) = -f(b-t)$;

(ii) $g$ is positive on $[a, b]$ and even at $a$ and $b$ which means that $g$ is the restriction of a function $g$ such that $g(a+t) = g(a-t)$, $g(b+t) = g(b-t)$;

The proof is similar to the description of the metric in polar coordinates (see [Be] Lemma 9.114 and Theorem 9.125).

Note that it is an easy exercise to prove that functions $f$, $g$ satisfying the ODE characterizing $\mathcal{AC}^\perp$-metrics together with the initial conditions (a), (b) extend to respectively odd and even (with respect to the points

$a$, $b$) real analytic functions around $a$, $b$. The fact that the conditions (a), (b) of Theorem 1 imply (i) and (ii) is based on the following elementary Lemma.

**Lemma 2.** *Let us assume that the function $Q$ is real analytic in a closed interval $[x_0, x_1]$, $Q$ is positive in $(x_0, x_1)$ and $Q(x_0) = Q(x_1) = 0$, $Q'(x_0) \neq 0$, $Q'(x_1) \neq 0$. Let a real function $\phi(t)$ satisfy an equation*

$$\phi'' = \frac{1}{2}Q'(\phi), \quad \phi(0) = x_1, \quad \phi'(0) = 0, \tag{1.6}$$

*Then $(\phi')^2 = Q(\phi)$ and $\phi$ is a real analytic and periodic function $\phi \in C^\omega(\mathbb{R})$ satisfying conditions $\phi(0) = x_0$, $\phi(l) = x_1$ where $l$ is the first positive real number such that $\phi(l) = x_1$. Also $\phi$ is an even function at $t = 0$ and $t = l$.*

**Proof.** The first part of Lemma is well known (see [B], p. 445, Lemma 16.37). If $\phi$ satisfies equation (1.6) then $(\phi')^2 = Q_0(\phi)$ where $Q_0$ is a primitive function of $Q'$. From the boundary conditions $Q_0 = Q$. It is enough to show that $\phi$ is even at $t = 0$ (for $t = l$ the proof is similar) which is equivalent to $\phi^{(2k-1)}(0) = 0$ for every $k \in \mathbb{N}$. We prove by induction with respect to $k$ that for every $F \in C^\omega(\mathbb{R})$ the function $\psi(t) = F(\phi(t))$ is even at $t = 0$. If $k = 1$ then $\psi'(0) = F'(\phi(0))\phi'(0) = 0$. Let us assume that the result holds for $2k - 1$. Let us write $\frac{1}{2}Q' = Q_1$. We get

$$\begin{aligned}
\psi^{(2k+1)} &= (\psi'')^{(2k-1)} = (F''(\phi)(\phi')^2 + F'(\phi)\phi'')^{(2k-1)} \\
&= (F''(\phi)Q(\phi) + F'(\phi)Q_1(\phi))^{(2k-1)} \\
&= \sum \frac{(2k-1)!}{p!(2k-1-p)!}\big((F''(\phi))^{(p)}Q(\phi)^{(2k-1-p)} \\
&\qquad\qquad + (F'(\phi))^{(p)}(Q_1(\phi))^{(2k-1-p)}\big).
\end{aligned}$$

Thus $\psi^{(2k+1)}(0) = 0$. It follows that $F(\phi)$ is even at 0 for every $F$. If $F = id$ then we get that $\phi$ is even at 0. □

Using this lemma one can show that in every case considered below in the paper function $g$ is even and $f$ is odd. Both functions are non-negative and $g$ is positive which follows from the formula on $g$.

## 2. Neutral bi-Hermitian Gray surfaces

In this section we shall construct bi-Hermitian neutral metrics on ruled surfaces $M_{k,g}$ of genus $g$. The ruled surface $(M_{k,g}, g)$ is locally of cohomogeneity 1 with respect to the group of all local isometries of $(M_{k,g}, g)$

and an open, dense submanifold $(U_{k,g}, g) \subset (M_{k,g}, g)$ is isometric to the manifold $(a,b) \times P_k$ where $(P_k, g_k)$ is a 3-dimensional semi-Riemannian $\mathcal{A}$-manifold (a principal circle bundle $p : P_k \to \Sigma_g$) over a Riemannian surface $(\Sigma_g, g_{can})$ of constant sectional curvature $K \in \{-4, 0, 4\}$ with a metric

$$g_{f,g} = dt^2 + f(t)^2\theta^2 - g(t)^2 p^* g_{can}, \tag{2.1}$$

where $g_k = \theta^2 - p^* g_{can}$ and $\theta$ is the connection form of $P_k$ such that $d\theta = 2\pi k\, p^*\omega$, where $\omega$ is proportional to the volume form of $(\Sigma_g, g_{can})$ and the de Rham cohomology class $[\omega] \in H^2(\Sigma_g, \mathbb{R})$ defined by the form $\omega$ is an integral class corresponding to the class $1 \in H^2(\Sigma_g, \mathbb{Z}) = \mathbb{Z}$ i.e. $\int_{\Sigma_g} \omega = 1$. Let $\theta^\sharp$ be a vector field dual to $\theta$ with respect to $g_k$. Let us consider a local orthonormal frame $\{X, Y\}$ on $(\Sigma_g, g_{can})$ and let $X^h$, $Y^h$ be horizontal lifts of $X$, $Y$ with respect to $p : M_{k,g} \to \Sigma_g$ (i.e. $dt(X^h) = \theta(X^h) = 0$ and $p(X^h) = X$) and let $H = \frac{\partial}{\partial t}$. Let us define two almost Hermitian structures $J, \bar{J}$ on $M$ as follows

$$JH = \frac{1}{f}\theta^\sharp, \quad JX^h = Y^h, \quad \bar{J}H = -\frac{1}{f}\theta^\sharp, \quad \bar{J}X^h = Y^h.$$

**Proposition 3.** *Let $\mathcal{D}$ be a distribution spanned by the fields $\{\theta^\sharp, H\}$. Then $\mathcal{D}$ is a non-degenerate totally geodesic foliation with respect to the metric $g_{f,g}$ where $g$ is a non constant function. Both structures $J$, $\bar{J}$ are Hermitian and $\mathcal{D}$ is contained in the nullity of $J$ and $\bar{J}$. The distribution $\mathcal{D}^\perp$ is umbilical with the mean curvature normal $\xi = -\nabla \ln g$. Let $\lambda$, $\mu$ be eigenvalues of the Ricci tensor $S$ of $g_{f,g}$ corresponding to eigendistributions $\mathcal{D}$, $\mathcal{D}^\perp$ respectively. Then the following conditions are equivalent:*

(a) *There exists $D \in \mathbb{R}$ such that $\lambda - \mu = Dg^2$,*

(b) *There exist $C, D \in \mathbb{R}$ such that $\mu = Dg^2 - C$,*

(c) *$\lambda - 2\mu$ is constant,*

(d) *$(U_{k,g}, g_{f,g})$ is a neutral bi-Hermitian Gray surface.*

**Proof.** The first assertion of Proposition 3 is a consequence similar to Proposition 3 in [J-2]. Note that $\nabla\lambda = H\lambda H$, $\nabla\mu = H\mu H$. Consequently $tr_g \nabla S = \frac{1}{2}\nabla\tau = (H\lambda + H\mu)H$. On the other hand one can easily check that $tr_g \nabla S = 2(\mu - \lambda)\xi + H\lambda H$. Thus

$$\nabla\mu = 2(\lambda - \mu)\nabla \ln g.$$

Now we prove that (a)$\Rightarrow$(b). If (a) holds then $\nabla\mu = 2Dg^2\frac{\nabla g}{g} = D\nabla g^2$. Thus $\nabla(\mu - Dg^2) = 0$ which implies (b).

(b)⇒(a). We have

$$-\frac{\nabla g}{g}2(\mu-\lambda)=\nabla\mu=2Dg\nabla g,$$

and consequently $\nabla g(\frac{(\mu-\lambda)}{g}+Dg)=0$ which is equivalent to (a).

(a)⇒(c). We have $\lambda-\mu=Dg^2$ and consequently $\nabla\mu=2Dg\nabla g=D\nabla g^2$. Thus $\nabla\lambda=\nabla(\mu+Dg^2)=2D\nabla g^2$ and $\nabla\lambda-2\nabla\mu=0$ which gives (c).

(c)⇒(a). If $\nabla\lambda=2\nabla\mu$ then $\nabla\lambda=4(\lambda-\mu)\frac{\nabla g}{g}$. Consequently $\nabla\lambda-\nabla\mu=2(\lambda-\mu)\frac{\nabla g}{g}$ and $\nabla\ln|\lambda-\mu|=2\frac{\nabla g}{g}=2\nabla\ln g$, which means that $\nabla\ln|\lambda-\mu|g^{-2}=0$ or $\lambda=\mu$ on the whole of $M$. In the last case we obtain an Einstein metric. It follows that $\ln\frac{|\lambda-\mu|}{g^2}=\ln D$ for some $D\in\mathbb{R}_+$ or $\lambda-\mu=0$ $(D=0)$, which is equivalent to (a).

(d)⇔(c). This equivalence follows from [J-3]. □

**Theorem 2.** *On any ruled surface $M_{k,g}$ of genus $g$ with $k>0$ there exists a one-parameter family of neutral bi-Hermitian $\mathcal{AC}^{\perp}$-metrics $\{g_x : x\in(0,\epsilon_s)\}$, where $\epsilon_s>0$ depends only on $g$ and $k$, which consists of neutral bi-Hermitian Gray metrics on $M_{k,g}$.*

**Proof.** Note that for the first Chern class $c_1(\Sigma_g)\in H^2(\Sigma_g,\mathbb{Z})$ of the complex curve $\Sigma_g$ we have the relation $c_1(\Sigma_g)=\chi\alpha$, where $\alpha\in H^2(\Sigma_g,\mathbb{Z})$ is an indivisible integral class and $\chi=2-2g$ is the Euler characteristic of $\Sigma_g$. Let us write $s=\frac{2k}{|\chi|}$ if $g\neq 1$ and $s=k$ if $g=1$. Then it is easy to show using O'Neill formulas for a semi-Riemannian submersion (see [ON], [B], [S]) that the manifold $(M_{k,g},g)$ with the metric $g$ given by (2.1) has the Ricci tensor with the following eigenvalues :

$$\lambda_0=-2\frac{g''}{g}-\frac{f''}{f}, \tag{2.2a}$$

$$\lambda_1=-\frac{f''}{f}-2\frac{f'g'}{fg}+2s^2\frac{f^2}{g^4}, \tag{2.2b}$$

$$\lambda_2=-\frac{g''}{g}-\frac{f'g'}{fg}-\left(\frac{g'}{g}\right)^2-2s^2\frac{f^2}{g^4}-\frac{K}{g^2}, \tag{2.2c}$$

where $\lambda_0$, $\lambda_1$, correspond to eigenfields $T=\frac{d}{dt}$, $\theta^\sharp$ and $\lambda_2$ corresponds to a two-dimensional eigendistribution orthogonal to $T$ and $\theta^\sharp$. Note that the equations we have got are practically the same as equations for Gray manifolds obtained in the Riemannian case. The only difference is that we now have $-K$ instead of $K$, so the cases $K>0$ and $K<0$ are now reversed, which give the different geometric meaning to the equations in Riemannian and semi-Riemannian neutral cases. We present here the method of solving equations (2.2) for the convenience of the reader and completeness, although the calculations are just the same as in [J-4]. If $(M,g) \in \mathcal{AC}^{\perp}$ is a bi-Hermitian Gray surface then $\lambda_0 = \lambda_1 = \lambda$, and if we denote $\mu = \lambda_2$, Proposition 3 (b) implies an equation

$$\mu = Dg^2 - C \tag{2.3}$$

for some $D, C \in \mathbb{R}$. Since $\lambda_0 = \lambda_1$ we get

$$f = \pm\frac{gg'}{\sqrt{s^2 + Ag^2}}. \tag{2.4}$$

Using a homothety of the metric we can assume that $A \in \{-1, 0, 1\}$. In the case $A = 0$ we get a neutral Kähler metric and these metrics on compact complex surfaces we shall describe in section 4. So we restrict our considerations to the case $A \in \{-1, 1\}$. Now we introduce a function $h$ such that $h^2 = s^2 + Ag^2$. Note that $\mathrm{im}h \subset (-s, s)$ if $A = -1$, and $\mathrm{im}h \subset (s, \infty)$ if $A = 1$. Then $g = \sqrt{|s^2 - h^2|}$. Let us introduce a function $z$ such that $h' = \sqrt{z(h)}$. Note that

$$f = h' \quad \text{and} \quad f' = \frac{1}{2}z'(h). \tag{2.5}$$

It follows that equation (2.3) is equivalent to

$$z'(h) - z(h)\frac{s^2 + h^2}{h(s^2 - h^2)} = \frac{4\epsilon}{h} + \frac{D(s^2 - h^2)^2}{h} - \frac{C(s^2 - h^2)}{h}, \tag{2.6}$$

where $\epsilon = -\mathrm{sgn}KA \in \{-1, 0, 1\}$. It follows that

$$\begin{aligned} z(h) = \left(1 - \left(\frac{h}{s}\right)^2\right)^{-1} &\left(-4\epsilon\left(\frac{h}{s}\right)^2 - \frac{Ds^4}{5}\left(\frac{h}{s}\right)^6\right. \\ &+ (Ds^4 - \frac{Cs^2}{3})\left(\frac{h}{s}\right)^4 + (2Cs^2 - 3Ds^4)\left(\frac{h}{s}\right)^2 \\ &\left. - 4\epsilon + Cs^2 - Ds^4 + \frac{E}{s}\frac{h}{s}\right). \end{aligned} \tag{2.7}$$

Let us denote again $C = Cs^2$, $D = Ds^4$, $E = \frac{E}{s}$ and let

$$\begin{aligned} z_0(t) = (1 - t^2)^{-1} &\left(-4\epsilon(1 + t^2) + D\left(-\frac{1}{5}t^6 + t^4 - 3t^2 - 1\right)\right. \\ &\left. + C\left(-\frac{1}{3}t^4 + 2t^2 + 1\right) + Et\right). \end{aligned} \tag{2.8}$$

Write

$$P(t) = -4\epsilon t^2 - \frac{D}{5}t^6 + \left(D - \frac{C}{3}\right)t^4 \tag{2.9}$$
$$+ (2C - 3D)t^2 + Et - 4\epsilon + C - D.$$

Then $z_0(t) = \frac{P(t)}{1-t^2}$. Note that $z(h) = z_0\left(\frac{h}{s}\right)$ and $z'(h) = \frac{1}{s}z_0'\left(\frac{h}{s}\right)$. In view of Th. 1. we are looking for real numbers $x, y \in \mathbb{R}$, $x > y$ such that

$$z_0(x) = 0, z_0'(x) = -2s, \tag{2.10a}$$

$$z_0(y) = 0, z_0'(y) = 2s, \tag{2.10b}$$

and $z_0(t) > 0$ for $t \in (y, x)$. Note that equations (2.10a) are equivalent to

$$-4\epsilon x^2 - \frac{D}{5}x^6 + \left(D - \frac{C}{3}\right)x^4 \tag{2.11a}$$
$$+ (2C - 3D)x^2 - 4\epsilon + C - D + Ex = 0$$

$$-8\epsilon x - \frac{6D}{5}x^5 + 4\left(D - \frac{C}{3}\right)x^3 \tag{2.11b}$$
$$+ 2(2C - 3D)x + E = -2s(1 - x^2).$$

Equations (2.11) yield

$$D = \frac{5(-3E - 6s - 24\epsilon x + 3Ex^2 - 12sx^2 - 8\epsilon x^3 + 2sx^4)}{2(-1 + x)x(1 + x)(15 + 10x^2 - x^4)}, \tag{2.12a}$$

(2.12b)

$$C = \frac{3(5E + 10s + 80\epsilon x + 30sx^2 - 10Ex^2 + 5Ex^4 - 10sx^4 - 16\epsilon x^5 + 2sx^6)}{2(-1 + x)x(1 + x)(-15 - 10x^2 + x^4)}$$

Solving in a similar way equations (2.10b) one can see that there exists a function $z_0$ satisfying the equations (2.10) if

$$(x + y)(-4\epsilon(-5x + x^3 + 5y + 2x^2y - 2xy^2 - y^3) \tag{2.13}$$
$$+ s(5 + 2x^3y + 2xy^3 + 3y^2 + 3x^2 + x^2y^2 - 16xy)) = 0,$$

where $x > y$, $x, y \in (-1, 1)$ in the case $A = -1$ and $x, y \in (1, \infty)$ in the case $A = 1$. Using standard methods one can check that in the case of the genus $g \leq 1$ (i.e. if $K = 4$ or $K = 0$) the only solutions of (2.13) giving a positive function $z$ are $x = -y \in (0, 1)$. In the case of genus $g \geq 2$ we have $K = -4$ and apart from the solutions with $x = -y$ (see [J-3]) there is an additional family of solutions with $\epsilon = -1$ for $s \in (0, \eta)$ on $M_{k,g}$ where $\eta = \frac{2}{17}(15 + 4\sqrt{13})\sqrt{\frac{1}{3}(13(10\sqrt{13} - 35)))} \simeq 2.05318...$ One can check that there exist families of solutions with $-1 < y < x < 1$ for $k \in \mathbb{N}$ such that

$k < (g-1)\eta$. In fact it is not difficult to check that if $g > 1$ then on the sets $F = \{(x,y) : -1 < y < x < 1\}$ and $H = \{(x,y) : 1 < y < x\}$ the function $G(x,y) = -4\epsilon(-5x + x^3 + 5y + 2x^2y - 2xy^2 - y^3) + s(5 + 2x^3y + 2xy^3 + 3y^2 + 3x^2 + x^2y^2 - 16xy)$ is negative and positive somewhere on the boundaries of both of $F$ and $H$ for $s < 2$. In the case $g \leq 1$ $G$ is positive on both $F$ and $H$. We have

$$G_x(x,y) = -4\epsilon(-5 + 3x^2 + 4xy - 2y^2) + 2s(3x^2y + y^2x + 3x - 8y + y^3),$$
$$G_y(x,y) = -4\epsilon(5 + 2x^2 - 4xy - 3y^2) + 2s(x^3 + 3y + x^2y - 8x + 3x^2y).$$

The equations $G_x = 0, G_y = 0$ are equivalent to

$$-4\epsilon(-5 + 3x^2 + 4xy - 2y^2) + 2s(3x^2y + y^2x + 3x - 8y + y^3) = 0,$$
$$(x+y)(10\epsilon(y-x) + s(-5 + x^2 + 3xy + y^2)) = 0.$$

Consequently for $\epsilon = -1$ and $s \in [2,\eta)$ $G$ attains its negative minimum in $F$ on the line $x = -y$. It follows that if $g \leq 1$ then $x = -y \in (0,1), E = 0$ and $\epsilon = 1$ or $\epsilon = 0$. These solutions are also valid for $g \geq 2$ but in this case we have also solutions with $x \neq -y$ both for $\epsilon = -1$ and $s \in (0,\eta)$ (then $-1 < y < x < 1$) and for $\epsilon = 1$ and $s \in (0,2)$ (in this case there exist solutions with $1 < y < x$). Now we give explicit formulas for the case $x = -y \in (0,1)$. Then we get

$$(2.14) \quad P(t) = \frac{1}{x(15 - 5x^2 - 11x^4 + x^6)}((t^2 - x^2)(s(-15 + 10x^2 - 3x^4) + t^2(10 + 12x^2 - 6x^4) + t^4(-3 - 6x^2 + x^4)) + 4\epsilon x(x^2(-5 + x^2) - t^4(3 + x^2) + t^2(5 + 2x^2 + x^4)))).$$

Thus

$$P(0) = \frac{-4\epsilon x^4(x^2 - 5) + sx(15 - 10x^2 + 3x^4)}{15 - 5x^2 - 11x^4 + x^6}$$

and, since $\lim_{x\to 0^+} \frac{P(0)}{x} = s > 0$, there exists $\epsilon_s > 0$ such that $P(t) > 0$ if $t \in (0,x)$ for all $x \in (0,\epsilon_s)$. In fact in the case $\epsilon = -1$ the real number $\epsilon_s$ is the first positive root of the polynomial $-4x^3(x^2-5) + s(-15 + 10x^2 - 3x^4)$ and $\epsilon_s = 1$ if $\epsilon = 1$. Note also that in both cases $\epsilon_s = 1$ if $s \geq 2$. Now the function $z_0(t) = \frac{1}{(1-t^2)}P(t)$ is positive on $(-x,x)$, $x \in (0,\epsilon_s)$. If $x \in (0,\epsilon_s)$ then there exists a solution $h : (-a,a) \to (-sx, sx)$, where

$$a = \lim_{t\to sx^-} \int_0^t \frac{dh}{\sqrt{z_0\left(\frac{h}{s}\right)}},$$

of an equation

$$h' = \sqrt{z_0\left(\frac{h}{s}\right)},$$

such that $h(-a) = -sx$, $h(a) = sx$, $h'(-a) = h'(a) = 0$, $h''(-a) = 1$, $h''(a) = -1$. It follows that functions $f = h'$, $g = \sqrt{s^2 - h^2}$ are smooth on $(-a, a)$ and satisfy the boundary conditions described in Th. 2. Consequently the metric

$$g_x = dt^2 + f(t)^2\theta^2 - g(t)^2 p^* g_{can},$$

on the manifold $(-a, a) \times P_k$ extends to the smooth metric on the compact ruled surface $M = P_k \times_{S^1} S^2$ which is a 2−sphere bundle over Riemannian surface $\Sigma_g$. Note that $g(-a) = g(a) = s\sqrt{1-x^2}$. □

**Theorem 3.** *On the surfaces* $M_{0,g} = \mathbb{CP}^1 \times \Sigma_g$ *where* $g \geq 2$ *there exists a one-parameter family* $\{g_\alpha : \alpha > 1\}$ *of bi-Hermitian* $\mathcal{AC}^{\perp}$*-metrics. The Ricci tensor* $\rho = \rho_\alpha$ *of* $(\Sigma_0, g_\alpha)$ *is bi-Hermitian and has two eigenvalues, which are everywhere different.*

**Proof.** Analogous to [J-3]. Let us write $g' = \sqrt{P(g)}$. Then

$$g^2 P''(g) - 2P'(g)g - 4P(g) + 16 + 6Cg^2 = 0. \tag{2.15}$$

Consequently

$$P(g) = \frac{A}{g} + Bg^4 + Cg^2 + 4, \tag{2.16}$$

where $A, B \in \mathbb{R}$ are arbitrary. Now let $D = 1$ and let us consider the equations (we are looking for unknown real numbers $A, B, C$ and $(x, y)$ where $0 < y < x$)

$$\begin{aligned} P(y) = 0, P(x) = 0,\\ P'(y) = 2, P'(x) = -2. \end{aligned} \tag{2.17}$$

Then $y = \frac{4(\alpha-1)(\alpha^2+3\alpha+1)}{\alpha(2\alpha^2+\alpha+2)}$, $x = \alpha y = \frac{4(\alpha-1)(\alpha^2+3\alpha+1)}{(2\alpha^2+\alpha+2)}$ where $\alpha > 1$. Note that $x, y > 0$ and $C > 0$, $A, B < 0$. Let us consider an equation ($P = P_\alpha$, $h = h_\alpha$ depend on the parameter $\alpha > 1$)

$$\frac{d^2h}{dt^2} = \frac{1}{2}P'(h), \quad h'(0) = 0, \quad h(0) = y = \frac{4(\alpha-1)(\alpha^2+3\alpha+1)}{\alpha(2\alpha^2+\alpha+2)}. \tag{2.18}$$

This equation is equivalent to (if $t \in D = \{t \geq 0 : h'(t) \geq 0\}$)

$$\frac{dh}{dt} = \sqrt{P(h)}, \quad h(0) = y = \frac{4(\alpha-1)(\alpha^2+3\alpha+1)}{\alpha(2\alpha^2+\alpha+2)}. \tag{2.19}$$

It follows that $P = P_\alpha$, where $\alpha > 1$, has exactly two positive roots $\{x, y\}$ and $P(t) > 0$ if $t \in (y, x)$. Note that equation (2.18) admits a smooth periodic solution $h$ defined on the whole of $\mathbb{R}$ and such that $\operatorname{im} h = [x, y]$. Now it is easy to check that $\lambda = -10Bh^2 - 3C$ and $\mu = -5Bh^2 - 3C$. The tensor $\rho - \frac{\tau}{3}g$ is a Killing tensor with eigenvalues $C$, $5Bh^2 + C$ corresponding to $\mathcal{D}$, $\mathcal{D}^\perp$ respectively. Then we obtain a one parameter family of bi-Hermitian $\mathcal{AC}^\perp$-metrics $\{g_\alpha : \alpha > 1\}$ on $M_{0,g}$. □

## 3. Einstein neutral bi-Hermitian surfaces

Let us note that the solutions $(M_{k,g}, g_x)$ with $D(x) = 0$ correspond to Einstein neutral surfaces. Let us recall that

$$D = \frac{5(-6s - 24\epsilon x - 12sx^2 - 8\epsilon x^3 + 2sx^4)}{2(-1+x)x(1+x)(15+10x^2-x^4)} \tag{3.1}$$

Consequently there exist Einstein metrics in the family of Gray metrics $(M_{k,g}, g_x)$ if and only if the equation

$$-6s - 24\epsilon x - 12sx^2 - 8\epsilon x^3 + 2sx^4 = 0 \tag{3.2}$$

has a real root $x \in (0,1)$. Now it is not difficult to check that in the case $g = 0$ we have $\epsilon = 1$ and $s \in \mathbb{N}$ and that equation (3.2) does not have any solution in $(0,1)$. Similarly in the case $g = 1$. In the case $g > 1$ we have $s = \frac{k}{g-1} > 0$ and $\epsilon = -1$. In that case equation (3.2) has a real root in $(0,1)$ if and only if $s \in (0,2)$. To show this, let us denote

$$Q(x) = -6s - 24\epsilon x - 12sx^2 - 8\epsilon x^3 + 2sx^4.$$

Then

$$Q''(x) = -24(s + 2\epsilon x - sx^2).$$

Consequently $Q''(x) < 0$ in $(0,1)$ for $\epsilon \geq 0$. In the case where $\epsilon = -1$, since $Q''(0) = -24s < 0$ and $Q''(1) = 48 > 0$, the equation $Q''(x) = 0$ has exactly real one root, say $\alpha_s$. Then, we have $Q'(\alpha_s) > 0$ if and only if $(0 <) s < (\sqrt{3}/2)^{\frac{1}{2}}(\sqrt{3}+1)$. In particular, for $0 < s < 2 (< (\sqrt{3}/2)^{\frac{1}{2}}(\sqrt{3}+1)$, $Q'(x) \geq Q'(\alpha_s) > 0$, and hence $Q(x)$ is monotone increasing, in $(0,1)$. Since $Q(0) = -6s < 0$ and $Q(1) = 32 - 16s$, it follows that $Q(x)$ has exactly one root in $(0,1)$ if $s \in (0,2)$. On the other hand, if $s \geq 2$, then, since $3+6x^2-x^4>0$ and $x^4<x^3$ in $(0,1)$, we see that

$$\begin{aligned} Q(x) &= 24x+8x^3-2s(3+6x^2-x^4) \leq 24x+8x^3-4(3+6x^2-x^4) \\ &< 24x+8x^3-12-24x^2+4x^3 = 12(x^3-2x^2+2x-1) \\ &= 12(x-1)(x^2-x+1) < 0. \end{aligned}$$

Thus the equation $Q(x)=0$ has exactly one real root $x \in (0,1)$ if and only if $s \in (0,2)$. Since $s = \frac{k}{g-1}$ it follows that the real root $x \in (0,1)$ exists if and only if $k \in \{1,2,3,\ldots,2g-3\}$. Note that for $D=0$ the polynomial $P$ is even and of degree 4 with exactly one positive root. Consequently we get an Einstein metric on $M_{k,g}$ for all $k \in \{1,2,\ldots,2g-3\}$. Thus we obtain on the ruled surfaces $M_{k,g}$ of genus $g>1$ constructed above exactly $2g-3$ different Einstein neutral bi-Hermitian metrics, exactly one Einstein metric on every surface $M_{k,g}$ for $k \in \{1,2,\ldots,2g-3\}$. Note that for $g=2$ we obtain only one metric corresponding to the Riemannian Einstein Bergery-Page metric on the first Hirzebruch surface $F_1$. Consequently we have proved

**Theorem 4.** *On a ruled surface $M_{k,g}$ of genus $g \geq 2$ for $k \in \{1,2,\ldots,2g-3\}$ there exists an Einstein bi-Hermitian non-Kähler neutral metric.*

Note that in the Riemannian case we have the Page metric which is only one Einstein co-homogeneity 1 Hermitian, non-Kähler metric (see [P], [B], [LeB], [S]).

## 4. Neutral Kähler Gray surfaces

The solutions with $A=0$ analogously as in the Riemannian case give neutral Kähler Gray surfaces. We shall prove

**Theorem 5.** *On a ruled surface $M_{k,g}$ of genus $g \geq 2$ for $k \in \{1,2,\ldots,2g-3\}$ there exists bi-Hermitian Kähler neutral metric.*

**Proof.** Since $A=0$ we get $f=\frac{gg'}{s}$. Consequently from (2.3) we get an equation

$$-2\frac{g''}{g} - 4\left(\frac{g'}{g}\right)^2 - \frac{K}{g^2} = Dg^2 + C. \tag{4.1}$$

Let us write $g' = \sqrt{P(g)}$. Then we can rewrite (4.1) as

$$P'(g) + \frac{4}{g}P(g) + \frac{K}{g} + Dg^3 + Cg = 0, \tag{4.2}$$

with the boundary conditions

$$\frac{1}{2}yP'(y) = s, \frac{1}{2}xP'(x) = -s, \tag{4.3}$$

where $g(a) = y < g(b) = x$, $P(x) = P(y) = 0$ and $P(t) > 0$ for $t \in (y, x)$.

Consequently $P(g) = -\frac{D}{8}g^4 - \frac{C}{6}g^2 + \frac{E}{g^4} - \frac{K}{4}$. Now exactly as in [J-2] we see that the solution satisfying the boundary conditions exists if $K = -4$ and $(s+2)y^4 + (s-2)x^4 = 0$. This condition implies that $s < 2$, $C = 0$, $D > 0$ and $y = \sqrt[4]{\frac{2(2-s)}{D}}$, $x = \sqrt[4]{\frac{2(2+s)}{D}}$ and $E = \frac{s^2-4}{2D} < 0$. In fact it is not difficult to check that a positive solution satisfying the boundary condition exists if and only if $s \in (0, 2)$ hence for $k < 2(g-1)$ and that the metrics corresponding to the same $s$ with different $D$ are homothetic. □

Note that the situation in the case of neutral 4-manifolds is again quite different from Riemannian case where we have only one irreducible non-Einstein compact Kähler Gray surface (see [A-C-G],[J-2]).

The author is very grateful to the referee for the valuable comments and pointing out a mistake in the first version of the paper.

## References

[A] V. Apostolov, *Generalized Goldberg-Sachs theorems for pseudo-Riemannian four-manifolds*, J. Geom. Phys. 27 (1998), 185–198.

[A-C-G] V. Apostolov, D. Calderbank and P. Gauduchon, *The geometry of weakly selfdual Kähler surfaces*, Compositio Math. 135 (2003),179–322.

[B] L. Bérard-Bergery, *Sur de nouvelles variétés riemanniennes d'Einstein*, Institut Élie Cartan, 6, Univ. Nancy, (1982), 1–60.

[Be] A. Besse, *Einstein manifolds*, Springer Verlag, 1987.

[D] M. Dunajski, *Anti-self-dual four manifolds with a parallel real spinor*, Proc. R. Soc. London. A(2002) 458, 1205–1222.

[G] A. Gray, *Einstein-like manifolds which are not Einstein*, Geom. Dedicata, 7 (1978), 259–280.

[J-1] W. Jelonek, *Compact Kähler surfaces with harmonic anti-self-dual Weyl tensor*, Diff. Geom. and Appl. 16 (2002), 267–276.

[J-2] W. Jelonek, *Extremal Kähler $\mathcal{AC}^{\perp}$-surfaces*, Bull. Belg. Math. Soc. – Simon Stevin, 9 (2002), 561–571.

[J-3] W. Jelonek, *Bi-Hermitian Gray surfaces*, (to appear in Pacific Math. J.)

[J-4] W. Jelonek, *Bi-Hermitian Gray surfaces II*, Preprint 2003.

[K] H. Kamada, *Self-dual Kähler metrics of neutral signature on complex surfaces*, Tohoku Mathematical Publications, September 2002.

[LeB] C. Lebrun, *Einstein metrics on complex surfaces*, in Geometry and Physics (Aarhus 1995) eds. J. Andersen, J. Dupont, H. Pedersen and A. Swann, Lect. notes in Pure Appl. Math., Marcel Dekker, 1996.

[M] Y. Matsushita, *Fields of 2-planes and two kinds of almost complex structures on compact 4-dimensional manifolds*, Math. Zeitschrift 207 (1991), 281–291.

[M-L] Y. Matsushita and P. Law, *Hitchin-Thorpe type inequalities for pseudo-Riemannian 4-manifolds of metric signature* $(++--)$, Geom. Dedicata 87 (2001), 65–89.

[M-S] B. Madsen, H. Pedersen, Y. Poon and A. Swann, *Compact Einstein-Weyl manifolds with large symmetry group*, Duke Math. J. 88 (1997), 407–434.

[ON] B. O'Neill, *Semi-Riemannian Geometry*, Academic Press, 1983.

[P] D. Page, *A compact rotating gravitational instanton*, Phys. Lett. 79 B (1978), 235–238.

[Pe] J. Petean, *Indefinite Kähler-Einstein Metrics on Compact Complex Surfaces*, Comm. Math. Phys. 189 (1997), 227–235.

[S] P. Sentenac, *Construction d'une métriques d'Einstein sur la somme de deux projectifs complexes de dimension 2*, Géométrie riemannienne en dimension 4 (Séminaire Arthur Besse 1978-1979) Cedic-Fernand Nathan, Paris (1981), pp. 292–307.

# ON THE DISCONTINUITY OF AFFINE MOTIONS AND AUSLANDER'S CONJECTURE

TATSUYA KAWABE

*Department of Mathematical Science*
*Graduate School of Science and Technology*
*Niigata Univ. Ikarashi, Niigata, 950-2181, JAPAN*
*E-mail: kawabe@sc.niigata-u.ac.jp*

For every affine crystallographic subgroup $\Gamma$ of $\mathrm{Aff}(n)$, L. S. Auslander conjectured that $\Gamma$ is virtually solvable. The purpose in the present paper is to find an algebraic condition on $\Gamma$ which leads the affirmative answer to the conjecture. The author considers the proper discontinuity of virtually torsionfree subgroups of $\mathrm{Aff}(n)$ by dividing into several types as for the Zariski closure of the linear part. This method leads a partial answer of Auslander's conjecture for $n \leqq 6$.

*Keywords*: discrete subgroup, proper discontinuity, virtual solvability

## Introduction

Let $\mathrm{Aff}(n)$ be the affine transformation group, i.e. the semidirect product $\mathrm{GL}(n, \mathbf{R}) \ltimes \mathbf{R}^n$ which acts on $\mathbf{R}^n$ as follows.

$$(g, s) \cdot x = gx + s \quad \text{for } g \in \mathrm{GL}(n, \mathbf{R}), \quad s,\ x \in \mathbf{R}^n.$$

$\mathbf{R}^n$ is identified with the group of all pure translations of $\mathbf{R}^n$.

A subgroup $\Gamma$ of $\mathrm{Aff}(n)$ is said to act *properly discontinuously* (resp. *uniformly*) on $\mathbf{R}^n$ if for every compact subset $K$ of $\mathbf{R}^n$, the set $\{\gamma \in \Gamma \mid \gamma K \cap K \neq \emptyset\}$ is finite (resp. $\mathbf{R}^n/\Gamma$ is compact). In addition, a subgroup $\Gamma$ of $\mathrm{Aff}(n)$ is said an *affine crystallographic group* if it acts properly discontinuously and uniformly on $\mathbf{R}^n$.

The following statement is a long standing conjecture about the affine crystallogrphic group by L. Auslander:

*Every affine crystallographic group is virtually solvable.*

*AMS(MOS) Subj. Class.*: 22E40

This conjecure has been proven in a number of the cases (see [1], [2], [4], [5], [6], [10]).

In this paper, our purpose is to find an algebraic condition on $\Gamma$ which leads the affirmative answer to the conjecture.

Let $\lambda$ : Aff($n$) $\rightarrow$ GL($n$, $\mathbf{R}$) be the natural projection to the linear part. Let us denote $G$ the Zariski closure of $\lambda(\Gamma)$ in $GL(n, \mathbf{R})$ and $S$ the Levi-factor of $G^o$. In section 1, we describe in detail the orbits of non-Euclidean affine action of dimension 2, and in section 2, we find out an abelian subgroup of Aff(6) which satisfies a contractive condition (in Lemma 2.2).

Making use of the results in sections 1 and 2, in section 3 we examine the proper discontinuity of virtually torsionfree subgroup $\Gamma$ of Aff($n$) which acts uniformly on $\mathbf{R}^6$ and satisfies $S = SO(3,2)$. Theorem A is obtained as a sequel. The method on the approach is based on the idea in [4] by D. Fried and W. Goldman, and [10] by G. Tomanov. But in details, it causes a technical difficulty; unfortunately, we can't similarly apply Margulis's sign condition as in [1], [2], [3] by H. Abels, G. A. Margulis, and G. A. Soifer because of dimension 6, and so need to use the other condition that $\mathbf{R}^6/\Gamma$ is compact.

The key of the proof of Theorem A is that $\Gamma$ with properly discontinuous and uniform action is isomorphic to a lattice of $SO(3,2)$ (this is caused by the uniformness of the action of $\Gamma$ on $\mathbf{R}^6$) and it contains two elements with certain contractive condition (see Lemma 3.6).

## 1. Affine crystallographic subgroup of Aff(2)

In this section, we characterize the action of an affine crystallographic subgroup of Aff(2), whose characterization will be used to the proof of Lemma 3.6 and Theorem A in later.

Let $\Gamma$ be a torsionfree affine crystallographic subgroup of Aff(2) the linear part of which is unipotent and let $\delta_1$, $\delta_2$ be generators of $\Gamma$ denoted as follows.

$$\delta_i = \left( \left[ \begin{array}{c|c} 1 & a_i \\ \hline 0 & 1 \end{array} \right], \left[ \begin{array}{c} c_i \\ \hline b_i \end{array} \right] \right) \qquad i = 1, 2.$$

We have the following lemmas by a direct calculation.

**Lemma 1.1.** *For every point $x = {}^t(s,t) \in \mathbf{R}^2$ and each $\delta_i$, the orbit $\{\delta_i^k x\}$*

*is placed on the one of the following curves:*

$$(1)\quad \mathbf{x}_2 = t, \quad (2)\quad \mathbf{x}_1 = \frac{c_i}{b_i}\mathbf{x}_2 + \alpha, \quad (3)\quad \mathbf{x}_1 = \frac{a_i}{2b_i}\left(\mathbf{x}_2 - \frac{a_ib_i - 2c_i}{2a_i}\right)^2 + \beta$$

*where variables* $\mathbf{x}_1$, $\mathbf{x}_2$ *correspond the coordinate* ${}^t(\mathbf{x}_1, \mathbf{x}_2)$, *and* $\alpha$, $\beta$ *are constant numbers related to* $s$, $t$.

Let us denote $C_i(x)$ the curve contains the orbit $\{\delta_i^k x\}$ of a point $x$ by the action of $\delta_i$ for each $i = 1, 2$.

**Lemma 1.2.** $C_1(x)$ *intersects* $C_2(x)$ *at only the point* $x \in \mathbf{R}^2$.

Hence the curves $C_1(\delta_2^k x)$, $C_2(\delta_1^j x)$ $(k, j \in \mathbf{Z})$ used a point $x \in \mathbf{R}^2$ are considered as the boundary curves of the fundamental domains of the action of the group $<\delta_1, \delta_2>$.

**Proposition 1.** $\delta_1$, $\delta_2$ *satisfy the one of the different* 3 *cases:*
(1) $b_1 = 0$ *and* $b_2 \neq 0$, *or* $b_2 = 0$ *and* $b_1 \neq 0$
(2) $a_1 = a_2 = 0$ *and* $(b_1, c_1) \neq (b_2, c_2)$
(3) $a_1/b_1 = a_2/b_2$ *and* $b_1/2 - c_1/a_1 \neq b_2/2 - c_2/a_2$.

The following lemma is a consequence considered about 3 cases of Proposition 1.

**Lemma 1.3.** $C_1(\delta_2^k x)$*'s (resp.* $C_2(\delta_1^k x)$*'s) are placed at equidistant intervals in direction parallel to the coordinate* $\mathbf{x}_1$ *(except the curve* $C_i$ *corresponding* $\delta_i$ *with* $b_i = 0$*).*

## 2. A Contractive affine action on a $\mathbf{R}^6$

Let $\Delta$ be a free abelian subgroup of rank 2 of Aff(6), which acts properly discontinuously and freely on $\mathbf{R}^6$. Let us assume that $\Delta$ is generated by $\delta_1$ and $\delta_2$, which is represented as follows:

$$(2.1)\quad \delta_i = \left(\left[\begin{array}{c|c} 1 & \mathbf{a}_i \\ \hline 0 & \mathbf{g}_i \end{array}\right], \left[\begin{array}{c} c_i \\ \hline \mathbf{b}_i \end{array}\right]\right), \quad \begin{array}{l}\text{where} \\ \mathbf{a}_i = (0, 0, a_i, 0, 0),\ \mathbf{b}_i = {}^t(0, 0, b_i, 0, 0), \\ \mathbf{g}_i = \operatorname{diag}\left[\lambda_i, \mu_i, 1, \mu_i^{-1}, \lambda_i^{-1}\right],\ 0 < \lambda_i,\ \mu_i < 1.\end{array}$$

Let $V$ be the 1-dimensional subspace $\{{}^t(*, 0, 0, 0, 0, 0)\}$. We need more notations to define some subspaces invariant for $\delta_i$ $(i = 1, 2)$. Let $\gamma \in \mathrm{Aff}(6)$ be an element whose the linear part has only real eigenvalues. For each eigenvalue $\mu$ of the linear part of $\gamma$, let $V_\mu(\gamma)$ be the eigenspace of the linear part corresponding to the eigenvalue $\mu$. We set $D^-(\gamma) = \bigoplus_{\mu \leqq 1} V_\mu(\gamma)$,

$W(\gamma) = V_1(\gamma)$, and $V^-(\gamma) = \bigoplus_{\mu<1} V_\mu(\gamma)$. If $\delta_i$'s are represented as (2.1), then $D^-(\delta_i)$ are written as the 4-dimensional subspace $\{{}^t(*,*,*,*,0,0)\}$, and $W(\delta_i)$ as the 2-dimensional subspace $\{{}^t(*,0,0,*,0,0)\}$, $V^-(\delta_i)$ as the 2-dimensional subspace $\{{}^t(0,*,*,0,0,0)\}$ for $i = 1, 2$. Let us describe simply $D^- = D^-(\delta_i)$, $W = W(\delta_i)$, $V^- = V^-(\delta_i)$ since they are common subspaces for $i = 1, 2$.

Every element $\delta \in \Delta$ acts affinely on $D^-$ and $W$, and each point in $D^-$ tends asymptotically to the subspace $W$ by the action of $\delta_i^n$ for positive integer $n$ i.e. for the Euclidean distance dist$(*,*)$, $\text{dist}(x, W) > \text{dist}(\delta^n x, W)$ for $x \in D^-$ $(i = 1,\ 2)$.

Since $\text{cd}\Delta = 2 = \dim W$. we see that $\Delta$ acts uniformly on $W$. Hence the subspace $W$ is divided into infinitely compact domains by the family of the curves defined in section 1 $\{C_1(\delta_2^m x_0),\ C_2(\delta_1^k x_0)\}_{k,m\in\mathbf{Z}}$ for fixed $x_0 \in W$. Let $K(i,j)$ denotes a compact domain encircled with the following boundary curves (see Fig.(K)):

$C_1(\delta_2^i x_0)$, $C_2(\delta_1^j x_0)$, $C_1(\delta_2^{i+1} x_0)$, $C_2(\delta_1^{j+1} x_0)$.

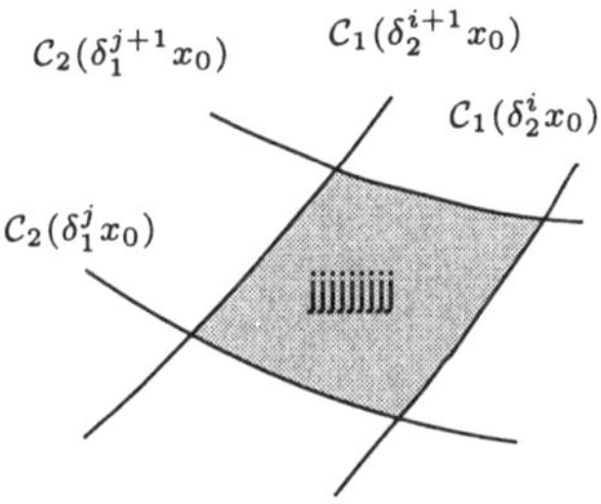


Fig. (K)

We have the following relation by the definition of $K(*,*)$:

$$\delta_1^i \delta_2^j K(r,s) = K(r+j, s+i) \quad \text{for every integer } i,\ j.$$

**Lemma 2.1.** *Let $Q = \{q_l\}$ be a sequence on the line $V$ and $Q' = \{\, q_l' := w_0 + q_l \,\}$ be a sequence on the line $w_0 + V$ for a point $w_0 \in W$. We denote $K(m_l, n_l)$ and $K(m_l', n_l')$ the domains containing $q_l$, $w_0 + q_l$ respectively for each $l$. Then $m_l - m_l'$ and $n_l - n_l'$ are constant numbers (not depend on $l$).*

This Lemma is immediately proved by Lemma 1.3.

Let us put $x_0$ as the origin 0 in the subspace $W$. Then we have the following

**Lemma 2.2.** *If we take suitable elements $\delta_1$, $\delta_2$ generating $\Delta$, there exist a sequence of points $\{\, q_l \in V \mid l = 1, 2, 3, \dots \,\}$, and an integral pair $(i,j)$ for each $q_l$ satisfying the following conditions:*

1. $\delta_1^i \delta_2^j q_l \in K(0,0)$
2. *The action of $\delta_1^i \delta_2^j$ on $D^-$ doesn't make any points of $D^-$ leave from $W$ i.e.* $\text{dist}(\delta_1^i \delta_2^j x, W) \leqq \text{dist}(x, W)$ *for any $x \in D^-$, where* dist$(*,*)$ *is the Euclidean distance.*

**Proof.** Let us take two generators $\delta_1$, $\delta_2$ of $\Delta$ represented as (2.1). We separate the proof into the following 3 cases by the sign of $b_i$ $(i = 1, 2)$:

$$\text{CASE 1: } [\,b_1 b_2 = 0\,], \quad \text{CASE 2: } [\,b_1 b_2 < 0\,], \quad \text{CASE 3: } [\,b_1 b_2 > 0\,].$$

In the case 1 and the case 2, let us divide $V$ into the segments $I_{r,s} = V \cap K(r,s)$. Then it is easy to see that $(\cup_{r,s \leqq 0} K(r,s)) \cap V$ is a ray with end point $0 \in V$, and hence there are infinite number of segments $I_{r,s}$ for non positive integers $r$, $s$ intersecting the ray of $V$. Let us take a point $q_{m(r,s)}$ on each segment $I_{r,s}$ for such $r$, $s$. It immediately follows that $\delta_1^s \delta_2^r q_{m(r,s)} \in \delta_1^{-s} \delta_2^{-r} K(r,s) = K(0,0)$. Since $-s$, $-r \geqq 0$, $\delta_1^{-s}\delta_2^{-r}$'s action on $D^-$ doesn't make any point of $D^-$ leave from $W$. This establishes the required construction.

Next we consider the case 3. It is sufficient to show the case that $b_1$, $b_2 > 0$. In this case, unfortunately, $(\cup_{r,s \leqq 0} K(r,s)) \cap V$ is not a ray but a segment $K(0,0) \cap V$. Let us consider an integral pair $(m,n)$ such that $\delta_1^m \delta_2^n$ satisfies the following 3 inequalities:

$$(A_1): \ \lambda_1^m \lambda_2^n < 1, \qquad (A_2): \ \mu_1^m \mu_2^n < 1, \qquad (A_3): \ m b_1 + n b_2 < 0.$$

By changing generators of $\Delta$ if necessary, there are infinitely many $\delta_1^m \delta_2^n$ satisfying $(A_1)$, $(A_2)$, $(A_3)$ in $\Delta$. We can change $\delta_1$, $\delta_2$ to such $\delta_1^m \delta_2^n$'s as mutually independent elements. Then it is arrived at the conclusion in the case 2. □

From now on, we may assume that $\delta_1$, $\delta_2 \in \Delta$ satisfy the statement of Lemma 2.2, and fix the corresponding subspace $D^-$. The following corollary is important to prove Lemma 3.6 in later.

**Corollary 1.** *Let $p$ be a point in $W$ and $x$ be a point in $D^-$. Then there exist a sequence of points $\{\, \widetilde{q_l} \in x + V \mid l = 1, 2, 3, \ldots \}$ and a compact neighborhood $K_p$ of $p$ such that for each index $l$, there exists an integral pair $(i,j)$ as follows:*

$$\delta_1^i \delta_2^j \widetilde{q_l} \in K_p.$$

**Proof.** Let us consider the decomposition of $D^-$ into a direct sum $W + V^-$. Write $x$ as $x = x_w + x^-$ using $x_w \in W$ and $x^- \in V^-$, they are uniquely determined. Let us take a sequence of points $\{\, q_l \in V \mid l = 1, 2, 3, \ldots \}$, and take exponents $i$, $j$ of $\delta_1^i \delta_2^j$ corresponding to the index $l$ as satisfying Lemma 2.2. We set $\widetilde{q_l} = x + q_l = x_w + q_l + x^-$ for $q_l \in V$ $(l = 1, 2, 3, \ldots)$. Let us divide the action of $\delta_i$ into the components $x_w + q_l \in W$ and $x^- \in V^-$ as follows:

$$\delta_1^i \delta_2^j \widetilde{q}_l = \delta_1^i \delta_2^j (x_w + q_l) + g_1^i g_2^j x^- .$$

Let us divide $W$ into the union of the domains $K(r, s)$ $(r, s \in \mathbf{Z})$ such as taken in the proof of Lemma 2.2. Let $K(j, i)$, $K(r_l, s_l)$ be the domains, each of which contains $q_l$, $x_w + q_l$ respectively. Then $\delta_1^i \delta_2^j K(-j, -i) = K(0, 0)$, and $\delta_1^i \delta_2^j$'s action on $D^-$ doesn't make any points of $D^-$ leave from $W$ for given $i$, $j$. Let us put $(i_0, j_0) = (r_l + j, s_l + i)$. Then we have that $\delta_1^{i_0} \delta_2^{j_0} K(-j, -i) = K(r_l, s_l)$. It immediately follows that

$$\delta_1^i \delta_2^j K(r_l, s_l) = \delta_1^{i+i_0} \delta_2^{j+j_0} K(-j, -i) = \delta_1^{i_0} \delta_2^{j_0} K(0, 0).$$

Note that $i_0$, $j_0$ are constant numbers (not depend on $l$) by Lemma 2.1. Since $\|g_1^i g_2^j x^-\| \leqq \|x^-\|$ for Euclidean norm $\| * \|$, if we take a sufficient large compact neighborhood $K_p$ of $p$ which contains $\delta_1^{i_0} \delta_2^{j_0} K(0, 0)$ and $p$, it follows that $\delta_1^i \delta_2^j \widetilde{q}_l \in K_p$ for each integral pair $(i, j, l)$ satisfying Lemma 2.2. □

**Remark.** In Lemma 2.2 and Corollary 1, each integral pair $(i, j)$ is uniquely determined for an index $l$ of the sequence $\{q_l\}$. This fact is important to prove Theorem A using Lemma 3.6 in later.

## 3. Levi factor SO(3,2)

**Theorem A.** *Let $\Gamma$ be a virtually torsionfree subgroup of* Aff(6). *We assume that a Levi-factor $S$ of the Zariski closure $G$ of $\lambda(\Gamma)$ in* $\mathrm{GL}(6, \mathbf{R})$ *is a simple algebraic group $SO(3, 2)$ and $\Gamma$ acts uniformly on $\mathbf{R}^6$. Then $\Gamma$ doesn't act properly discontinuously on $\mathbf{R}^6$.*

**Notation.** Let $\gamma$ be an element in Aff(6) such that each eigenvalue $\mu$ of $\lambda(\gamma)$ is real number. let $V_\mu(\gamma)$ denotes the eigenspace of $\lambda(\gamma)$ corresponding to the eigenvalue $\mu$. We set the subspace $E^-(\gamma)$ of $\mathbf{R}^6$ as follows:

$$E^-(\gamma) = \bigoplus_{\mu \in \Omega^-} V_\mu(\gamma),$$

where $\Omega^-$ is the set of the eigenvalues of $\lambda(\gamma)$ not greater than 1.

$E^-(\gamma)$ is decomposed as a direct sum of two subspaces as follows;

$$E^-(\gamma) = V^-(\gamma) \oplus V^0(\gamma), \text{ where } V^-(\gamma) := \bigoplus_{\mu \in \Omega^- \setminus \{1\}} V_\mu(\gamma), \quad V^0(\gamma) := V_1(\gamma).$$

As $\lambda(\gamma)$ has an eigenvalue equal to 1, there exists a unique $\gamma$-invariant affine subspace $W(\gamma)$ with the same dimension as $V_1(\gamma)$ and parallel to $V_1(\gamma)$ on which $\gamma$ acts as a translation. In addition, there is a unique $\gamma$-invariant affine subspace $D^-(\gamma)$ with same dimension as $E^-(\gamma)$ and parallel to $E^-(\gamma)$ containing $W(\gamma)$.

**Remark.** $W(\gamma)$ is an affine subspace of same dimension and parallel to $V^0(\gamma)$, and on which $\gamma$'s action is of parabolic type. In general, $\gamma$ doesn't act as a translation on $W(\gamma)$ (see section 2). $\gamma^k$ acts on $D^-(\gamma)$ as a contraction to $W(\gamma)$ for positive integer $k$.

**Proof of Theorem A.** Suppose that $\Gamma$ acts properly discontinuously on $\mathbf{R}^6$. Since $\Gamma$ is virtually torsionfree group, if necessary taking $\Gamma$ as a torsionfree subgroup of finite index, we may assume that $\Gamma$ acts freely on $\mathbf{R}^6$.

Making use of Proposition 1 in [6], deforming the linear part by the conjugation of some element in $GL(6,\mathbf{R})$, the possible representation of $G^o$ is considered to be the one of the following forms:

$$(a): \quad G^0 = \left\{ \left[\begin{array}{c|c} g & * \\ \hline 0 & 1 \end{array}\right] : g \in S \right\}, \quad (b): \quad G^0 = \left\{ \left[\begin{array}{c|c} 1 & * \\ \hline 0 & g \end{array}\right] : g \in S \right\}.$$

We separate the proof into the two cases corresponding (a), (b).

CASE (a): Let $\pi$ be the natural projection from $G \cdot \mathbf{R}^6 \subset \mathrm{Aff}(6)$ to $S = SO(3,2)$. We set $\Gamma_1 = [\Gamma, \Gamma]$. Since $\pi(\Gamma)$ is Zariski dense in $S$, $\pi(\Gamma_1)$ is Zariski dense in $S$. $\Gamma_1$ has the following form:

$$\left\{ \left( \left[\begin{array}{c|c} g & * \\ \hline 0 & 1 \end{array}\right], \left[\begin{array}{c} s \\ \hline 0 \end{array}\right] \right) : g \in S \right\}.$$

Let $\widetilde{V}$ denotes the subspace $\{{}^t(*,*,*,*,*,0)\}$ of $\mathbf{R}^6$, where the coordinate of which corresponds to the translational part of $\Gamma$. Clearly $\Gamma_1$ acts affinely and freely on $\widetilde{V}$. Since the linear part of $\Gamma_1$ is $\pi(\Gamma_1)$ which is Zariski dense in $SO(3,2)$, it is immediately follows from Theorem A in [1] by H. Abels, G. A. Margulis and G. A. Soifer that $\Gamma_1$ doesn't act properly discontinuously on $\widetilde{V}$. This completes the proof of case (a).

CASE (b): Let $\rho : \Gamma \to SO(3,2)$ be the projection.

**Lemma 3.1.** *$\rho$ is injective and* $\operatorname{img}(\rho)$ *is a discrete subgroup of $SO(3,2)$.*

**Proof.** Let $\widetilde{\rho} : \Gamma \to SO(3,2) \ltimes \mathbf{R}^5 (= \operatorname{Aff}(\mathbf{R}^6/V))$ be the projection as follows:

$$\left( \left[ \begin{array}{c|c} 1 & * \\ \hline 0 & A \end{array} \right], \left[ \begin{array}{c} * \\ \hline x \end{array} \right] \right) \mapsto (A, x),$$

and $V$ the 1 dimensional subspace $\{{}^t(*,0,0,0,0,0)\}$ of $\mathbf{R}^6$, where the coordinate of which corresponds the translational part of $\Gamma$. If $\ker(\widetilde{\rho}) \neq 1$, then $\widetilde{\rho}(\Gamma)$ acts properly discontinuously and uniformly on $\mathbf{R}^5 (= \mathbf{R}^6/V)$ by Lemma 2.1 in [5]. Since $\widetilde{\rho}(\Gamma)$ is not solvable, this contradicts that Auslander's conjecture is correct in dimension 5. Hence $\widetilde{\rho}$ is injective.

Since the linear part $\lambda(\widetilde{\rho}(\Gamma))$ is Zariski dense in $SO(3,2)$, the pure translational subgroup $\widetilde{\rho}(\Gamma) \cap \mathbf{R}^5$ is trivial. Thus we have the following isomorphism:

$$\Gamma \cong \widetilde{\rho}(\Gamma) \cong \lambda(\widetilde{\rho}(\Gamma)) = \rho(\Gamma).$$

Hence $\rho$ is injective.

Next, we assume that $\rho(\Gamma)$ is not discrete. Let $(\overline{\rho(\Gamma)})^0$ denotes as the connected component of the closure of $\rho(\Gamma)$ in the Euclidean topology of $SO(3,2)$. Then $(\overline{\rho(\Gamma)})^0$ is nontrivial and it is known that $(\overline{\rho(\Gamma)})^0$ is solvable (see [8] Theorem 8.24). Since $\rho(\Gamma)$ normalizes $(\overline{\rho(\Gamma)})^0$ and $\rho(\Gamma)$ is Zariski dense in $SO(3,2)$, $(\overline{\rho(\Gamma)})^0$ is normal subgroup of $SO(3,2)$. This contradicts that $SO(3,2)$ has an irreducible representation. Hence $\rho(\Gamma)$ is discrete subgroup of $SO(3,2)$. □

From this lemma, we have the following inequality.

$$6 = \dim(SO(3,2)/S(O(3) \times O(2))) \geq \operatorname{cd}(\rho(\Gamma)) = \operatorname{cd}(\Gamma) = 6.$$

Thus $\operatorname{cd}(\rho(\Gamma)) = \dim(SO(3,2)/S(O(3) \times O(2)))$, i.e. $\rho(\Gamma)$ is an uniform lattice of $SO(3,2)$. Making use of Theorems 2.5, 2.7, 2.8 in [7], we have the following

**Lemma 3.2.** *There exist two element $\delta_1$, $\delta_2$ in $\Gamma$ satisfying that*

1. *$\rho(\delta_i)$ has pairwise distinct real positive eigenvalues for each $i = 1, 2$ and they generate a lattice of a maximal real torus $T_{\mathbf{R}}$ of $SO(3,2)$,*
2. *$E^-(\delta_1) = E^-(\delta_2)$.*

**Lemma 3.3.** *$D^-(\delta_1) = D^-(\delta_2)$ and $W(\delta_1) = W(\delta_2)$.*

**Proof.** Since $\delta_1$ and $\delta_2$ are commutative, we have

$$\delta_1\delta_2 D^-(\delta_1) = \delta_2 D^-(\delta_1), \qquad \delta_1\delta_2 W(\delta_1) = \delta_2 W(\delta_1).$$

The $\delta_i$-invariant flag $W(\delta_i) \subset D^-(\delta_i)$ with the contractive property of the $\delta_i$-action is uniquely determined by $\delta_i$ respectively for $i = 1, 2$. Hence

$$\delta_2 D^-(\delta_1) = D^-(\delta_1), \qquad \delta_2 W(\delta_1) = W(\delta_1),$$

which implies that $D^-(\delta_1) = D^-(\delta_2)$, $W(\delta_1) = W(\delta_2)$. □

If necessary, after taking a suitable conjugate by an element of Aff(6), we can assume that $\delta_i$ is represented as follows:

$$\delta_i = \left( \left[\begin{array}{c|c} 1 & \mathbf{a}_i \\ \hline 0 & \mathbf{g}_i \end{array}\right], \left[\begin{array}{c} c_i \\ \hline \mathbf{b}_i \end{array}\right] \right), \quad \begin{array}{l} \text{where} \\ \mathbf{a}_i = (0,0,a_i,0,0),\ \mathbf{b}_i = {}^t(0,0,b_i,0,0), \\ \mathbf{g}_i = \mathrm{diag}\left[\lambda_i, \mu_i, 1, \mu_i^{-1}, \lambda_i^{-1}\right],\ 0 < \lambda_i,\ \mu_i < 1. \end{array}$$

Then $D^-(\delta_i)$ (resp. $W(\delta_i)$) is written as the form $\{{}^t(*,*,*,*,0,0)\}$, and (resp. $\{{}^t(*,0,0,*,0,0)\}$). Let $V$ be the 1-dimensional subspace $\{{}^t(*,0,0,0,0,0)\}$, and $V^c$ be the 5-dimensional subspace $\{{}^t(0,*,*,*,*,*)\}$. Let us denote $V^\rho$ as $E^-(\delta_i) \cap V^c = \{{}^t(0,*,*,*,0,0)\}$. We may assume that $\delta_i$'s satisfy the statement of Lemma 2.2, and Corollary 1.

From the Zariski-density of $\rho(\Gamma)$ in $SO(3,2)$, we have following

**Lemma 3.4.** *Let $T_1$ be the maximal real torus of $SO(3,2)$ which contains $\rho(\delta_1)$, $\rho(\delta_2)$ satisfying Lemma 3.2. Then there is an element $\gamma \in \Gamma$ such that*

1. *$T_1$ and $\rho(\gamma)T_1\rho(\gamma)^{-1}$ don't have common invariant subspace of $V^c$, and*
2. *$V^\rho + \rho(\gamma)V^\rho = V^c$.*

Let us denote $\gamma_i = \gamma\delta_i\gamma^{-1}$, $(i = 1, 2)$. Then $\rho(\gamma_i)$ has pairwise distinct real positive eigenvalues for each $i = 1, 2$ and they generate a lattice of the maximal real torus $\rho(\gamma)\, T_1\, \rho(\gamma)^{-1}$.

Now we consider about the action of $\gamma_i$, $\delta_j$ $(i, j \in \{1,2\})$. From Lemma 3.2, Lemma 3.3, Lemma 3.4, they satisfy the following properties:

- $D^-(\gamma_1) = D^-(\gamma_2)$, $D^-(\delta_1) = D^-(\delta_2)$,
- $W(\gamma_1) = W(\gamma_2)$, $W(\delta_1) = W(\delta_2)$,
- the group generated by $\gamma_i$'s (resp. $\delta_i$'s) act uniformly on $W(\gamma_1)$ (resp. $W(\delta_1)$),
- $D^-(\delta_1) \cap D^-(\gamma_1) \neq \varnothing$.

**Lemma 3.5.** *If $\delta_1^k \gamma_1^l \delta_2^m \gamma_2^n = 1$, then $(k,l,m,n) = (0,0,0,0)$.*

**Proof.** From the equality of the condition, we have that $\rho(\delta_1)^k \rho(\delta_2)^m = \rho(\gamma_1)^{-l} \rho(\gamma_2)^{-n}$. Since $T_1 \cap \rho(\gamma) T_1 \rho(\gamma)^{-1} = 1$ by Lemma 3.4, it follows that $\rho(\delta_1)^k \rho(\delta_2)^m = I$ and $\rho(\gamma_1)^{-l} \rho(\gamma_2)^{-n} = I$. Using the fact that $\rho(\delta_i)$'s (resp. $\rho(\gamma_i)$'s) are mutually independent in the real torus $T_1$ (resp. $\rho(\gamma) T_1 \rho(\gamma)^{-1}$), we have $(k,m) = (0,0)$ and $(l,n) = (0,0)$. □

The following lemma is most important for the proof of Theorem A.

**Lemma 3.6.** *There are two points $p_1 \in W(\delta_1)$, $p_2 \in W(\gamma_1)$ and sufficient large compact neighborhoods $K_{p_1}$, $K_{p_2}$ respectively of them and a sequence of points $q_l \in D^-(\delta_1) \cap D^-(\gamma_1)$ satisfying the following condition for each index $l$.*

**($\mathfrak{C}$):** *There exists an integral pair $(i,j)$ as follows:*

1. $\delta_1^i \delta_2^j q_l \in K_{p_1}$,
2. $\gamma_1^i \gamma_2^j q_l \in K_{p_2}$.

**Proof.** Let $p_1 \in W(\gamma_1)$, and $x \in D^-(\delta_1) \cap D^-(\gamma_1)$. Then $D^-(\delta_1) \cap D^-(\gamma_1)$ includes $x + V$. About $\delta_1$, $\delta_2$, it had been already shown in Corollary 1 in section 2 that there exist a sequence $\{q_l\} \subset x + V$ and indices $i$, $j$ corresponding each $q_l$ which satisfy the condition 1.

For each $q_l$, we take such an integral pair $(i,j)$. To show this lemma, we investigate the behaviour of the orbit $\gamma_1^i \gamma_2^j q_l$ for the integers $i$, $j$. Let us put $p_2$ as $\gamma p_1$, which belongs to $W(\gamma_1)$. We claim that there is a sufficient large compact neighborhood $K_{p_2}$ of $p_2$ such that $\gamma_1^i \gamma_2^j q_l \in K_{p_2}$. For each $q_l$, we have $\gamma^{-1} q_l \in D^{-1}(\delta_1)$. Let us denote $q_l = q_l' + q'$ ($q_l' \in W(\delta_1)$, $q' \in V^-(\delta_1)$), and $\gamma^{-1} q_l = q_l'' + q''$ ($q_l'' \in W(\delta_1)$, $q'' \in V^-(\delta_1)$). Let $W(\delta_1)$ divides into the union of the domains $K(r,s)$ ($r, s \in \mathbf{Z}$) such as taken in Lemma 2.2, and we set $p_1 \in K(r_0, s_0)$, and $q_l' \in K(r_l, s_l)$, $q_l'' \in K(r_l', s_l')$ ($l = 1, 2, \ldots$). Note that $\bar{r} := r_l' - r_l$ and $\bar{s} := s_l' - s_l$ are constant numbers (not depend on index $l$) since we can denote that $q_l'' = q_l' + w_0$ ($w_0 \in W$) by looking at $\gamma^{-1}$-action on the sequence $\{q_l'\}$ in $V$, and appliable to Lemma 2.1. Then $\delta_1^i \delta_2^j q_l' \in K(r_l + j, s_l + i)$, and it follows that $r_0 = r_l + j$ and $s_0 = s_l + i$. Hence we have

$$\delta_1^i \delta_2^j q_l'' \in K(r_l' + j, s_l' + i) = K(r_l' + r_0 - r_l, s_l' + s_0 - s_l) = K(\bar{r} + r_0, \bar{s} + s_0).$$

In view of the action by dividing into $W(\delta_1)$ and $V^-(\delta_1)$, we have

$$\delta_1^i \delta_2^j \gamma^{-1} q_l = \delta_1^i \delta_2^j q_l'' + \mathbf{g}_1^i \mathbf{g}_2^j q'',$$

and $\|\mathbf{g}_1^i\mathbf{g}_2^j q''\| \leqq \|q''\|$ as for Euclidean norm $\| * \|$. From Corollary 1 we see that there is a large compact neighborhood $K'_{p_1}$ of $p_1$ such that $\delta_1^i\delta_2^j\gamma^{-1}q_l \in K'_{p_1}$. Hence we have

$$\gamma_1^i\gamma_2^j q_l = \gamma\delta_1^i\delta_2^j\gamma^{-1}q_l \in \gamma K'_{p_1}.$$

Taking $K_{p_2}$ as $\gamma K'_{p_1}$, it immediately follows that our claim is the case. Thus the sequence $\{q_l\}$ and the compact neighborhood $K_{p_2}$ of $p_2$ satisfy the condition 2. This completes the proof of Lemma 3.6. □

By Lemma 3.6, it implies that there are infinitely many integral pairs $(i,j)$ such that $\delta_1^i\delta_2^j\gamma_1^{-i}\gamma_2^{-j}K_{p_2} \cap K_{p_1} \neq \varnothing$. Thus by Lemma 3.5, there are infinitely many $\tilde{\gamma} \in \Gamma$ such that $\tilde{\gamma}K_{p_2} \cap K_{p_1} \neq \varnothing$, i.e. $\Gamma$ doesn't act properly discontinuously on $\mathbf{R}^6$.

In consequence, we have completely proven Theorem A. □

## 4. A remark on the Auslander's conjecture

Let us consider the Auslander's conjecture for dimension 6. Let $\Gamma$ be an affine crystallographic subgroup of Aff(6). Applying Selberg's lemma ([9]) to $\Gamma$, we may assume that $\Gamma$ is torsion free. Let $G$ be the Zariski closure of $\lambda(\Gamma)$ in $GL(6,R)$ and $S$ the Levi-factor of $G^0$. Applying the following property (F) about the linear part of $\Gamma$, it is easy to see that $S$ does not have the irreducible representation.

**(F)**: *Let $\Gamma$ be asubgroup of* Aff$(n)$ *which acts freely on $\mathbf{R}^n$ and $G$ be the Zariski closure of $\lambda(\Gamma)$ in $GL(n,R)$. For every $\gamma \in \Gamma$,the linear part $\lambda(\gamma)$ has at least one eigenvalue equal to* 1. *Moreover every element $g$ in $G$ has at least one eigenvalue equal to* 1.

This property is proved in [6], for example.

G. Tommanov has proved the Auslander's conjecture for the case of $\mathrm{rank}_R(S_i) \leq 1$ for each simple factors $S_i$ of $S$ ([10]).

Hence we have to consider only the following cases, where $S$ is;

$$\begin{array}{llll} \mathcal{S}_3, & \mathcal{S}_4, & \mathcal{S}_5, & SO(3,2), \\ SL(3,\mathbf{R}) \times SL(2,\mathbf{R}), & SL(3,\mathbf{R}) \times SO(2,1), & \mathcal{S}_3 \times SO(2), & \end{array}$$

where $\mathcal{S}_i$ is a simple real algebraic group which has an irreducible reprensentation of dimension $i$,

If we can show that all non-trivial types of $S$ does not arise, Auslander's conjecture for $n = 6$ is proved to be true.

Theorem A shows that $S = SO(3,2)$ cannot arise. Making use of Proposition 1 and main Theorem in [6], it can be easily shown that $S = \mathcal{S}_i$ $(i = 3,4,5)$ are impossible to arise. The cases where

$$S = SL(3,\mathbf{R}) \times SL(2,\mathbf{R}), \quad SL(3,\mathbf{R}) \times SO(2,1), \quad \mathcal{S}_3 \times SO(2)$$

remain unsolvable.

## References

1. H. Abels and G.A. Margulis and G.A. Soifer, *On the Zariski closure of the linear part of a properly discontinuous group of affine transformations*, J. Differential Geom. **60**(2002), no. 2, 315–344.
2. H. Abels and G.A. Margulis and G.A. Soifer, *Properly discontinuous groups of affine transformations with orthogonal linear part*, Comptes Rendus Acad. Sci. Paris **324 I**(1997), 253–258.
3. H. Abels and G.A. Margulis and G.A. Soifer, *Properly Discontinuous Groups of Affine Transformations*, SFB-Preprint 97–083.
4. D. Fried and W. Goldman, *Three dimensional affine crystallographic groups*, Adv. in Math. **47**(1983), 1–49.
5. F. Grunewald and G. Margulis, *Transitive and quasitransitive affine actions preserving a generalized Lorentz structure*, J. G. P. (4) **5**(1988), 493–531.
6. T. Kawabe, *On The Properly Discontinuous Subgroups of Affine Motions* Manuscripta. Math. **95**(1998), 529–543.
7. G. Prasad and M.S. Raghunathan, *Cartan subgroups and lattices in semi-simple groups*, Annals of Math. **96**(1972), 296–317.
8. M.S. Raghunathan, *Discrete subgroup of Lie groups*, Ergebnisse Math. u. i. Grenzgeb., Vol. 68, Springer-Verlag (1972).
9. A. Selberg, *On discontinuous groups in higher dimensional symmetric spaces*, Contributions to Function Theory (Int. Colloq. Bombay, 1960), Tata Inst. Fund. Res., Bombay (1960), 147–164.
10. G.M. Tomanov, *The virtually solvability of the fundamental group of a generalized Lorentz space form*, J. Diff. Geom. **32**(1990), 539–547.

# REAL HYPERSURFACES IN NON-FLAT COMPLEX SPACE FORMS CONCERNED WITH THE STRUCTURE JACOBI OPERATOR AND RICCI TENSOR

U-HANG KI

*Department of Mathematics*
*Kyungpook National University*
*Daegu 702-701, Korea*

SETSUO NAGAI

*Department of Education*
*Toyama University*
*Toyamashi 930-8555, Japan*
*E-mai: snagai@edu.toyama-u.ac.jp*

RYOICHI TAKAGI

*Department of Mathematics and Informatics*
*Chiba University*
*Chibashi 263-8522, Japan*
*E-mai: takagi@math.s.chiba-u.ac.jp*

Let $M$ be a real hypersurface with almost contact metric structure $(\phi, \xi, \eta, g)$ in a non-flat complex space form $M_n(c)$. In this paper we investigate real hypersurfaces of $M_n(c)$ whose structure Jacobi operator $R_\xi$ commutes with both the structure tensor $\phi$ and the Ricci tensor $S$ of $M$. We characterize Hopf hypersurfaces of $M_n(c)$.

*Keywords*: real hypersurfaces, Ricci tensor, the structure Jacobi operator.

## 0. Introduction

Let $M$ be a real hypersurface of a complex space form $M_n(c)$. Then $M$ has an almost contact metric structure $(\phi, \xi, \eta, g)$ induced from the complex structure $J$ and the Kähler metric of $M_n(c)$ (for details see §1). The struc-

2000 *Mathematics subject classification*: 53C40, 53C15.

ture vector $\xi$ is said to be principal if $A\xi = \alpha\xi$ is satisfied, where $A$ is the shape operator of $M$ and $\alpha = \eta(A\xi)$. A real hypersurface is said to be a Hopf hypersurface if the structure vector $\xi$ of $M$ is principal.

In a complex projective space $\mathbb{C}P_n$ Hopf hypersurfaces with constant principal curvatures are just the homogeneous real hypersurfaces (cf. [5]). Further, Hopf hypersurfaces with constant principal curvatures in a non-flat complex space forms were completely classified as follows:

**Theorem T** ([6]). *Let $M$ be a homogeneous real hypersurface of $\mathbb{C}P_n$. Then $M$ is a tube of radius $r$ over one of the following Kähler submanifolds:*

($A_1$) *a hyperplane $\mathbb{C}P_{n-1}$, where $0 < r < \frac{\pi}{2}$,*

($A_2$) *a totally geodesic $\mathbb{C}P_k$ $(1 \leq k \leq n-2)$, where $0 < r < \frac{\pi}{2}$,*

($B$) *a complex quadric $Q_{n-1}$, where $0 < r < \frac{\pi}{4}$,*

($C$) *$\mathbb{C}P_1 \times \mathbb{C}P_{\frac{n-1}{2}}$, where $0 < r < \frac{\pi}{4}$ and $n \geq 5$ is odd,*

($D$) *a complex Grassmann $G_{2,5}\mathbb{C}$, where $0 < r < \frac{\pi}{4}$ and $n = 9$,*

($E$) *a Hermitian symmetric space $\mathrm{SO}(10)/\mathrm{U}(5)$, where $0 < r < \frac{\pi}{4}$ and $n = 15$.*

**Theorem B** ([1]). *Let $M$ be a real hypersurface of a complex hyperbolic space $\mathbb{C}H_n$. Then $M$ has constant principal curvatures and $\xi$ is principal if and only if $M$ is locally congruent to one of the following:*

($A_0$) *a self-tube, that is, a horosphere,*

($A_1$) *a geodesic hypersphere, or a tube over a hyperplane $\mathbb{C}H_{n-1}$,*

($A_2$) *a tube over a totally geodesic $\mathbb{C}H_k$ $(1 \leq k \leq n-2)$,*

($B$) *a tube over a totally real hyperbolic space $\mathbb{R}H_n$.*

Let $M$ be a real hypersurface of $M_n(c)$ and S is the Ricci tensor of $M$. Then the classification of $M$ with the commutativity condition $\phi S = S\phi$ is still open and very important problem. The condition $\phi S = S\phi$ leads the fact that the structure vector $\xi$ is an eigenvector of $S$, that is, $S\xi = \sigma\xi$ for some scalar $\sigma$. Further, the condition $S\xi = \sigma\xi$ leads the condition $R_\xi S = SR_\xi$, where $R_\xi$ denotes the structure Jacobi operator of $M$ (for the definition see §1). So it is a very important problem to investigate real hypersurfaces satisfying the condition $R_\xi S = SR_\xi$.

The purpose of this paper is to eatablish the following theorems.

**Theorem 4.1.** *Let $M$ be a real hypersurface with $R_\xi\phi = \phi R_\xi$ and $R_\xi S = SR_\xi$ in a complex space form $M_n(c)$, $c \neq 0$. If $\theta = 3\left\{(\rho-\lambda)^2 - \frac{c}{4}\right\} \neq 0$, then $M$ is a Hopf hypersurface.* (for the definitions of $\rho$ and $\lambda$ see §2.)

**Theorem 4.2.** *Let $M$ be a real hypersurface in a complex hyperbolic space $\mathbb{C}H_n$. If it satisfies $R_\xi \phi = \phi R_\xi$ and at the same time $R_\xi S = SR_\xi$, then $M$ is locally congruent to one of $(A_0)$, $(A_1)$ or $(A_2)$-type provided that $\eta(A\xi) \neq 0$.*

The authors would like to express their sincere gratitude to the referee for his valuable suggestions and comments.

## 1. Fundamental facts of real hypersurfaces

Let $M$ be a real hypersurface of a complex space form $M_n(c)$ and $N$ be a unit normal vector field on $M$. By $\tilde{\nabla}$ we denote the Levi-Civita connection with respect to the Fubini-Study metric $\tilde{g}$ of $M_n(c)$. Then the Gauss and Weingarten formulas are given respectively by

$$\tilde{\nabla}_Y X = \nabla_Y X + g(AY, X)N, \ \tilde{\nabla}_X N = -AX$$

for any vector fields $X$ and $Y$ on $M$, where $g$ denotes the Riemannian metric of $M$ induced from $\tilde{g}$ and $A$ is the shape operator of $M$ in $M_n(c)$. For any vector fields $X$ tangent to $M$, we put

$$JX = \phi X + \eta(X)N, \ JN = -\xi.$$

Then we may see that the structure $(\phi, \xi, \eta, g)$ is an almost contact metric structure on $M$, that is, we have

$$\phi^2 X = -X + \eta(X)\xi, \ g(\phi X, \phi Y) = g(X, Y) - \eta(X)\eta(Y),$$
$$\eta(\xi) = 1, \ \phi\xi = 0, \ \eta(X) = g(X, \xi)$$

for any vector fields $X$ and $Y$ on $M$.

From the fact $\tilde{\nabla} J = 0$ and making use of the Gauss and Weingarten formulas, we obtain

$$(\nabla_X \phi)Y = \eta(Y)AX - g(AX, Y)\xi, \ \nabla_X \xi = \phi AX. \tag{1.1}$$

Since the ambient space is of constant holomorphic sectional curvature $c$, we have the following Gauss and Codazzi equations:

$$\begin{aligned} R(X,Y)Z = \frac{c}{4}\{&g(Y,Z)X - g(X,Z)Y \\ &+ g(\phi Y, Z)\phi X - g(\phi X, Z)\phi Y - 2g(\phi X, Y)\phi Z\} \\ &+ g(AY,Z)AX - g(AX,Z)AY, \end{aligned} \tag{1.2}$$

$$(\nabla_X A)Y - (\nabla_Y A)X = \frac{c}{4}\{\eta(X)\phi Y - \eta(Y)\phi X - 2g(\phi X, Y)\xi\} \tag{1.3}$$

for any vector fields $X$, $Y$ and $Z$ on $M$, where $R$ denotes the Riemannian curvature tensor of $M$.

In what follows, to write our formulas in convention forms, we denote by $\alpha = g(A\xi,\xi)$, $\beta = g(A^2\xi,\xi)$, $\gamma = g(A^3\xi,\xi)$, $h = \operatorname{Tr} A$ and $h_{(2)} = \operatorname{Tr} {}^tAA$ and for a function $f$ we denote by $\nabla f$ the gradient vector field of $f$.

If we put $U = \nabla_\xi\xi$, then $U$ is orthogonal to the structure vector $\xi$. We get

$$\phi U = -A\xi + \alpha\xi, \tag{1.4}$$

which shows that $g(U,U) = \beta - \alpha^2$. Thus we easily see that $\xi$ is a principal curvature vector, that is, $A\xi = \alpha\xi$ if and only if $\beta - \alpha^2 = 0$. From the Gauss structure equation (1.2), the Ricci tensor $S$ of $M$ is given by

$$S = \frac{c}{4}\{(2n+1)I - 3\eta\otimes\xi\} + hA - A^2, \tag{1.5}$$

where $I$ is the identity map, which implies

$$S\xi = \frac{c}{2}(n-1)\xi + hA\xi - A^2\xi. \tag{1.6}$$

We put

$$A\xi = \alpha\xi + \mu W, \tag{1.7}$$

where $W$ is a unit vector field orthogonal to $\xi$. Then we have $U = \mu\phi W$, and $W$ is also orthogonal to $U$. Further we have $\mu^2 = \beta - \alpha^2$.

By the definition of $U$ and the second equation of (1.1) and (1.7), it is verified that

$$g(\nabla_X\xi, U) = \mu g(AW, X). \tag{1.8}$$

Now, differentiating (1.4) covariantly along $M$ and using (1.1), we find

$$\begin{aligned}&\eta(X)g(AU + \nabla\alpha, Y) + g(\phi X, \nabla_Y U)\\ &= g((\nabla_Y A)X,\xi) - g(A\phi AX, Y) + \alpha g(A\phi X, Y),\end{aligned} \tag{1.9}$$

which shows that

$$(\nabla_\xi A)\xi = 2AU + \nabla\alpha \tag{1.10}$$

because of (1.3) and (1.8). Since $W$ is orthogonal to $U$, we see, using (1.1), that

$$\mu g(\nabla_X W, \xi) = g(AU, X). \tag{1.11}$$

Because of (1.2), the structure Jacobi operator $R_\xi$ is given by

$$R_\xi X = R(X,\xi)\xi = \frac{c}{4}(X - \eta(X)\xi) + \alpha AX - \eta(AX)A\xi \tag{1.12}$$

for any vector field $X$ on $M$.

## 2. The Jacobi operator of real hypersurfaces

Let $M$ be a real hypersurface of a complex space form $M_n(c)$, $c \neq 0$ satisfying $R_\xi \phi = \phi R_\xi$. Then by (1.12) we have

$$\alpha(\phi AX - A\phi X) = g(A\xi, X)U + g(U, X)A\xi. \tag{2.1}$$

We set $\Omega = \{p \in M | \mu(p) \neq 0\}$, and suppose that $\Omega$ is nonvoid, that is, $\xi$ is not a principal curvature vector on $M$. In the sequel we discuss our arguments on the open set $\Omega$ of $M$ unless otherwise stated. Then, it is, using (2.1), clear that $\alpha \neq 0$ on $\Omega$. So a function $\lambda$ given by $\beta = \alpha\lambda$ is defined. Thus, replacing $X$ by $U$ in (2.1) and using (1.4), we find

$$\phi AU = \lambda A\xi - A^2\xi. \tag{2.2}$$

Now, suppose that the Ricci tensor $S$ of type (1,1) and the Jacobi operator $R_\xi$ with respect to the structure vector $\xi$ commute to each other, that is, $R_\xi S = SR_\xi$. Then we have

$$\alpha A^3\xi = (\alpha h - \frac{c}{4})A^2\xi + (\gamma - \beta h + \frac{c}{4})A\xi + \frac{c}{4}(\beta - h\alpha)\xi, \tag{2.3}$$

$$\mu\{g(A^2\xi, Y)w(X) - g(A^2\xi, X)w(Y)\} = \beta\{\eta(Y)g(A\xi, X) - \eta(X)g(A\xi, Y)\},$$

where a 1-form $w$ is defined by $w(X) = g(W, X)$.

Using these equations, we have

$$A^2\xi = \rho A\xi + \sigma\xi \tag{2.4}$$

for certain scalars $\rho$ and $\sigma$ on $M$. Then we have

$$\sigma = \alpha(\lambda - \rho). \tag{2.5}$$

Combining (2.4) with (2.2), it is seen that

$$AU = (\rho - \lambda)U. \tag{2.6}$$

From (1.7) and (2.4) we also have

$$AW = \mu\xi + (\rho - \alpha)W \tag{2.7}$$

and hence

$$A^2W = \rho AW + \alpha(\lambda - \rho)W \tag{2.8}$$

by virtue of $\mu \neq 0$.

Differentiating (2.4) covariantly along $\Omega$ and taking account of (1.1), we find

$$\begin{aligned} &g((\nabla_X A)A\xi, Y) + g(A(\nabla_X A)\xi, Y) + g(A^2\phi AX, Y) - \rho g(A\phi AX, Y) \\ &= g(\nabla\rho, X)g(A\xi, Y) + \rho g((\nabla_X A)\xi, Y) + g(\nabla\sigma, X)\eta(Y) \\ &\quad + \alpha(\lambda - \rho)g(\phi AX, Y) \end{aligned} \tag{2.9}$$

for any vector fields $X$ and $Y$ on $M$, which together with (1.3) and (1.10) implies that

$$(\nabla_\xi A)A\xi = \rho AU - \frac{c}{4}U + \frac{1}{2}\nabla\beta.$$

We put $X = \xi$ in (2.9) and use (1.10), (2.6) and the last equation. Then we obtain

$$\begin{aligned}\frac{1}{2}\nabla\beta &= -A\nabla\alpha + \rho\nabla\alpha + g(\nabla\rho,\xi)A\xi + g(\nabla\sigma,\xi)\xi \\ &\quad - \left\{(\rho-\lambda)(\rho+\alpha-3\lambda) - \frac{c}{4}\right\}U,\end{aligned} \tag{2.10}$$

which connected to (2.4) and (2.6) gives

$$\begin{aligned}&\frac{1}{2}(A\nabla\beta - \rho\nabla\beta) = -A^2\nabla\alpha + 2\rho A\nabla\alpha - \rho^2\nabla\alpha + g(\nabla\sigma,\xi)A\xi \\ &+ g(\sigma\nabla\rho - \rho\nabla\sigma,\xi)\xi + \lambda\left\{(\rho-\lambda)(\rho+\alpha-3\lambda) - \frac{c}{4}\right\}U.\end{aligned} \tag{2.11}$$

Because of (2.5), we see, using (2.10), that

$$\frac{1}{2}d\beta(\xi) = \alpha d\alpha(\xi) + \mu d\alpha(W),$$

where $d$ denotes the exterior differential operator, which together with $\beta = \alpha\lambda$ implies that

$$\alpha d\lambda(\xi) = (2\alpha - \lambda)d\alpha(\xi) + 2\mu d\alpha(W). \tag{2.12}$$

We verify also, making use of (2.7) and (2.10), that

$$\frac{1}{2}d\beta(W) - \alpha d\alpha(W) = \mu(d\rho(\xi) - d\alpha(\xi)),$$

which enables us to obtain

$$\alpha d\lambda(W) = (2\alpha - \lambda)d\alpha(W) + 2\mu(d\rho(\xi) - d\alpha(\xi)). \tag{2.13}$$

**Remark 2.1.** We notice here that $\rho - \lambda \neq 0$ on $\Omega$. For the proof of this fact, see [4], §3.

## 3. Lemmas

We will continue now our arguments under the same hypotheses $R_\xi\phi = \phi R_\xi$ and at the same time $R_\xi S = SR_\xi$ as in section 2.

First of all, we give the following two lemmas without proofs. For the proofs see [4].

**Lemma 3.1.** *Let $M$ be a real hypersurface satisfying $R_\xi\phi = \phi R_\xi$ and $R_\xi S = SR_\xi$ in $M_n(c)$. Then we have*

$$\alpha(\nabla\rho - \nabla\lambda) = \theta U \tag{3.1}$$

*on $\Omega$, where $\theta$ is given by*

$$\theta = 3(\rho-\lambda)^2 - \frac{3}{4}c. \tag{3.2}$$

**Lemma 3.2.** *Let $M$ be a real hypersurface of $M_n(c)$, $c \neq 0$ such that $R_\xi\phi = \phi R_\xi$ and $R_\xi S = SR_\xi$ hold. If it satisfies $du = 0$, then we have*

$$\nabla\alpha = (3\lambda - 2\rho)U, \tag{3.3}$$

$$\alpha\nabla\lambda = \left\{2(\rho-\lambda)(3\lambda-\alpha-\rho) + \lambda(3\lambda-2\rho) + \frac{c}{2}\right\}U, \tag{3.4}$$

$$\alpha\nabla\rho = (\rho^2 - 2\rho\alpha + 2\alpha\lambda - \frac{c}{4})U. \tag{3.5}$$

In the proof of Lemma 3.1 [4], we showed the following two equations. We give them here for later use.

$$\begin{aligned}&\theta(\lambda d\alpha(\xi) - \alpha d\lambda(\xi))A\xi\\ &= \theta\alpha(\lambda\nabla\alpha - \alpha\nabla\lambda) + \theta\alpha\left\{2(\rho-\lambda)(3\lambda-\alpha-\rho) + \frac{c}{2}\right\}U,\end{aligned} \tag{3.6}$$

$$\theta\left\{\mu\lambda d\alpha(\xi) - \mu\alpha d\lambda(\xi) - \alpha\lambda d\alpha(W) + \alpha^2 d\lambda(W)\right\} = 0, \tag{3.7}$$

where a 1-form $u$ is defined by $u(X) = g(U, X)$.

Now, we prove

**Lemma 3.3.** *Under the same hypotheses as those stated in Lemma 3.2, we have*

$$\begin{aligned}(\nabla_U A)X =&(\lambda-\alpha)\left\{A^2X - \frac{c}{4}(X - \eta(X)\xi)\right\}\\ &+ \frac{2}{\alpha}\left\{(\rho-\lambda)^2 - \frac{c}{4}\right\}u(X)U + \frac{c}{4}\left\{\eta(X)W + w(X)\xi\right\}.\end{aligned} \tag{3.8}$$

**Proof.** Since we have $A\xi = \alpha\xi + \mu W$, (2.1) becomes

$$(\phi A - A\phi)X = \eta(X)U + u(X)\xi + \tau\left\{w(X)U + u(X)W\right\}, \tag{3.9}$$

where we put

$$\alpha\tau = \mu, \quad \lambda - \alpha = \mu\tau. \tag{3.10}$$

From this we have

$$(\lambda-\alpha)\nabla\alpha + \mu\alpha\nabla\tau = \mu\nabla\mu.$$

Further, from (3.3) and (3.4) we are led to

$$\mu\nabla\mu = \Big\{(\rho-\alpha)(2\lambda-\rho)+\frac{c}{4}\Big\}U. \tag{3.11}$$

Combining last two equations, we obtain

$$\mu\alpha\nabla\tau = \Big\{(\rho-\lambda)(3\lambda-\alpha-\rho)+\frac{c}{4}\Big\}U. \tag{3.12}$$

We notice here that

$$d\alpha(\xi)=0,\ d\alpha(W)=0,\ d\lambda(\xi)=0,\ d\lambda(W)=0,\ d\rho(W)=0 \tag{3.13}$$

by virtue of Lemma 3.2.

In the following, let $\{e_1,\ldots,e_{2n-1}\}$ be an orthonormal frame field of $M$. And the coefficients of tensors are defined by $\xi^i=g(\xi,e_i)$, $A_i{}^j=g(Ae_i,e_j)$, $\phi_i{}^j = g(\phi e_i,e_j)$, $\nabla_i U^j = g(\nabla_{e_i}U,e_j)$ and so on. Furthermore, we use Einstein rule, that is, we take the summation for the same indices in the equation.

Differentiating (3.9) covariantly and using (1.1), we find

$$\begin{aligned}&(\nabla_k A_{jr})\phi_i{}^r + (\nabla_k A_{ir})\phi_j{}^r + A^2_{kj}\xi_i - (A_{jr}\xi^r)A_{ki} + A^2_{ki}\xi_j\\ &- (A_{ir}\xi^r)A_{kj} + (\xi_i+\tau W_i)\nabla_k U_j + (\xi_j+\tau W_j)\nabla_k U_i\\ &+ U_j\nabla_k\xi_i + U_i\nabla_k\xi_j + \tau_k(U_jW_i+U_iW_j)\\ &+ \tau(U_j\nabla_k W_i + U_i\nabla_k W_j) = 0,\end{aligned} \tag{3.14}$$

where $\tau_k=\partial_k\tau$. From this, taking the skew-symmetric part with respect to indices $k$ and $j$, and making use of (1.3) and $\nabla_jU_i-\nabla_iU_j=0$, we have

$$\begin{aligned}&(\nabla_k A_{ir})\phi_j{}^r - (\nabla_j A_{ir})\phi_k{}^r + \frac{c}{4}(\xi_k g_{ji} - \xi_j g_{ki})\\ &+ \xi_j A^2_{ki} - \xi_k A^2_{ji} - (A_{jr}\xi^r)A_{ki} + (A_{kr}\xi^r)A_{ji} + (\xi_j+\tau W_j)\nabla_k U_i\\ &- (\xi_k+\tau W_k)\nabla_j U_i + U_j\nabla_k\xi_i - U_k\nabla_j\xi_i + U_i(\nabla_k\xi_j - \nabla_j\xi_k)\\ &+ \tau_k(U_jW_i+U_iW_j) - \tau_j(U_kW_i+U_iW_k)\\ &+ \tau\{U_j\nabla_k W_i - U_k\nabla_j W_i + U_i(\nabla_k W_j - \nabla_j W_k)\} = 0.\end{aligned}$$

Thus, interchanging $k$ and $i$, we find

$$\begin{aligned}&(\nabla_k A_{ir})\phi_j{}^r - (\nabla_k A_{jr})\phi_i{}^r + \frac{c}{2}(\xi_i g_{jk} - \xi_j g_{ik}) + \xi_j A^2_{ki} - \xi_i A^2_{jk}\\ &- (A_{jr}\xi^r)A_{ik} + (A_{ir}\xi^r)A_{jk} + (\xi_j+\tau W_j)\nabla_i U_k - (\xi_i+\tau W_i)\nabla_j U_k\\ &+ U_j\nabla_i\xi_k - U_i\nabla_j\xi_k + U_k(\nabla_i\xi_j - \nabla_j\xi_i) + \tau_i(U_jW_k+U_kW_j)\\ &- \tau_j(U_iW_k+U_kW_i) + \tau\{U_k(\nabla_i W_j - \nabla_j W_i) + U_j\nabla_i W_k - U_i\nabla_j W_k\} = 0,\end{aligned}$$

where we have used (1.3). Adding this to (3.14) and taking account of (3.12), we find

$$\begin{aligned}&(\nabla_k A_{ir})\phi_j{}^r + \frac{c}{4}(\xi_i g_{jk} - \xi_j g_{ik}) + \xi_j A^2_{ik} - (A_{jr}\xi^r)A_{ki}\\&+\frac{1}{\mu\alpha}\left\{(\rho-\lambda)(3\lambda-\alpha-\rho)+\frac{c}{4}\right\}U_iU_kW_j + (\xi_j+\tau W_j)\nabla_k U_i\\&+\frac{1}{2}\{U_k(\nabla_i\xi_j - \nabla_j\xi_i) + U_i(\nabla_k\xi_j - \nabla_j\xi_k) + U_j(\nabla_k\xi_i + \nabla_i\xi_k)\}\\&+\frac{1}{2}\tau\{U_k(\nabla_i W_j - \nabla_j W_i) + U_i(\nabla_k W_j - \nabla_j W_k)\\&+U_j(\nabla_k W_i + \nabla_i W_k)\} = 0.\end{aligned} \tag{3.15}$$

On the other hand, differentiating (1.7) covariantly and using (1.1), we obtain

$$(\nabla_k A_{jr})\xi^r - A_{jr}A_{ks}\phi^{rs} = \alpha_k\xi_j - \alpha A_{kr}\phi_j{}^r + \mu_k W_j + \mu\nabla_k W_j,$$

from which, taking the skew-symmetric part and making use of (1.3), (3.3) and (3.11),

$$\begin{aligned}&-\frac{c}{2}\phi_{kj} - 2A_{jr}A_{ks}\phi^{rs} = (3\lambda-2\rho)(U_k\xi_j - U_j\xi_k) - \alpha(A_{kr}\phi_j{}^r - A_{jr}\phi_k{}^r)\\&+\frac{1}{\mu}\left\{(\rho-\alpha)(2\lambda-\rho)+\frac{c}{4}\right\}(U_kW_j - U_jW_k) + \mu(\nabla_k W_j - \nabla_j W_k).\end{aligned}$$

Putting $e_j = \xi$ in this and using (2.6), we have

$$\mu\xi^r(\nabla_k W_r - \nabla_r W_k) = (\alpha-\lambda)U_k. \tag{3.16}$$

Replacing $e_j$ by $\xi$ in (3.15) and taking account of (3.16), we get

$$\nabla_k U_i = \alpha A_{ki} - A^2_{ki} + \frac{c}{4}(g_{ki} - \xi_k\xi_i) + \frac{\lambda}{\alpha}U_kU_i. \tag{3.17}$$

Differentiating (2.6) covariantly and using (3.1) and (3.17), we get

$$\begin{aligned}(\nabla_k A_{jr})U^r = {}& A^3_{jk} + (\lambda-\rho-\alpha)A^2_{jk} + \left\{\alpha(\rho-\lambda) - \frac{c}{4}\right\}A_{jk}\\&+\frac{c}{4}\xi_k(A_{jr}\xi^r) + \frac{\theta}{\alpha}U_kU_j + \frac{c}{4}(\rho-\lambda)(g_{jk} - \xi_j\xi_k).\end{aligned} \tag{3.18}$$

Differentiating $\phi U = -\mu W$ covariantly and using (1.1), we find

$$\mu_k W_j + \mu\nabla_k W_j = (\rho-\lambda)U_k\xi_j + \phi_{jr}\nabla_k U^r,$$

or, using (3.11) and (3.17), we obtain

$$\begin{aligned}&\left\{(\rho-\alpha)(2\lambda-\rho)+\frac{c}{4}\right\}U_kW_j + \mu^2\nabla_k W_j\\&= \mu(\rho-\lambda)U_k\xi_j + \mu\left\{\alpha A_{kr}\phi_j{}^r + \frac{c}{4}\phi_{jk} - A^2_{kr}\phi_j{}^r\right\} + \lambda(\lambda-\alpha)U_kW_j.\end{aligned}$$

Putting $e_k = W$ in this, and using (2.7) and (2.8), we have

$$\mu^2 \nabla_W W = \left\{ (\rho-\alpha)^2 + \alpha(\lambda-\rho) - \frac{c}{4} \right\} U. \tag{3.19}$$

Now, if we take the inner product (3.15) with $\mu W_j$, and use (2.7), (3.10), (3.17) and (3.19), then we obtain

$$(\nabla_k A_{jr})U^r = (\lambda-\alpha)\left\{ A^2_{jk} - \frac{c}{4}(g_{jk} - \xi_j\xi_k) \right\} + \frac{2}{\alpha}\left\{ (\rho-\lambda)^2 - \frac{c}{4} \right\} U_j U_k$$
$$- \frac{c}{4}\mu W_k \xi_j.$$

Comparing this with (3.18), we obtain

$$\begin{aligned} &A^3_{ji} - \rho A^2_{ji} + \left\{ \alpha(\rho-\lambda) - \frac{c}{4} \right\} A_{ji} + \frac{c}{4}(\rho-\alpha)(g_{ji} - \xi_j\xi_i) \\ &+ \frac{c}{4}\xi_j(A_{ir}\xi^r) + \frac{c}{4}\mu W_j \xi_i + \frac{1}{\alpha}\left\{ (\rho-\lambda)^2 - \frac{c}{4} \right\} U_j U_i = 0. \end{aligned} \tag{3.20}$$

Combining this with (3.18), we, using (1.3), verify that

$$U^r \nabla_r A_{ji} = (\lambda-\alpha)\left\{ A^2_{ji} - \frac{c}{4}(g_{ji} - \xi_j\xi_i) \right\} + \frac{2}{\alpha}\left\{ (\rho-\lambda)^2 - \frac{c}{4} \right\} U_j U_i$$
$$+ \frac{c}{4}\mu(\xi_j W_i + \xi_i W_j).$$

This completes the proof. □

Since $\mu^2 = \alpha(\lambda-\alpha)$, taking the trace of (3.8), we find

$$dh(U) = (\lambda-\alpha)\left\{ h_{(2)} + 2(\rho-\lambda)^2 - \frac{c}{2}n \right\}. \tag{3.21}$$

From (3.8) we also have

$$\sum_{j,i} A_{ji}\nabla_U A_{ji} = (\lambda-\alpha)\left\{ h_{(3)} - \frac{c}{4}(h-\alpha) + 2(\rho-\lambda)^3 - \frac{c}{2}(\rho-\lambda) + \frac{c}{2}\alpha \right\},$$

where $h_{(3)} = \operatorname{Tr} A^3$. This shows that

$$\frac{1}{2}\nabla_U h_{(2)} = (\lambda-\alpha)\left\{ h_{(3)} + 2(\rho-\lambda)^3 - \frac{c}{4}(h-\alpha) + \frac{c}{2}(\alpha+\lambda-\rho) \right\} \tag{3.22}$$

because of $\nabla h_{(2)} = 2\sum_{j,i} A_{ji}\nabla A_{ji}$.

On the other hand, the trace of (3.20) is given by

$$h_{(3)} = \rho h_{(2)} - h\alpha(\rho-\lambda) + \frac{c}{4}h - \frac{c}{2}(n-1)(\rho-\alpha)$$
$$- \frac{c}{4}\alpha + (\lambda-\alpha)\left\{ \frac{c}{4} - (\rho-\lambda)^2 \right\}.$$

Substituting this into (3.22), we find

$$\begin{aligned}\frac{1}{2}\nabla_U h_{(2)} = (\lambda-\alpha)\Big\{\rho h_{(2)} - h\alpha(\rho-\lambda) - (\lambda-\alpha)(\rho-\lambda)^2 \\ + 2(\rho-\lambda)^3 - \frac{c}{2}n(\rho-\alpha) + \frac{c}{4}(3\lambda-\alpha)\Big\}.\end{aligned} \tag{3.23}$$

Now, we demonstrate the following.

**Lemma 3.4.** *Under the same hypotheses as those stated in Lemma* 3.2, *we have* $\Omega$ *is void if* $\rho = h$.

**Proof.** Since $\rho = h$ is assumed, (3.5) becomes

$$\alpha\nabla h = (\rho^2 - 2\rho\alpha + 2\alpha\lambda - \frac{c}{4})U,$$

which shows that

$$\alpha dh(U) = \alpha(\lambda-\alpha)(\rho^2 - 2\rho\alpha + 2\alpha\lambda - \frac{c}{4}).$$

From this and (3.21) we have

$$h_{(2)} + 2(\rho-\lambda)^2 - \frac{c}{2}n = \rho^2 - 2\rho\alpha + 2\alpha\lambda - \frac{c}{4}. \tag{3.24}$$

Differentiating this covariantly, we find

$$\frac{1}{2}\alpha\nabla h_{(2)} = (2\lambda - 2\rho - \alpha)\alpha(\nabla\rho - \nabla\lambda) + \rho\alpha\nabla\rho + (\lambda-\rho)\alpha\nabla\alpha, \tag{3.25}$$

which together with (3.1), (3.3) and (3.5) yields

$$\frac{1}{2}\alpha\nabla h_{(2)} = fU$$

for certain function $f$ on $\Omega$. Because of (3.23) and $h = \rho$, it follows that

$$\begin{aligned}\frac{1}{2}\alpha\nabla h_{(2)} = \Big\{\rho h_{(2)} + 2(\rho-\lambda)^3 - (\lambda-\alpha)(\rho-\lambda)^2 - \rho\alpha(\rho-\lambda) \\ -\frac{c}{2}n(\rho-\alpha) + \frac{c}{4}(3\lambda-\alpha)\Big\}U.\end{aligned} \tag{3.26}$$

Substituting (3.1), (3.3), (3.5) and (3.26) into (3.25), we find that

$$8(\rho-\lambda)^3 + (\alpha - \lambda - 2\rho)(\rho-\lambda)^2 + \frac{c}{2}(n-2)\alpha + \frac{3}{4}c(3\lambda - 2\rho) = 0,$$

where we have used (3.2) and (3.24), or equivalently

$$\theta(3\lambda - 2\rho) = \alpha\left\{(\rho-\lambda)^2 + \frac{c}{2}(n-2)\right\}. \tag{3.27}$$

We define the function $\nu$ by $\nu = \rho - \lambda$. Then (3.27) becomes

$$3(\nu^2 - \frac{c}{4})(\lambda - 2\nu) = \alpha\left\{\nu^2 + \frac{c}{2}(n-2)\right\}. \tag{3.28}$$

From (3.1) and (3.5), we have respectively

$$\alpha\nabla\nu = 3(\nu^2 - \frac{c}{4})U, \tag{3.29}$$

$$\alpha\nabla\lambda = \left\{-2\nu^2 + 2(\lambda-\alpha)\nu + \lambda^2 + \frac{c}{2}\right\}U. \tag{3.30}$$

If $\lambda \equiv 2\nu$, then by virtue of (3.28), we are led to

$$\nabla\nu = 0,\ \nabla\lambda = 2\nabla\nu = 0,\ \nabla\rho = \nabla(\nu+\lambda) = 0.$$

If $\lambda \neq 2\nu$, then from (3.28) and our assumption $c \neq 0$, we obtain

$$\nu^2 - \frac{c}{4} \neq 0.$$

From (3.3) and (3.29), we have

$$\nabla\alpha = \frac{\alpha(\lambda - 2\nu)}{3(\nu^2 - \frac{c}{4})}\nabla\nu. \tag{3.31}$$

According to (3.29) and (3.30), we obtain

$$3(\nu^2 - \frac{c}{4})\nabla\lambda = \left\{-2\nu^2 + 2(\lambda-\alpha)\nu + \lambda^2 + \frac{c}{2}\right\}\nabla\nu. \tag{3.32}$$

Differentiating (3.28), we have

$$\left[3(\nu^2 - \frac{c}{4})\left\{6\nu\lambda - 18\nu^2 + \frac{3}{2}c - 2\alpha\nu\right\} - \alpha(\lambda-2\nu)\left\{\nu^2 + \frac{c}{2}(n-2)\right\}\right]\nabla\nu$$
$$+ 9(\nu^2 - \frac{c}{4})^2\nabla\lambda = 0. \tag{3.33}$$

Making use of (3.32) and (3.33), we deduce

$$3\left\{(\nabla\lambda)^2 + (\nabla\nu)^2\right\}(\nu^2 - \frac{c}{4})\left[3(\nu^2 - \frac{c}{4})\left\{-2\nu^2 + 2(\lambda-\alpha)\nu + \lambda^2 + \frac{c}{2}\right\}\right.$$
$$\left. + 3(\nu^2 - \frac{c}{4})\left\{6\nu\lambda - 18\nu^2 + \frac{3}{2}c - 2\alpha\nu\right\} - \alpha(\lambda - 2\nu)\left\{\nu^2 + \frac{c}{2}(n-2)\right\}\right] = 0.$$

If $(\nabla\lambda)^2 + (\nabla\nu)^2 \neq 0$, then we have

$$\begin{aligned} &\left\{-10\nu^3 - \lambda\nu^2 + (n+1)c\nu - \frac{1}{2}(n-2)c\lambda\right\}\alpha \\ &- 60\nu^4 + 24\nu^3\lambda + (21c + 3\lambda^2)\nu^2 - 6c\lambda\nu - \frac{3}{4}c\lambda^2 - \frac{3}{2}c^2 = 0. \end{aligned} \tag{3.34}$$

Using (3.28) and (3.34), we have

$$4(3n-5)\nu^2 - 3(2n-3)\lambda\nu - (n-2)c = 0, \tag{3.35}$$

since $\nu^2 \neq \frac{c}{4}$.

Differentiating (3.35), we obtain

$$\{8(3n-5)-3(2n-3)\lambda\}\nabla\nu-3(2n-3)\nabla\lambda=0. \tag{3.36}$$

From (3.32) and (3.36), we are led to

$$\begin{aligned}&2(2n-3)\nu^2\alpha+2(2n-3)\nu^3+\{5(3-2n)\lambda+8(3n-5)\}\nu^2\\&+\left\{(3-2n)\lambda^2+\frac{c}{2}(3-2n)\right\}\nu+\frac{3}{4}(2n-3)c\lambda+2(5-3n)c=0.\end{aligned} \tag{3.37}$$

Making use of (3.28) and (3.37), we have

$$\begin{aligned}&10(2n-3)\nu^5+\{(3-2n)\lambda+8(5-3n)\}\nu^4\\&+\left\{(2n-3)\lambda^2+(-2n^2+2n+\frac{3}{2})c\right\}\nu^3\\&+\left\{(5n^2-16n+\frac{51}{4})c\lambda+(-12n^2+50n-50)c\right\}\nu^2\\&+\left\{(n^2-\frac{7}{2}n+3)c\lambda+(\frac{1}{2}n^2-\frac{7}{4}n+\frac{3}{2})c^2\right\}\nu\\&+(-\frac{3}{4}n^2+\frac{21}{8}n-\frac{9}{4})c^2\lambda+(3n^2-11n+10)c^2=0.\end{aligned} \tag{3.38}$$

Caluculating the resultant of (3.35) and (3.38) with respect to $\nu$, we obtain

$$a_0(c,n)\lambda^6+a_1(c,n)\lambda^5+\cdots+a_6(c,n)=0, \tag{3.39}$$

where $a_i(c,n)$ are polynomials of $c$, $n$ and

$$a_0(c,n)=-\frac{9}{4}c^2(n-2)(22n-35)(2n-3)^4(216n^2-666n+515).$$

(We use a computer to calculate this.)

When $n>2$, (3.39) is a non-trivial algebraic equation of $\lambda$. So we have

$$\nabla\lambda=0.$$

Furthermore, using (3.35), we obtain

$$\nabla\nu=0.$$

These contradict to oue assumption $(\nabla\lambda)^2+(\nabla\nu)^2\neq 0$. So it must be

$$\nabla\lambda=\nabla\nu=0.$$

When $n=2$, (3.35) is reduced to

$$\nu(4\nu-3\lambda)=0. \tag{3.40}$$

On the set of points of $\nu\neq 0$, we have

$$4\nu=3\lambda. \tag{3.41}$$

Differentiating (3.41), and making use of (3.32), we deduce that

$$(\frac{14}{9}\nu^2 + 2\alpha\nu - \frac{3}{2}c)\nabla\nu = 0. \tag{3.42}$$

If $\nabla\nu \neq 0$, then we obtain

$$\frac{14}{9}\nu^2 + 2\alpha\nu - \frac{3}{2}c = 0. \tag{3.43}$$

According to (3.28), (3.41) and (3.43), we are led to

$$\frac{22}{9}\nu^2 + \frac{c}{2} = 0.$$

This contradicts to the fact that $\nabla\nu \neq 0$.

By all the above areguments, we have

$$\nabla\nu = \nabla\lambda = \nabla\rho = 0.$$

Thus, by (3.1) we verify that $\theta = 0$, which together with (3.27) will produce a contradiction. Thus, $\Omega$ is void. □

Finally, we prove

**Lemma 3.5.** *Let $M$ be a real hypersurface in $M_n(c)$, $c \neq 0$ such that $R_\xi\phi = \phi R_\xi$ and $R_\xi S = SR_\xi$. If $du = 0$, then $\Omega$ is void.*

**Proof.** From (2.4) we have

$$A^3\xi = (\rho^2 - \beta - \rho\alpha)A\xi + \rho(\beta - \rho\alpha)\xi.$$

Comparing this with (2.3), we verify that

$$\mu(h-\rho)(\beta - \rho\alpha - \frac{c}{4}) = 0. \tag{3.44}$$

In the previous paper, we prove (cf. [3], [2])

**Theorem KL** ([4]). *Let $M$ be a real hypersurface of $M_n(c)$, $c \neq 0$ satisfying $R_\xi\phi = \phi R_\xi$ and $AR_\xi = R_\xi A$. Then $\Omega$ is void.*

**Theorem KKL** ([4]). *Let $M$ be a real hypersurface of $M_n(c)$, $c \neq 0$ satisfying $R_\xi\phi = \phi R_\xi$, $R_\xi S = SR_\xi$ and $g(S\xi, \xi) = const.$, then $\Omega$ is void.*

Let $\Omega_0 := \{p \in \Omega | (h-\rho)(p) \neq 0\} \subset \Omega \neq \emptyset$. Then by (3.44), we have $\beta - \rho\alpha = \frac{c}{4}$ on $\Omega_0$ and thus $AR_\xi = R_\xi A$ (cf. Rcmark 4 of [4]). According to Theorem KL, we see that $\Omega_0$ is void. Therefore we have $h - \rho = 0$ on $\Omega$. Owing to Lemma 3.4, we see that $\Omega$ is empty. □

**Lemma 3.6.** *Let $M$ be a real hypersurface of $M_n(c)$, $c \neq 0$ which satisfies $R_\xi\phi = \phi R_\xi$ and $S\xi = g(S\xi, \xi)\xi$. If $du = 0$, then $\Omega$ is void.*

**Proof.** From the assumption $S\xi = g(S\xi, \xi)\xi$ and (1.6) we have

$$A^2\xi = hA\xi + (\beta - h\alpha)\xi,$$

which together with (2.2) implies that

$$AU = (h - \lambda)U.$$

Thus, all formulas developed above are valid replacing $\rho$ by $h$. Owing to Lemma 3.4, we see that $\Omega = \emptyset$. □

## 4. Theorems

Let $M$ be a real hypersurface of $M_n(c)$, $c \neq 0$ satisfying $R_\xi\phi = \phi R_\xi$ and $R_\xi S = SR_\xi$.

Now, suppose that $\theta \neq 0$ in (3.2). Then (3.7) implies

$$\mu\left\{\lambda d\alpha(\xi) - \alpha d\lambda(\xi)\right\} = \alpha\left\{\lambda d\alpha(W) - \alpha d\lambda(W)\right\},$$

or using (2.13)

$$\lambda d\alpha(\xi) - \alpha d\lambda(\xi) = 2\mu d\alpha(W) - 2\alpha(d\rho(\xi) - d\alpha(\xi)). \tag{4.1}$$

From (3.1) and (3.2) we have

$$\alpha\nabla\theta = 6(\rho - \lambda)\theta U. \tag{4.2}$$

Differentiating (3.1) covariantly and taking the skew-symmetric part obtained one, and using (4.2), we find

$$d\alpha(Y)d(\rho - \lambda)(X) - d\alpha(X)d(\rho - \lambda)(Y) = \theta du(Y, X),$$

which together with (3.1) and $\theta \neq 0$ gives

$$d\alpha(Y)u(X) - d\alpha(X)u(Y) = \alpha du(Y, X), \tag{4.3}$$

which apply by $\sum_{j,i}\phi_{ji}$, we have

$$\alpha\sum_{j,i}\phi_{ji}\nabla_j U_i = \mu d\alpha(W). \tag{4.4}$$

On the other hand, taking the trace of (1.9), and making use of (1.3), (1.4) and (2.6), we find

$$\sum_{j,i}\phi_{ji}\nabla_j U_i = dh(\xi) - d\alpha(\xi),$$

which together with (4.4) yields

$$\mu d\alpha(W) = \alpha d(h - \alpha)(\xi).$$

Combining this with (4.1), we verify that

$$\lambda d\alpha(\xi) - \alpha d\lambda(\xi) = 2\alpha d(h-\rho)(\xi). \tag{4.5}$$

From this and (3.6) we obtain

$$\lambda\nabla\alpha - \alpha\nabla\lambda = 2\left\{(\rho-\lambda)^2 + (\rho-\lambda)(\alpha-2\lambda) - \frac{c}{4}\right\}U + 2\alpha d(h-\rho)(\xi)A\xi.$$

As is already shown in the proof of Lemma 3.5, we may consider the case where $h-\rho=0$ on $\Omega$. Thus, the last equation becomes

$$\lambda\nabla\alpha - \alpha\nabla\lambda = 2\left\{(\rho-\lambda)^2 + (\rho-\lambda)(\alpha-2\lambda) - \frac{c}{4}\right\}U. \tag{4.6}$$

Using the same method as that used to derive (4.3) from (3.1), we can derive from (4.6) the following:

$$\begin{aligned} &d\lambda(Y)d\alpha(X) - d\lambda(X)d\alpha(Y) \\ &= (\rho-\lambda)\{d(\alpha-2\lambda)(Y)u(X) - d(\alpha-2\lambda)(X)u(Y)\} \\ &\quad + \left\{(\rho-\lambda)^2 + (\rho-\lambda)(\alpha-2\lambda) - \frac{c}{4}\right\}du(Y,X), \end{aligned}$$

where we have used (3.1). Substituting (4.3) and (4.6) into this, we find

$$\left\{(\rho-\lambda)^2 - \frac{c}{4}\right\}\{d\alpha(Y)u(X) - d\alpha(X)u(Y)\} = 0,$$

which together with (4.3) implies that $du = 0$. Thus, by Lemma 3.5, we have

**Theorem 4.1.** *Let $M$ be a real hypersurface with $R_\xi\phi = \phi R_\xi$ and $R_\xi S = SR_\xi$ in $M_n(c)$, $c \neq 0$. If $\theta = 3\left\{(\rho-\lambda)^2 - \frac{c}{4}\right\} \neq 0$, then $M$ is a Hopf hypersurface.*

From Lemma 3.6 and Theorem 4.1 we have

**Corollary 4.1.** *Let $M$ be a real hypersurface with $R_\xi\phi = \phi R_\xi$, and $S\xi = g(S\xi,\xi)\xi$ in $M_n(c)$, $c \neq 0$. If $\theta \neq 0$, then $M$ is a Hopf hypersurface.*

On a real hypersurface of a complex hyperbolic space $H_n\mathbb{C}$, we see that $\theta \neq 0$. Thus we have by (2.1) and Theorem 4.1

**Theorem 4.2.** *Let $M$ be a real hypersurface in a complex hyperbolic space. If it satisfies $R_\xi\phi = \phi R_\xi$ and at the same time $R_\xi S = SR_\xi$, then $M$ is locally congruent to one of $(A_0)$, $(A_1)$ or $(A_2)$-type provided that $\eta(A\xi) \neq 0$.*

**Corollary 4.2.** *Let $M$ be a real hypersurface in a complex hyperbolic space $H_n\mathbb{C}$. If it satisfies $R_\xi\phi = \phi R_\xi$, and $S\xi = g(S\xi,\xi)\xi$, then $M$ is the same type as that of Theorem 4.2.*

## References

1. J. Berndt, *Real hypersurfaces with constant principal curvatures in complex hyperbolic space*, J. Reine Angew. Math. **395**(1989), 132–141.
2. J.T. Cho and U-H. Ki, *Real hypersurfaces of a complex projective space in terms of the Jacobi operators*, Acta Math. Hungar **80**(1998), 155–167.
3. U-H. Ki, S.-J. Kim and S.-B. Lee, *Some characterizations of real hypersurfaces of type A*, Kyungpook Math. J. **31**(1991), 73–82.
4. U-H. Ki, S.J. Kim and S.-B. Lee, *The structure Jacobi operator on real hypersurfaces in a nonflat complex space form*, to appear in Bull. Korean Math. Soc.
5. M. Kimura, *Real hypersurfaces and complex submanifolds in complex projective space*, Trans. Amer. Math. Soc. **296**(1986), 137–149.
6. R. Takagi, *On homogeneous real hypersurfaces in a complex projective space*, Osaka J. Math. **10**(1973), 495–506.

# REAL HYPERSURFACES IN A NONFLAT COMPLEX SPACE FORM FROM THE VIEWPOINT OF THEIR GEODESICS

SADAHIRO MAEDA*

*Department of Mathematics, Shimane University,*
*Matsue, Shimane, 690-8504, JAPAN*
*E-mail: smaeda@math.shimane-u.ac.jp*

TOSHIAKI ADACHI†

*Department of Mathematics, Nagoya Institute of Technology*
*Gokiso, Nagoya, 466-8555, JAPAN*
*E-mail: adachi@nitech.ac.jp*

In this paper we give a survey on characterizations of totally $\eta$-umbilic real hypersurfaces, ruled real hypersurfaces and Hopf hypersurfaces with constant principal curvatures in a complex projective or complex hyperbolic space by the extrinsic shape of geodesics on these real hypersurfaces.

*Keywords*: nonflat complex space forms, totally $\eta$-umbilic real hypersurfaces, Hopf hypersurfaces, constant principal curvatures, ruled real hypersurfaces, geodesics, circles

## 1. Introduction

When we study submanifolds, it gives information on their shapes to study how their geodesics are seen in the ambient space. It is well-known that a hypersurface in a real space form $M^n(c;\mathbb{R})$ is totally umbilic if and only if every geodesic on this hypersurface is a circle in the ambient space $M^n(c;\mathbb{R})$. In a nonflat complex space form $M^n(c;\mathbb{C})$, there does not exist a real hypersurface all of whose geodesics are circles in the ambient space,

*The first author is partially supported by Grant-in-Aid for Scientific Research (C) (No.14540080), Ministry of Education, Science, Sports, Culture and Technology.
†The second author is partially supported by Grant-in-Aid for Scientific Research (C) (No. 14540075), Ministry of Education, Science, Sports, Culture and Technology.

which is equivalent to the fact that $M^n(c;\mathbb{C})$ does not admit a totally umbilic real hypersurface. However, there exist real hypersurfaces some of whose geodesics *orthogonal to the characteristic vector field* are circles in a nonflat complex space form. In this paper we summarize characterization theorems for totally $\eta$-umbilic real hypersurfaces, ruled real hypersurfaces and Hopf hypersurfaces with constant principal curvatures in a nonflat complex space form from this point of view. We shall call these real hypersurfaces *standard examples* of real hypersurfaces in a nonflat complex space form in this paper.

## 2. Real hypersurfaces in a nonflat complex space form

In this section, we list some important examples of real hypersurfaces in a nonflat complex space form $M^n(c;\mathbb{C}), c \neq 0$, which is either a complex projective space $\mathbb{C}P^n(c)$ of constant holomorphic sectionl curvature $c$ $(> 0)$ or a complex hyperbolic space $\mathbb{C}H^n(c)$ of constant holomorphic sectional curvature $c$ $(< 0)$. For a real hypersurface $M$ in a Kähler manifold $\widetilde{M}$ with complex structure $J$, we denote by $\mathcal{N}$ a unit normal vector field on $M$ and define the characteristic vector field $\xi$ by $-J\mathcal{N}$, a function $\eta : TM \to \mathbb{R}$ on the tangent bunble $TM$ by $\eta(v) = \langle v, \xi\rangle = \langle Jv, \mathcal{N}\rangle$ and a $(1,1)$-tensor $\phi$ by $\phi(v) = J(v - \eta(v)\xi)$. We then have $\phi^2(v) = -v + \eta(v)\xi$ and $\phi(\xi) = 0$. These $(\phi, \xi, \eta, \langle\ ,\ \rangle)$ form an almost contact metric structure on $M$. We denote by $T^0M$ the holomorphic distribution $\{v \in TM \mid v \perp \xi_{\varpi(v)}\}$, where $\varpi : TM \to M$ is a canonical projection.

A real hypersurface in a nonflat $M^n(c;\mathbb{C})$ is called a *Hopf hypersurface* if its characteristic vector field $\xi$ is principal. We denote by $\alpha$ the principal curvature corresponding to $\xi$.

**Lemma 2.1.** *For a Hopf hypersurface in a nonflat complex space form $M^n(c;\mathbb{C})$ the following hold:*

(1) *$\alpha$ is locally constant.*
(2) *If $Av = \lambda v$ for $v \in T^0M$, then $(2\lambda - \alpha)A\phi v = (\alpha\lambda + (c/2))\phi v$ holds. In particular, $A\phi v = \dfrac{\alpha\lambda + (c/2)}{2\lambda - \alpha}\phi v$ when $c > 0$.*

Hopf real hypersurfaces with *constant principal curvatures* are completely classified. A Hopf hypersurface in $\mathbb{C}P^n(c)$ $(n \geqq 2)$ with constant principal curvatures is locally congruent to one of the following:

($A_1$) a geodesic sphere of radius $r$, where $0 < r < \pi/\sqrt{c}$;

($A_2$) a tube of radius $r$ around totally geodesic $\mathbb{C}P^k(c)$ ($1 \leqq k \leqq n-2$), where $0 < r < \pi/\sqrt{c}$;
(B) a tube of radius $r$ around complex hyperquadric $\mathbb{C}Q^{n-1}$, where $0 < r < \pi/(2\sqrt{c})$,
(C) a tube of radius $r$ around $\mathbb{C}P^1(c) \times \mathbb{C}P^{(n-1)/2}(c)$, where $0 < r < \pi/(2\sqrt{c})$ and $n$ ($\geqq 5$) is odd;
(D) a tube of radius $r$ around complex Grassmann $\mathbb{C}G_{2,5}$, where $0 < r < \pi/(2\sqrt{c})$ and $n = 9$;
(E) a tube of radius $r$ around Hermitian symmetric space $SO(10)/U(5)$, where $0 < r < \pi/(2\sqrt{c})$ and $n = 15$.

These real hypersurfaces are said to be of type $A_1$, $A_2$, B, C, D and E. Summing up real hypersurfaces of type $A_1$ and $A_2$, we call them real hypersurfaces of type A. The numbers of distinct principal curvatures of these real hypersurfaces are $2, 3, 3, 5, 5, 5$, respectively.

| | ($A_1$) | ($A_2$) | (B) | (C, D, E) |
|---|---|---|---|---|
| $\lambda_1$ | $\frac{\sqrt{c}}{2}\cot\frac{\sqrt{c}r}{2}$ | $\frac{\sqrt{c}}{2}\cot\frac{\sqrt{c}r}{2}$ | $\frac{\sqrt{c}}{2}\cot\left(\frac{\sqrt{c}}{2}(r-\frac{\pi}{4})\right)$ | $\frac{\sqrt{c}}{2}\cot\left(\frac{\sqrt{c}}{2}(r-\frac{\pi}{4})\right)$ |
| $\lambda_2$ | — | $-\frac{\sqrt{c}}{2}\tan\frac{\sqrt{c}r}{2}$ | $\frac{\sqrt{c}}{2}\cot\left(\frac{\sqrt{c}}{2}(r+\frac{\pi}{4})\right)$ | $\frac{\sqrt{c}}{2}\cot\left(\frac{\sqrt{c}}{2}(r+\frac{\pi}{4})\right)$ |
| $\lambda_3$ | — | — | — | $\frac{\sqrt{c}}{2}\cot\frac{\sqrt{c}r}{2}$ |
| $\lambda_4$ | — | — | — | $-\frac{\sqrt{c}}{2}\tan\frac{\sqrt{c}r}{2}$ |
| $\alpha$ | $\sqrt{c}\cot(\sqrt{c}r)$ | $\sqrt{c}\cot(\sqrt{c}r)$ | $\sqrt{c}\cot(\sqrt{c}r)$ | $\sqrt{c}\cot(\sqrt{c}r)$ |

Here $\lambda_i$ denotes a principal curvature of $A|_{T^oM}$. It should note that a geodesic sphere of radius $r$ ($0 < r < \pi/\sqrt{c}$) in $\mathbb{C}P^n(c)$ is congruent to a tube of radius $\left((\pi/\sqrt{c}) - r\right)$ over a totally geodesic hyperplane $\mathbb{C}P^{n-1}(c)$.

A Hopf hypersurface $M$ in $\mathbb{C}H^n(c)$ ($n \geqq 2$) with constant principal curvatures is locally congruent to one of the following:

($A_0$) a horosphere in $\mathbb{C}H^n(c)$,
($A_{1,0}$) a geodesic sphere of radius $r$ ($0 < r < \infty$),
($A_{1,1}$) a tube of radius $r$ around totally geodesic $\mathbb{C}H^{n-1}(c)$, where $0 < r < \infty$,
($A_2$) a tube of radius $r$ around totally geodesic $\mathbb{C}H^k$ ($1 \leqq k \leqq n-2$), where $0 < r < \infty$,
(B) a tube of radius $r$ around totally real totally geodesic $\mathbb{R}H^n(c/4)$, where $0 < r < \infty$.

These real hypersurfaces are said to be of type $\mathrm{A}_0$, $\mathrm{A}_1$, $\mathrm{A}_1$, $\mathrm{A}_2$ and B. Here, type $\mathrm{A}_1$ means either type $\mathrm{A}_{1,0}$ or type $\mathrm{A}_{1,1}$. Summing up real hypersurfaces of type $\mathrm{A}_0$, $\mathrm{A}_1$ and $\mathrm{A}_2$, we call them real hypersurfaces of type A. A real hypersurface of type B with radius $r = (1/\sqrt{|c|})\log_e(2+\sqrt{3})$ has two distinct constant principal curvatures. Except this, the numbers of distinct principal curvatures of these real hypersurfaces are $2, 2, 2, 3, 3$, respectively. There is an error in p. 150 in [4].

| | $\lambda_1$ | $\lambda_2$ | $\alpha$ |
|---|---|---|---|
| $(\mathrm{A}_0)$ | $\frac{\sqrt{\vert c\vert}}{2}$ | — | $\sqrt{\vert c\vert}$ |
| $(\mathrm{A}_{1,0})$ | $\frac{\sqrt{\vert c\vert}}{2}\coth\frac{\sqrt{\vert c\vert}r}{2}$ | — | $\sqrt{\vert c\vert}\coth(\sqrt{\vert c\vert}r)$ |
| $(\mathrm{A}_{1,1})$ | $\frac{\sqrt{\vert c\vert}}{2}\tanh\frac{\sqrt{\vert c\vert}r}{2}$ | — | $\sqrt{\vert c\vert}\coth(\sqrt{\vert c\vert}r)$ |
| $(\mathrm{A}_2)$ | $\frac{\sqrt{\vert c\vert}}{2}\coth\frac{\sqrt{\vert c\vert}r}{2}$ | $\frac{\sqrt{\vert c\vert}}{2}\tanh\frac{\sqrt{\vert c\vert}r}{2}$ | $\sqrt{\vert c\vert}\coth(\sqrt{\vert c\vert}r)$ |
| (B) | $\frac{\sqrt{\vert c\vert}}{2}\coth\frac{\sqrt{\vert c\vert}r}{2}$ | $\frac{\sqrt{\vert c\vert}}{2}\tanh\frac{\sqrt{\vert c\vert}r}{2}$ | $\sqrt{\vert c\vert}\tanh(\sqrt{\vert c\vert}r)$ |

If we denote by $V_{\lambda_i}$ the principal vector subbundle associated with $\lambda_i$, it satisfies the following:

(1) $V_{\lambda_i}$ is invariant under the action of $\phi$ for a hypersurface of type A.
(2) $\phi(V_{\lambda_1}) = V_{\lambda_2}$, $\phi(V_{\lambda_2}) = V_{\lambda_1}$ for a hypersurface of type B.
(3) $\phi(V_{\lambda_1}) = V_{\lambda_2}$, $\phi(V_{\lambda_2}) = V_{\lambda_1}$, and $V_{\lambda_3}, V_{\lambda_4}$ are invariant for a hypersurface of type C, D and E.

Among Hopf hypersurfaces of constant principal curvatures, we have a nice class of real hypersufaces. A real hypersurface $M$ in $M^n(c;\mathbb{C})$ $(n \geqq 2)$ is called *totally $\eta$-umbilic* if there are smooth functions $\beta$ and $\delta$ on $M$ such that its shape operator $A$ is of the form $A = \beta I + \delta\eta\otimes\xi$, that is,

$$Av = \beta(\varpi(v))v + \delta(\varpi(v))\eta(v)\xi.$$

It is known that these two functions $\beta$ and $\delta$ are automatically constant. It is clear that a real hypersurface $M$ is totally $\eta$-umbilic if and only if every $v \in T^0M$ is principal. A totally $\eta$-umbilic real hypersurface in a nonflat complex space form $M^n(c;\mathbb{C})$ $(n \geqq 2)$ is locally congruent to one of the following:

$(\mathrm{A}_1)$ a geodesic sphere of radius $r$ $(0 < r < \pi/\sqrt{c})$ in $\mathbb{C}P^n(c)$, which satisfies the condition with $\beta = (\sqrt{c}/2)\tan(\sqrt{c}r/2)$ and $\delta = -1/\beta$,

($\mathrm{A}_0$) horospheres in $\mathbb{C}H^n(c)$, which satisfy the condition with $\beta = \delta = \sqrt{|c|}/2$,

($\mathrm{A}_{1,0}$) a geodesic sphere of radius $r$ $(0 < r < \infty)$ in $\mathbb{C}H^n(c)$, which satisfies the condition with $\beta = (\sqrt{c}/2)\coth(\sqrt{c}r/2)$ and $\delta = 1/\beta$,

($\mathrm{A}_{1,1}$) a tube of radius $r$ $(0 < r < \infty)$ around totally geodesic complex hyperplane $\mathbb{C}H^{n-1}(c)$ in $\mathbb{C}H^n(c)$, which satisfies the condition with $\beta = (\sqrt{c}/2)\tanh(\sqrt{c}\,r/2)$ and $\delta = 1/\beta$.

As important real hypersurfaces which are not Hopf hypersurafaces we have ruled real hypersurfaces. A real hypersurface $M$ in a nonflat complex space form $M^n(c;\mathbb{C})$ $(n \geqq 2)$ is said to be a *ruled real hypersurface* if the holomorphic distribution $T^0M$ is integrable and each of its integral manifold is a totally geodesic complex hypersurface $M^{n-1}(c;\mathbb{C})$ of $M^n(c;\mathbb{C})$. A ruled real hypersurface is constructed in the following manner. Given an arbitrary regular curve $\gamma : I \to M^n(c;\mathbb{C})$ on an interval $I$ we attach at each point $\gamma(t)$ $(t \in I)$ a totally geodesic complex hypersurface $M_t^{n-1} \cong M^{n-1}(c;\mathbb{C})$ which is orthogonal to the plane spanned by $\dot\gamma(t)$ and $J\dot\gamma(t)$. We then see that $M = \bigcup_{t\in I} M_t^{n-1}$ is a ruled real hypersurface in $M^n(c;\mathbb{C})$. We have the following characterization of a ruled real hypersurface in terms of its shape operator.

**Lemma 2.2.** *For a real hypersurface $M$ in a nonflat complex space form, the following conditions are mutually equivallent:*

(1) *$M$ is a ruled real hypersurface.*

(2) *The shape operator $A$ of $M$ is of the form*

$$A\xi = \mu\xi + \nu U, \quad AU = \nu\xi, \quad AX = 0 \quad X \perp \xi, U$$

*with a unit vector field $U$ orthogonal to $\xi$ and differentiable functions $\mu, \nu : M \to \mathbb{R}$ satisfying $\nu \neq 0$.*

(3) *$\langle Av, w\rangle = 0$ for arbitrary $v, w \in T_x^0 M$ at each point $x \in M$.*

## 3. Extrinsic shape of geodesics on standard examples

For a curve $\gamma$ on a real hypersurface $M$ in a nonflat complex space form $M^n(c;\mathbb{C})$, we consider $\gamma$ as a curve on $M^n(c;\mathbb{C})$ and call it the extrinsic shape of $\gamma$. In this section we study the extrinsic shape of geodesics whose initial vector is orthogonal to the characteristic vector field $\xi$ on those standard examples of real hypersurfaces in $M^n(c;\mathbb{C})$ in section 2. A smooth curve $\gamma$ parametrized by its arclength is said to be a *circle* if it satisfies $\nabla_{\dot\gamma}\nabla_{\dot\gamma}\dot\gamma = -\kappa^2\dot\gamma$ with some nonnegative constant $\kappa$. This condition is equivalent to the condition that there exist a nonnegative constant $\kappa$ and a field of unit vectors $Y$ along this curve which satisfy $\nabla_{\dot\gamma}\dot\gamma = \kappa Y$ and $\nabla_{\dot\gamma}Y = -\kappa\dot\gamma$. We call this constant $\kappa$ the *curvature* of $\gamma$. We note that geodesics are treated as circles of null curvature.

We denote by $\widetilde\nabla$ and $\nabla$ the Riemannian connections of a nonflat complex space form $M^n(c;\mathbb{C})$ and a real hypersurface $M$ in $M^n(c;\mathbb{C})$, respectively. For vector fields $X, Y$ on $M$, the following formulas of Gauss and Weingarten show the relationship of connections:

$$\widetilde\nabla_X Y = \nabla_X Y + \langle AX, Y\rangle\mathcal{N}, \quad \widetilde\nabla_X\mathcal{N} = -AX, \tag{1}$$

where $\langle\ ,\ \rangle$ is the Riemannian metric on $M$ induced from the standard metric $\langle\ ,\ \rangle$ on $M^n(c;\mathbb{C})$. These lead us to

$$(\nabla_X\phi)Y = \eta(Y)AX - \langle AX, Y\rangle\xi, \quad \nabla_X\xi = \phi AX. \tag{2}$$

The Codazzi equation is given as

$$(\nabla_X A)Y - (\nabla_Y A)X = (c/4)\{\eta(X)\phi Y - \eta(Y)\phi X - 2\langle\phi X, Y\rangle\xi\}. \tag{3}$$

For a geodesic $\gamma$ on a hypersurface $M$, we have $\widetilde\nabla_{\dot\gamma}\dot\gamma = \langle A\dot\gamma, \dot\gamma\rangle\mathcal{N}$ by (1). Thus we see

**Lemma 3.1.** *Let $\gamma$ be a geodesic on a hypersurface $M$ in $\widetilde M$.*

(1) *The extrinsic shape of $\gamma$ is a geodesic if and only if $\langle A\dot\gamma, \dot\gamma\rangle \equiv 0$,*
(2) *The extrinsic shape of $\gamma$ is a circle of positive curvature if and only if $\dot\gamma$ is everywhere principal and $\langle A\dot\gamma, \dot\gamma\rangle$ is nonzero constant along $\gamma$.*

For a geodesic $\gamma$ on a hypersurface $M$ in $M^n(c;\mathbb{C})$, we also have

$$\nabla_{\dot\gamma}\langle\dot\gamma, \xi\rangle = \langle\dot\gamma, \nabla_{\dot\gamma}\xi\rangle = \langle\dot\gamma, \phi A\dot\gamma\rangle.$$

by (2). When $M$ is of type A, as it satisfies $\phi A = A\phi$, we see $\langle\dot\gamma, \xi\rangle$ is constant along $\gamma$, hence we obtain the following:

**Proposition 3.1.** *On an $\eta$-umbilic real hypersurface $M$ whose shape operator is of the form $A = \beta I + \delta\eta\otimes\xi$ the extrinsic shape of each geodesic $\gamma$ with $\dot\gamma(0) \in T^0M$ is a circle of positive curvature $|\beta|$ in $M^n(c;\mathbb{C})$. It satisfies $\dot\gamma(s) \in T^0M$ at each $s$.*

Next we study geodesics on hypersurfaces of type $A_2$. It is well-known that a hypersurface in $M^n(c;\mathbb{C})$ is of type A if and only if its shape operator satisfies $(\nabla_X A)Y = -(c/4)\{\langle\phi X, Y\rangle\xi + \eta(Y)\phi X\}$ for arbitrary vector fields $X, Y$ (see for example [6]). We hence find for a geodesic $\gamma$ on a real hypersurface of type A that $\langle A\dot{\gamma}, \dot{\gamma}\rangle$ is constant along $\gamma$.

**Proposition 3.2.** *Let $M$ be a real hypersurface of type $\mathrm{A}_2$ in $M^n(c;\mathbb{C})$.*

(1) *The extrinsic shape of each geodesic $\gamma$ with $\dot{\gamma}(0) \in V_{\lambda_\ell}$ is a circle of positive curvature $|\lambda_\ell|$ in $M^n(c;\mathbb{C})$.*
(2) *When $c > 0$, the extrinsic shape of each geodesic $\gamma$ with $\dot{\gamma}(0) = (v_1 + \lambda_1 v_2)/\sqrt{\lambda_1^2+1}$ for arbitrary unit tangent vectors $v_\ell \in V_{\lambda_\ell}$ is a geodesic in $M^n(c;\mathbb{C})$.*
(3) *The extrinsic shape of other geodesics with initial vector contained in $T^0M$ are not circles.*

When $M$ is a ruled real hypersurface, as the integral manifold for $T^0M$ through an arbitrary point is totally geodesic, we see every geodesic $\gamma$ with $\dot{\gamma}(0) \in T^0M$ lies on some integral manifold. We hence have the following:

**Proposition 3.3.** *On a ruled real hypersurface $M$ the extrinsic shape of each geodesic $\gamma$ with $\dot{\gamma}(0) \in T^0M$ is a geodesic in $M^n(c;\mathbb{C})$. It satisfies $\dot{\gamma}(s) \in T^0M$ at each $s$.*

We now make mention of the extrinsic shape of some geodesics on Hopf hypersurfaces with constant principal curvatures.

**Proposition 3.4.** *Let $M$ be a Hopf hypersurface with constant principal curvatures in a nonflat $M^n(c;\mathbb{C})$. The extrinsic shape of a geodesic $\gamma$ on $M$ whose initial vector $\dot{\gamma}(0)$ is a principal curvature vector with $\dot{\gamma}(0) \in T^0M$ is a circle of positive curvature in $M^n(c;\mathbb{C})$. If the principal curvature of $\dot{\gamma}(0)$ is $\lambda$ (i.e. $A\dot{\gamma}(0) = \lambda\dot{\gamma}(0)$), its curvature is $|\lambda|$, and it satisfies $A\dot{\gamma}(s) = \lambda\dot{\gamma}(s)$ and $\dot{\gamma}(s) \in T^0M$ at each $s$.*

**Proof.** For a hypersurface of type A we already see this result. We hence consider hypersurfaces of type B, C, D and E. The key point is that for these real hypersurfaces we have the following integrability on the principal distributions $V_{\lambda_\ell}$:

i) $V_{\lambda_\ell}$ is integrable if it is not invariant under $\phi$.
ii) $\widetilde{V}_{\lambda_\ell} = V_{\lambda_\ell} \oplus \mathbb{R}\xi$ is integrable if $V_{\lambda_\ell}$ is invariant under $\phi$. In this case $V_{\lambda_\ell}$ is not integrable.

For sections $X, Y$ of $V_{\lambda_\ell}$ we have

$$A\nabla_X Y = \nabla_X(AY) - (\nabla_X A)Y = \lambda_\ell \nabla_X Y - (\nabla_X A)Y. \tag{4}$$

When the principal distribution $V_{\lambda_\ell}$ is not invariant under $\phi$, Codazzi equation shows for every vector field $Z$ on $M$ that

$$\begin{aligned}\langle(\nabla_X A)Y, Z\rangle &= \langle(\nabla_X A)Z, Y\rangle\\ &= \langle(\nabla_Z A)X + (c/4)\{\eta(X)\phi Z - \eta(Z)\phi X - 2\langle\phi X, Z\rangle\xi\}, Y\rangle\\ &= \langle(\nabla_Z A)X, Y\rangle = \langle\nabla_Z(AX) - A\nabla_Z X, Y\rangle\\ &= \langle(Z\lambda_\ell)X + (\lambda_\ell I - A)\nabla_Z X, Y\rangle = 0.\end{aligned}$$

Hence we find $\nabla_X Y \in V_{\lambda_\ell}$, so that every leaf $L$ of $V_{\lambda_\ell}$ is a totally geodesic submanifold in $M$. The Gauss formula (1) guarantees that $L$ is a totally umbilic submanifold in $M^n(c;\mathbb{C})$. If $\gamma$ is a geodesic on $M$ with $\dot\gamma(0) \in V_{\lambda_\ell}$, it lies on some leaf $L$, hence its extrinsic shape is a circle of curvature $|\lambda_\ell|$. We here note that each leaf $L$ is a totally umbilic hypersurface of constant curvature $\lambda_\ell^2 + (c/4)$ in $M^m(c/4;\mathbb{R})$ which is a totally real totally geodesic submanifold of $M^m(c;\mathbb{C})$, where $m = \dim_{\mathbb{R}}(V_{\lambda_\ell}) + 1$.

When the principal distribution $V_{\lambda_\ell}$ is invariant under $\phi$, for a section $X$ of $V_{\lambda_\ell}$ we obtain by (3)

$$\begin{aligned}(c/4)\phi X &= (\nabla_\xi A)X - (\nabla_X A)\xi = \nabla_\xi(AX) - A\nabla_\xi X - \nabla_X(A\xi) + A\nabla_X\xi\\ &= (\lambda_\ell I - A)\nabla_\xi X - \alpha\phi AX + A\phi AX\\ &= (\lambda_\ell I - A)\nabla_\xi X + \lambda_\ell(\lambda_\ell - \alpha)\phi X.\end{aligned}$$

This shows that $\langle(\lambda_\ell I - A)\nabla_\xi X, Z\rangle = 0$ for an arbitrary section $Z$ of $T^0M$ orthogonal to $V_{\lambda_\ell}$, so that $\nabla_\xi X$ is a section of $\widetilde{V}_{\lambda_\ell}$. In fact, for each $Z \in T^0M$ with $AZ = \lambda Z$ and $\lambda \neq \lambda_\ell$, we have $0 = \langle(\lambda_\ell I - A)\nabla_\xi X, Z\rangle = (\lambda_\ell - \lambda)\langle\nabla_\xi X, Z\rangle$. Next, for sections $X, Y$ of $V_{\lambda_\ell}$ and an arbitrary section $Z$ of $T^0M$, we find by (3) that

$$\begin{aligned}\langle(\nabla_X A)Y, Z\rangle &= \langle(\nabla_X A)Z, Y\rangle = \langle(\nabla_Z A)X, Y\rangle\\ &= \langle\nabla_Z(AX) - A\nabla_Z X, Y\rangle = \langle(\lambda_\ell I - A)\nabla_Z X, Y\rangle = 0.\end{aligned}$$

If $Z$ is a section of other principal distribution $V_{\lambda_j}$, by using (4) we see $\langle(\nabla_X A)Y, Z\rangle = (\lambda_\ell - \lambda_j)\langle\nabla_X Y, Z\rangle$. Hence $\nabla_X Y$ is also a section of $\widetilde{V}_{\lambda_\ell}$. Since for a section $X$ of $V_{\lambda_\ell}$ we see $\nabla_X\xi = \phi AX = \lambda_\ell\phi X$ is a section of $V_{\lambda_\ell}$, we find $\widetilde{V}_{\lambda_\ell}$ is integrable. Then every leaf $L$ of $\widetilde{V}_{\lambda_\ell}$ is a totally geodesic submanifold in $M$. In view of the principal curvature, when $\ell = 3$, we see $L$ is locally congruent to a geodesic sphere of radius $r$ $(0 < r < \pi/\sqrt{c})$ in $\mathbb{C}P^{m+1}(c)$ which is a totally geodesic complex submanifold of $\mathbb{C}P^n(c)$, where $2m = \dim_{\mathbb{R}} V_{\lambda_3}$, and when $\ell = 4$, we see $L$ is locally congruent to a geodesic sphere of radius $((\pi/\sqrt{c}) - r)$ in $\mathbb{C}P^{m+1}(c)$ which is a totally geodesic complex submanifold of $\mathbb{C}P^n(c)$, where $2m = \dim_{\mathbb{R}} V_{\lambda_4}$. If $\gamma$ is a geodesic on $M$ with $\dot{\gamma}(0) \in V_{\lambda_\ell}$ ($\ell = 3, 4$), it lies on some leaf $L$, hence we find by Proposition 3.1 that its extrinsic shape is a circle of positive curvature $|\lambda_\ell|$. □

At the end of this section we remark that every geodesic $\gamma$ with $\dot{\gamma}(0) = \xi_{\gamma(0)}$ on *any Hopf hypersurface* $M$ of $M^n(c; \mathbb{C}), c \neq 0$ is an integral curve generated by the characteristic vector field $\xi$ of $M$ because $\nabla_\xi \xi = \phi A \xi = 0$.

## 4. Characterizations of standard real hypersurfaces

In this section we give characterizations of standard examples of real hypersurfaces in a nonflat complex space form $M^n(c; \mathbb{C})$ from the viewpoint of the extrinsic shape of geodesics.

**Theorem 4.1.** *A real hypersurface $M$ in $M^n(c; \mathbb{C}), c \neq 0$ is totally $\eta$-umbilic if and only if at each point $x \in M$ there exist orthonormal vectors $v_1, v_2, \ldots, v_{2n-2} \in T^0_x M$ such that all geodesics on $M$ with initial direction $v_i + v_j$ $(1 \leqq i \leqq j \leqq 2n-2)$ are circles of positive curvature in $M^n(c; \mathbb{C})$.*

**Proof.** By Proposition 3.1 we need to show the "if" part. Since the extrinsic shape of the geodesic $\gamma_i$ with $\dot{\gamma}_i(0) = v_i$ is a circle of curvature $\kappa_i$ $(> 0)$ we find by (1) that

$$-\kappa_i^2 \dot{\gamma}_i = \widetilde{\nabla}_{\dot{\gamma}_i}(\widetilde{\nabla}_{\dot{\gamma}_i} \dot{\gamma}_i) = \langle (\nabla_{\dot{\gamma}_i} A)\dot{\gamma}_i, \dot{\gamma}_i \rangle \mathcal{N} - \langle A\dot{\gamma}_i, \dot{\gamma}_i \rangle A\dot{\gamma}_i.$$

Comparing the tangential components of both hand-sides, we obtain $\kappa_i^2 \dot{\gamma}_i = \langle A\dot{\gamma}_i, \dot{\gamma}_i \rangle A\dot{\gamma}_i$, in particular, $\langle Av_i, v_i \rangle Av_i = \kappa_i^2 v_i$. We therefore find that $v_i$ is a principal vector with $|\langle Av_i, v_i \rangle| = \kappa_i > 0$ and hence $\langle Av_i, v_j \rangle = 0$ for $1 \leqq i < j \leqq 2n-2$. Applying the same argument as above to the geodesic $\gamma_{ij}$ with $\dot{\gamma}_{ij}(0) = (v_i + v_j)/\sqrt{2}$ we get

$$\langle A(v_i + v_j), (v_i + v_j) \rangle A(v_i + v_j) = 2\kappa_{ij}^2 (v_i + v_j) \tag{5}$$

for some positive $\kappa_{ij}$. Taking the inner product of both hand-sides with $v_i - v_j$ we see that $\kappa_i = \kappa_j$ for $1 \leqq i, j \leqq 2n-2$. Since $\kappa_{ij} > 0$, (5) shows that $\langle Av_i, v_i \rangle + \langle Av_j, v_j \rangle \neq 0$. We can hence conclude that there is a positive constant $\kappa_x$ with $Av = \kappa_x v$ for all $v \in T^0_x M$ at arbitrary $x \in M$, and find that $M$ is totally $\eta$-umbilic in $M^n(c; \mathbb{C})$ □

The assumption on positivity on curvatures of circles in the direction $v_i + v_j$ $(1 \leqq i < j \leqq 2n-2)$ is important. The following shows that Theorem 4.1 does not hold without the assumption of positivity.

**Theorem 4.2.** *A real hypersurface $M$ in $M^n(c;\mathbb{C}), c \neq 0$ is locally congruent to either a totally $\eta$-umbilic hypersurface or a hypersurface of type $A_2$ of radius $\pi/(2\sqrt{c})$ in $\mathbb{C}P^n(c)$ if and only if at each point $x \in M$ there exist orthonormal vectors $v_1, v_2, \dots, v_{2n-2} \in T_x^0 M$ such that all geodesics $\gamma_{ij}$ on $M$ with initial direction $v_i + v_j$ $(1 \leqq i \leqq j \leqq 2n-2)$ are circles in $M^n(c;\mathbb{C})$ and the curvature of $\gamma_{ii}$ $(1 \leqq i \leqq 2n-2)$ is positive.*

**Proof.** By Propositions 3.1, 3.2 we are enough to show the "if" part. Since $\gamma_{ii}$ is a circle of positive curvature, the proof of Theorem 4.1 shows that $v_i$ is principal. We denote its principal curvature by $\kappa_i$. As $\gamma_{ij}$ $(i < j)$ is a circle, we also find the proof of Theorem 4.1 that $(\kappa_i + \kappa_j)(\kappa_i - \kappa_j) = 0$. Thus we obtain either $Av_i = \kappa_x v_i$ or $Av_i = -\kappa_x v_i$ holds for $1 \leqq i \leqq 2n-2$ with some constant $\kappa_x$. Hence $\xi$ is principal, because $\langle A\xi, v_i\rangle = \langle \xi, Av_i\rangle = 0$ for $1 \leqq i \leqq 2n-2$. Thus $\alpha$ is locally constant and $M$ is a real hypersurface with at most three distinct principal curvatures $\kappa_x, -\kappa_x$ and $\alpha = \langle A\xi, \xi\rangle$ at each point $x \in M$.

When the case $c > 0$ or the case $c < 0$ and $2\kappa_x \neq \alpha_x$, it follows from Lemma 2.1 that both of $\kappa$ and $\alpha$ are constant locally. When $c < 0$ and $2\kappa_x = \alpha$, the above discussion and the continuity of principal curvature functions guarantee that $2\kappa \equiv \alpha$ holds, so that $\kappa$ is also locally constant. Thus $M$ is locally congruent to one of real hypersurfaces of type A and B in $M^n(c;\mathbb{C})$. By the list of principal curvatures for these real hypersurfaces we obtain the desirable result. □

**Remark 4.1.** In Theorem 4.2, if the extrinsic shape of some geodesic $\gamma_{ij}$ $(i < j)$ is a geodesic, then $M$ is locally congruent to a hypersurface of type $A_2$ of radius $\pi/(2\sqrt{c})$ in $\mathbb{C}P^n(c)$.

The proof of Theorem 4.2 also shows the following:

**Theorem 4.3.** *A real hypersurface $M$ in $M^n(c;\mathbb{C}), c \neq 0$ is locally congruent to either a totally $\eta$-umbilic hypersurface or a hypersurface of type $A_2$ of radius $\pi/(2\sqrt{c})$ in $\mathbb{C}P^n(c)$ if and only if at each point $x \in M$ there exist orthonormal vectors $v_1, v_2, \dots, v_{2n-2} \in T_x^0 M$ such that all geodesics of $M$ with initial vector $v_i$ $(1 \leqq i \leqq 2n-2)$ are circles of the same positive curvature $\kappa_x$ in $M^n(c;\mathbb{C})$.*

Refining Remark 4.1, we obtain the following characterization of a real hypersurface of type $A_2$ with radius $\pi/(2\sqrt{c})$ in $\mathbb{C}P^n(c)$.

**Theorem 4.4.** *A real hypersurface $M$ in $M^n(c;\mathbb{C})$ $(c \neq 0,\ n \geqq 3)$ is locally congruent to a hypersurface of type $A_2$ with radius $\pi/(2\sqrt{c})$ in $\mathbb{C}P^n(c)$ if and only if there are a function $d : M \to \mathbb{N}$ and orthonormal vectors $v_1, v_2, \dots, v_{2n-2} \in T_x^0 M$ at each point $x \in M$ satisfying the following two conditions:*

i) *All geodesics of $M$ with initial vector $v_i$ $(1 \leqq i \leqq 2n-2)$ are circles of positive curvature in $M^n(c;\mathbb{C})$.*
ii) *All geodesics $\gamma_{ij}$ on $M$ with initial direction $v_i + v_j$ $(1 \leqq i \leqq d_x < j \leqq 2n-2)$ are geodesics in $M^n(c;\mathbb{C})$.*

*In this case $d$ is a locally constant even number, and $M$ is locally congruent to a tube of radius $\pi/(2\sqrt{c})$ around totally geodesic $\mathbb{C}P^{d/2}(c)$ or $\mathbb{C}P^{(2n-2-d)/2}(c)$ in $\mathbb{C}P^n(c)$.*

**Proof.** We find that each $v_i$ is principal by the first condition and that $\langle A(v_i+v_j), v_i+v_j\rangle = 0$ for $1 \leqq i \leqq d_x < j \leqq 2n-2$ by the second condition. Denoting by $\lambda_i$ the principal curvature for $v_i$, we see $\lambda_i + \lambda_j = 0$. Hence $M$ has at most three distinct principal curvatures at each point. The proof of Theorem 4.2 leads us to our assertion. □

We now characterize ruled hypersurfaces in a nonflat complex space form by the extrinsic shape of geodesics.

**Theorem 4.5.** *A real hypersurface $M$ in $M^n(c;\mathbb{C})$ $(c \neq 0, n \geqq 2)$ is ruled if and only if at each point $x \in M$ there exist orthonormal vectors $v_1, v_2, \dots, v_{2n-2} \in T_x^0 M$ such that all geodesics of $M$ with initial direction $v_i + v_j$ $(1 \leqq i \leqq j \leqq 2n-2)$ are geodesics in $M^n(c;\mathbb{C})$.*

**Proof.** We are enough to show the "if" part. Taking geodesics $\gamma_i$ and $\gamma_{ij}$ with $\dot\gamma_i(0) = v_i, \dot\gamma_{ij}(0) = (v_i + v_j)/\sqrt{2}$ for $1 \leqq i < j \leqq 2n-2$, we find by the condition that $\langle A\dot\gamma_i, \dot\gamma_i\rangle = \langle A\dot\gamma_{ij}, \dot\gamma_{ij}\rangle \equiv 0$. In particular, we have $\langle Av_i, v_i\rangle = \langle A(v_i + v_j), v_i + v_j\rangle = 0$, hence $\langle Av_i, v_j\rangle = 0$ for every $1 \leqq i < j \leqq 2n-2$. We therefore obtain $\langle Av, w\rangle = 0$ for every $v, w \in T_x^0 M$ at each point $x \in M$, and get the conclusion by Lemma 2.2. □

**Theorem 4.6.** *A real hypersurface $M$ in $M^n(c;\mathbb{C})$ $(c \neq 0, n \geqq 2)$ is locally congruent to a totally $\eta$-umbilic real hypersurface or a ruled real hypersurface if and only if every geodesic $\gamma$ on $M$ with $\dot\gamma(0) \in T^0 M$ is a circle in $M^n(c;\mathbb{C})$.*

**Proof.** We show the "if" part. We define a function $f : U_x^0M \to \mathbb{R}$ on $U_x^0M = \{u \in T_x^0M \mid \|u\| = 1\} \cong S^{2n-3}$ by $f(u) = \langle Au, u\rangle^2$. By the condition, Lemma 3.1 guarantees that either $u$ is principal or $\langle Au, u\rangle = 0$ holds for $u \in T^0M$. In particular, we see $v(f)_u = 4\langle Au, u\rangle\langle Au, v\rangle = 0$ for every tangent vector $v \in TS^{2n-3}$ at arbitrary $u \in S^{2n-3}$, because $v$ is identified with a tangent vector orthogonal to $u$. Thus we find that $f$ is a constant function and that there is a nonnegative function $k$ with $|\langle Au, u\rangle| = k(x)$ for every $u \in T_x^0M$.

When $k(x) \neq 0$, we see every $u \in T_x^0M$ is principal with principal curvature $\pm k(x)$. We hence obtain $Au = k(x)u$ for every $u \in T_x^0M$ or $Au = -k(x)u$ for every $u \in T_x^0M$ and find $x$ is an $\eta$-umbilic point. Hence $M$ is locally congruent to a totally $\eta$-umbilic real hypersurface. When $k(x) = 0$, the above argument guarantees that $k \equiv 0$ on a neighborhood $U$ of $x$. Since $A$ is symmetric, we find that $\langle Au, v\rangle = 0$ for all $u, v \in T^0M|_U$, and hence that $M$ is locally congruent to a ruled real hypersurface by Lemma 2.2. □

Finally we give a characterization of all Hopf hypersurfaces all of whose principal curvatures are constant in a nonflat complex space form. For proof see [2] and [4].

**Theorem 4.7.** *A real hypersurface $M$ in $M^n(c;\mathbb{C})$ $(c \neq 0)$ is locally congruent to a Hopf hypersurface all of whose principal curvatures are constant on $M$ if and only if at each point $x \in M$ there exist orthonormal vectors $v_1, v_2, \dots, v_{2n-2} \in T_x^0M$ such that all geodesics with initial vector $v_i$ $(1 \leqq i \leqq 2n-2)$ are circles of positive curvature in $M^n(c;\mathbb{C})$.*

Every circle $\gamma$ in a nonflat complex space form $M^n(c;\mathbb{C})$ which is the extrinsic shape of a geodesic in Theorem 4.7 is a simple curve lying on some totally real totally geodesic $\mathbb{R}P^2(c/4)$ or $\mathbb{R}H^2(c/4)$. Following the result on circles on $M^n(c;\mathbb{C})$ we find the following(see [1]):

i) $\gamma$ is closed if $c > 0$, or $c < 0$ and the corresponding principal curvature $\lambda$ for $\gamma$ satisfies $\lambda > \sqrt{|c|}/2$.
ii) $\gamma$ is not bounded if $c < 0$ and $\lambda \leqq \sqrt{|c|}/2$.

For about tangent vectors $v_1, v_2, \dots, v_{2n-2}$ in Theorem 4.7 we do not need to assume they form a local field of orthonormal frames. However, one can take them as a local field of orthonormal frames on $M$.

## References

1. T. Adachi and S. Maeda, "Global behaviours of circles in a complex hyperbolic space", *Tsukuba J. Math.* **21**(1997), 29–42.
2. T. Adachi, M. Kimura and S. Maeda, "A characterization of all homogeneous real hypersurfaces in a complex projective space by observing the extrinsic shape of geodesics", *Arch. Math.(Basel)* **73** (1999), 303–310.
3. ______, "Real hypersurfaces some of whose geodesics are plane curves in non-flat complex space forms", to appear in Tohoku Math. J..
4. B.-Y. Chen and S. Maeda, "Hopf hypersurfaces with constant principal curvatures in complex projective or complex hyperbolic spaces" *Tokyo J. Math.* **24** (2001), 133–152.
5. S. Maeda and K. Ogiue, "Characterizations of geodesic hyperspheres in a complex projective space by observing the extrinsic shape of geodesics" *Math Z.* **225** (1997), 537–542.
6. R. Niebergall and P. J. Ryan, "Real hypersurfaces in complex space forms" *Tight and Taut Submanifolds* T. E. Cecil and S. S. Chern, eds. 233–305 Cambridge University Press (1998).

# TANGENT BUNDLES WITH SASAKI METRIC AND ALMOST HYPERCOMPLEX PSEUDO-HERMITIAN STRUCTURE

MANCHO MANEV

*University of Plovdiv,*
*Faculty of Mathematics and Informatics,*
*236 Bulgaria blvd., Plovdiv 4003, BULGARIA*
*E-mail: mmanev@pu.acad.bg, mmanev@yahoo.com*

The tangent bundle as a $4n$-manifold is equipped with an almost hypercomplex pseudo-Hermitian structure and it is characterized with respect to the relevant classifications. A number of 8-dimensional examples of the considered type of manifold are received from the known explicit examples in that manner.

## Introduction

The geometry of the almost hypercomplex manifolds with Hermitian metric is known (e.g. [1]). A parallel direction including indefinite metrics is the developing of the geometry of the almost hypercomplex manifolds with pseudo-Hermitian metric structure. It has a natural origination from the geometry of the $n$-dimensional quaternionic Euclidean space.

The beginning was put by our joint works with K. Gribachev and S. Dimiev in [4] and [5]. More precisely we have combined the ordinary Hermitian metrics with the so-called by us *skew-Hermitian metrics* with respect to the almost complex structures of a hypercomplex structure.

First, we have ascertained four types of bilinear forms compatible with a hypercomplex structure: three metrics and one Kähler form which are skew-Hermitian or Hermitian with respect to the different almost complex structures, i. e. we have constructed a *pseudo-Hermitian structure.*

2000 *Mathematics Subject Classification.* Primary 55R10, 53C15; Secondary 32Q60, 53C26, 53C50, 53C55

Second, we have developed the notion of an *almost hypercomplex pseudo-Hermitian manifold.*

Third, we have considered mainly the (integrable) hypercomplex pseudo-Hermitian manifolds and its special subclass the class of pseudo-hyper-Kähler manifolds.

After that in [8] we have continued the investigations as we have constructed and characterized to a certain extent several 4-dimensional examples of the considered manifolds.

The aim of the present work is to generate the considered structure on a tangent bundle, to characterize it in terms of the corresponding manifolds and to construct 8-dimensional examples relevant to the known 4-dimensional examples.

In § 1 we recall the notions of the almost complex manifolds with Hermitian metric or skew-Hermitian metrics and the almost hypercomplex manifolds with pseudo-Hermitian metric structure.

In § 2 we remind the definitions and properties of the prolongations of vector fields to tangent bundles known as their vertical and horizontal lifts.

The main part of this paper is given in § 3, where we consider an almost skew-Hermitian manifold as a base manifold and we generate its tangent bundle with a metric, which is a prolongation of the base metric, known as the Sasaki metric. In that way we get the tangent bundle with almost hypercomplex pseudo-Hermitian structure. We characterize differential-geometrically this manifold having in mind also the relevant classifications of the received manifolds.

The results of § 3 and the known examples from [4] and [8] are used in the last § 4 to construct corresponding 8-dimensional examples in the discussed manner.

## 1. Some Kinds of Differentiable Manifolds with Almost Complex Structures

### 1.1. *Almost Complex Manifolds with Hermitian Metric or Skew-Hermitian Metric*

The notion of the *almost complex manifold* $(M^{2n}, J)$ is well-known. There exists a possibility it to be equipped with two different kinds of metrics. When $J$ acts as an isometry on each tangent space then the manifold is an *almost Hermitian manifold.* But, in the case when $J$ acts as an anti-isometry on each tangent space, the notion of the so-called *almost skew-Hermitian manifold* is available. Let us consider more precisely the last one.

Every $n$-dimensional complex Riemannian manifold induces a real $2n$-dimensional manifold $(M^{2n}, J, g, \tilde{g})$ with a complex structure $J$, a metric $g$ and an associated metric $\tilde{g} = g(\cdot, J\cdot)$. Both metrics are indefinite of signature $(n, n)$. This manifold we call an *almost skew-Hermitian manifold.* Such metric is known also as a B-metric or a Norden metric because it is introduced by A. P. Norden in [9]. An almost skew-Hermitian manifold is a *skew-Kähler manifold* if $J$ is parallel with respect to the Levi-Civita connection $\nabla$ of the metric $g$. The class of these manifolds is contained in every other class of the almost skew-Hermitian manifolds. A classification with respect to $\nabla J$ consisting of three basic classes is given in [2]. The class $\mathcal{W}_1$ is the main class, i. e. the class where $\nabla J$ has an explicit expression in terms of the metric tensors and the Lie form.

### 1.2. *Almost Hypercomplex Manifolds with Pseudo-Hermitian Structure*

Let us recall the notion of the almost hypercomplex structure $H$ on the manifold $M^{4n}$. It is the triple $H = (J_\alpha)$ $(\alpha = 1, 2, 3)$ of anticommuting almost complex structures satisfying the property $J_3 = J_1 \circ J_2$ ([1], [13]).

The introduction and the beginning of the study of the pseudo-Hermitian structures on an almost hypercomplex manifold is given in [4,5]. A pseudo-Riemannian metric $g$ of signature $(2n, 2n)$ on $(M^{4n}, H)$ is introduced as follows

$$g(\cdot,\cdot) = g(J_1\cdot, J_1\cdot) = -g(J_2\cdot, J_2\cdot) = -g(J_3\cdot, J_3\cdot). \tag{1.1}$$

We called such metric a *pseudo-Hermitian metric.* It generates a Kähler 2-form $\Phi$ and two pseudo-Hermitian metrics $g_2$ and $g_3$ by the following way

$$\Phi := g(J_1\cdot,\cdot), \qquad g_2 := g(J_2\cdot,\cdot), \qquad g_3 := g(J_3\cdot,\cdot). \tag{1.2}$$

Let us note that $g$ ($g_2$, $g_3$, respectively) has an Hermitian compatibility with respect to $J_1$ ($J_3$, $J_2$, respectively) and a skew-Hermitian compatibility with respect to $J_2$ and $J_3$ ($J_1$ and $J_2$, $J_1$ and $J_3$, respectively).

On the other hand, a quaternionic inner product $<\cdot,\cdot>$ in $\mathbb{H}$ generates in a natural way the bilinear forms $g$, $\Phi$, $g_2$ and $g_3$ by the following decomposition: $<\cdot,\cdot> = -g + i\Phi + jg_2 + kg_3$.

We called the structure $(H, G) := (J_1, J_2, J_3; g, \Phi, g_2, g_3)$ a *hypercomplex pseudo-Hermitian structure* on $M^{4n}$ or shortly a $(H, G)$*-structure* on $M^{4n}$. We called the manifold $(M, H, G)$ an *almost hypercomplex pseudo-Hermitian manifold* or shortly an *almost* $(H, G)$*-manifold.*

It is well known, that the almost hypercomplex structure $H=(J_\alpha)$ is a *hypercomplex structure* if the Nijenhuis tensors $N_\alpha(\cdot,\cdot)=[\cdot,\cdot]+J_\alpha[\cdot,J_\alpha\cdot]+J_\alpha[J_\alpha\cdot,\cdot]-[J_\alpha\cdot,J_\alpha\cdot]$ vanish for each $\alpha=1,2,3$. Moreover, one $H$ is hypercomplex iff two of $N_\alpha$ vanish.

We introduced *structural* $(0,3)$*-tensors* of the almost $(H,G)$-manifold by

$$F_\alpha(x,y,z)=g\big((\nabla_x J_\alpha)\,y,z\big)=\big(\nabla_x g_\alpha\big)\,(y,z) \quad (\alpha=1,2,3),$$

where $\nabla$ is the Levi-Civita connection generated by $g$ and $x,y,z\in T_pM$ at any $p\in M$.

There are valid relations between $F_\alpha$'s, e.g. $F_1(\cdot,\cdot,\cdot)=F_2(\cdot,J_3\cdot,\cdot)+F_3(\cdot,\cdot,J_2\cdot)$.

Since $g$ is a Hermitian metric with respect to $J_1$, according to Gray-Hervella [3] the basic subclass $\mathcal{W}_4$ of the Hermitian manifolds is determined by

$$\begin{aligned}F_1(x,y,z)=\tfrac{1}{2(2n-1)}[&g(x,y)\theta_1(z)-g(x,z)\theta_1(y)\\&-g(x,J_1y)\theta_1(J_1z)+g(x,J_1z)\theta_1(J_1y)],\end{aligned}$$

where $\theta_1(\cdot)=g^{ij}F_1(e_i,e_j,\cdot)=\delta\Phi(\cdot)$ is the *Lie form* for any basis $\{e_i\}_{i=1}^{4n}$, and $\delta$ – the coderivative.

On other side, the metric $g$ is a skew-Hermitian one with respect to $J_2$ and $J_3$. According to the classification of the almost complex manifolds with skew-Hermitian metric (Norden metric or B-metric) given in [2] the basic classes are defined as follows:

$$\begin{aligned}&\mathcal{W}_1: F_\alpha(x,y,z)=\tfrac{1}{4n}[g(x,y)\theta_\alpha(z)+g(x,z)\theta_\alpha(y)\\&\qquad\qquad\qquad+g(x,J_\alpha y)\theta_\alpha(J_\alpha z)+g(x,J_\alpha z)\theta_\alpha(J_\alpha y)],\\&\mathcal{W}_2: \underset{x,y,z}{\sigma} F_\alpha(x,y,J_\alpha z)=0, \qquad \mathcal{W}_3: \underset{x,y,z}{\sigma} F_\alpha(x,y,z)=0,\end{aligned} \tag{1.3}$$

where $\theta_\alpha(\cdot)=g^{ij}F_\alpha(e_i,e_j,\cdot)$, $\alpha=2,3$, are the corresponding Lie forms for an arbitrary basis $\{e_i\}_{i=1}^{4n}$.

Here, we denote the main subclasses of the respective complex manifolds by $\mathcal{W}(J_\alpha)$, where $\mathcal{W}(J_1):=\mathcal{W}_4(J_1)$ in [3], and $\mathcal{W}(J_\alpha):=\mathcal{W}_1(J_\alpha)$ for $\alpha=2,3$ in [2].

We obtained a sufficient condition an almost $(H,G)$-manifold to be an integrable one:

**Theorem 1.1 ([4]).** *Let $(M,H,G)$ belongs to $\mathcal{W}(J_\alpha)\bigcap\mathcal{W}(J_\beta)$. Then $(M,H,G)$ is of class $\mathcal{W}(J_\gamma)$ for all cyclic permutations $(\alpha,\beta,\gamma)$ of $(1,2,3)$.*

We say that a pseudo-Hermitian manifold is a *pseudo hyper-Kähler manifold* and denote $(M, H, G) \in \mathcal{K}$, if $F_\alpha = 0$ for every $\alpha = 1, 2, 3$, i.e. the manifold is of Kählerian type with respect to each $J_\alpha$.

**Theorem 1.2** ([4]). *If* $(M, H, G) \in \mathcal{K}(J_\alpha) \bigcap \mathcal{W}(J_\beta)$ $(\alpha \neq \beta \in \{1, 2, 3\})$ *then* $(M, H, G) \in \mathcal{K}$ .

We gave a geometric characteristic of the pseudo-hyper-Kähler manifolds according to the curvature tensor $R = [\nabla, \nabla] - \nabla_{[\ ,\ ]}$ induced by the Levi-Civita connection.

**Theorem 1.3** ([5]). *Each pseudo-hyper-Kähler manifold is a flat pseudo-Riemannian manifold with signature* $(2n, 2n)$.

## 2. The Tangent Bundle

Let us consider a $2n$-dimensional differentiable manifold $M$ and then $T_pM$ is the tangent space at a point $p$ of $M$. It is known that the set

$$TM = \bigcup_{p \in M} T_pM$$

is called the *tangent bundle* over the manifold $M$ ([14]).

For any point $u$ of $TM$ such that $u \in T_pM$ the correspondence $u \to p$ defines the bundle projection $\pi : TM \to M$ by $\pi(u) = p$. The set $\pi^{-1}(p)$, i.e. $T_pM$, is called the *fiber* of $TM$ over $p \in M$ and $M$ — the *base manifold* of $TM$. The base space $M$ is a submanifold differentiably imbedded in $TM$.

Let $(x^i)$ are the local coordinates of a point $p$ in a neighborhood $U \subset M$. Then a point $u \in T_pM$ with coordinates $(y^i)$ in $T_pM$ with respect to the natural basis $\left\{\frac{\partial}{\partial x^i}\right\}$ is represented by the ordered pair $(p, u)$. Therefore an arbitrary point $\tilde{p}$ in $TM$ has local coordinates of the kind $(x^i, y^i)$ in the open set $\pi^{-1}(U) \subset TM$. These coordinates is called the *induced coordinates* in $\pi^{-1}(U)$ from $(x^i)$.

Obviously, the tangent bundle of a base manifold $M$ with almost complex structure is of dimension $4n$.

Let us denote by $\mathcal{T}_s^r(M)$ the set of all differentiable tensor fields of type $(r, s)$ in $M$ and let $\mathcal{T}(M) = \sum_{r,s=0}^{\infty} \mathcal{T}_s^r(M)$ be the set of all tensor fields in $M$. Similarly, let us denote by $\mathcal{T}_s^r(TM)$ and $\mathcal{T}(TM)$ respectively the corresponding sets of tensor fields in the tangent bundle $TM$.

### 2.1. *Vertical Lifts*

The *vertical lift* $f^V$ *of a function* $f$ in $M$ is called the composition of $\pi : TM \to M$ and $f : M \to \mathbb{R}$, i.e. $f^V = f \circ \pi$, where $f \in \mathcal{T}_0^0(M)$ and

$f^V \in \mathcal{T}_0^0(TM)$.

Since a correspondence $TM \to \mathbb{R}$ is possible to be considered as the action of a 1-form in $M$ or a function in $TM$, then if $\omega$ is a 1-form in $M$, it is regarded, in a natural way, as a function in $TM$, denoted by $\imath\omega$.

Let $\tilde{X} \in \mathcal{T}_0^1(TM)$ be such that $\tilde{X} f^V = 0$ for all $f \in \mathcal{T}_0^0(M)$. Then $\tilde{X}$ is called a *vertical vector field* in $TM$. The vector field $X^V$ in $TM$ defined by $X^V(\imath\omega) = (\omega(X))^V$ for an arbitrary 1-form $\omega$ in $M$, is called the *vertical lift of vector field* $X$ in $M$ to $TM$. If $X$ has components $X^i$ in $M$, then $X^V$ has components

$$X^V : \begin{pmatrix} 0 \\ X^i \end{pmatrix} \tag{2.1}$$

with respect to the induced coordinates in $TM$. Consequently, it is clear, that $\left(\frac{\partial}{\partial x^i}\right)^V = \frac{\partial}{\partial y^i}$.

There are valid the following formulas for any $X, Y \in \mathcal{T}_0^1(M)$ and $f \in \mathcal{T}_0^0(M)$.

$$X^V f^V = 0,\ (X+Y)^V = X^V + Y^V,\ (fX)^V = f^V X^V,\ \left[X^V, Y^V\right] = 0.$$

## 2.2. *Horizontal Lifts*

Let $\nabla$ be the Levi-Civita connection in $M$ induced by the pseudo-Riemannian metric $g$. *Horizontal lift* $f^H$ *of a function* $f$ in $M$ to $TM$ is defined by $f^H = \imath(df) - \gamma(\nabla f)$, where $\gamma(\nabla f) = y^k \nabla_k f$. There follows immediately that $f^H = 0$ for any $f \in \mathcal{T}_0^0(M)$.

Suppose that $X$ and $\nabla$ have local components $X^k$ and $\Gamma_{ij}^k$, respectively, in $M$. *Horizontal lift* $X^H$ *of a vector field* $X$ in $M$ to $TM$ is called the vector field in $TM$ with components

$$X^H : \begin{pmatrix} X^k \\ -y^i \Gamma_{ij}^k X^j \end{pmatrix} \tag{2.2}$$

with respect to the induced coordinates in $TM$.

For the action of $X^H$ we have that $X^H f^V = (Xf)^V$, $\omega^V(X^H) = \{\omega(X)\}^V$, $J^V X^H = (JX)^V$.

## 2.3. *Adapted Frames*

In each coordinate neighborhood $\{U, x^k\}$ of $M$, $\dim M = 2n$, we denote $e_{(i)} = \frac{\partial}{\partial x^i}$. Then $4n$ local vector fields $e_{(i)}^H$ and $e_{(i)}^V$ form a basis of the tangent space $T_{\tilde{p}}(TM)$ at each point $\tilde{p} \in \pi^{-1}(p)$ and their components are given by

$$e^H_{(i)} : \begin{pmatrix} \delta^k_i \\ -y^j \Gamma^k_{ij} \end{pmatrix}, \qquad e^V_{(i)} : \begin{pmatrix} 0 \\ \delta^k_i \end{pmatrix}$$

with respect to the induced coordinates $(x^k, y^k)$ in $TM$. The set $\left\{e^H_{(i)}, e^V_{(i)}\right\}$ is called the *frame adapted to the connection* $\nabla$ in $\pi^{-1}(U)$. On putting $\tilde{e}_{(i)} = e^H_{(i)}$, $\tilde{e}_{(\bar{\imath})} = e^V_{(i)}$ we write the adapted frame as $\{\tilde{e}_{(A)}\} = \{\tilde{e}_{(i)}, \tilde{e}_{(\bar{\imath})}\}$.

Then the horizontal and vertical lifts of $X$ with local components $X^k$ are components

$$X^H : \begin{pmatrix} X^k \\ 0 \end{pmatrix}, \qquad X^V : \begin{pmatrix} 0 \\ X^k \end{pmatrix}$$

with respect to the adapted frame $\{\tilde{e}_{(A)}\}$ in $TM$.

## 3. The Tangent Bundle with an Almost Hypercomplex Pseudo-Hermitian Structure

Our purpose is a determination of an almost hypercomplex pseudo-Hermitian structure $(H, G)$ on $TM$ when the base manifold $M$ has an almost skew-Hermitian structure $(J, g, \tilde{g})$.

We will use the horizontal and vertical lifts of the vector fields on $M$ to get the corresponding components of the considered tensor fields on $TM$. These components are sufficient to describe the characteristic tensor fields on $TM$ in general.

### 3.1. *An Almost Hypercomplex Structure on the Tangent Bundle*

As it is known [15], for any affine connection in $M$, the induced horizontal and vertical distributions in $TM$ are mutually complementary. Then we define tensor field $J_1$, $J_2$ and $J_3$ in $TM$ by their action over the horizontal and vertical lifts of an arbitrary vector field in $M$ as follows:

$$\begin{aligned} &J_1 : X^H \to X^V,\ X^V \to -X^H; \\ &J_2 : X^H \to (JX)^V,\ X^V \to (JX)^H; \quad J_3 := J_1 \circ J_2, \end{aligned} \tag{3.1}$$

where $J$ is the given almost complex structure in $M$.

By direct computations we get the following

**Proposition 3.1.** *There exists an almost hypercomplex structure* $H = (J_1, J_2, J_3)$, *defined by* (3.1) *in* $TM$ *over an almost complex manifold* $(M, J)$ *with an affine connection* $\nabla$. *The received* $4n$*-dimensional manifold is an almost hypercomplex manifold* $(TM, H)$.

Let $N_\alpha$ denotes the Nijenhuis tensor regarding $J_\alpha$ for each $\alpha = 1,2,3$ and $\tilde{X}, \tilde{Y} \in \mathcal{T}_1^0(TM)$, i.e.

$$N_\alpha(\tilde{X},\tilde{Y}) = [\tilde{X},\tilde{Y}] + J_\alpha[J_\alpha\tilde{X},\tilde{Y}] + J_\alpha[\tilde{X},J_\alpha\tilde{Y}] - [J_\alpha\tilde{X},J_\alpha\tilde{Y}]. \tag{3.2}$$

Having in mind (2.1) and (2.2) we receive by direct computations the following

**Lemma 3.2.** *Let* $\nabla$ *be a torsion-free affine connection in* $M$ *and* $R$ *be its curvature tensor. Then for any* $X, Y \in \mathcal{T}_1^0(M)$ *at* $u \in T_pM$ *we have*

$$[X^H,Y^H] = [X,Y]^H - \{R(X,Y)u\}^V, \qquad [X^H,Y^V] = (\nabla_X Y)^V,$$
$$[X^V,Y^V] = 0, \qquad [X^V,Y^H] = -(\nabla_Y X)^V.$$

Using the definitions (3.1), (3.2) and Lemma 3.2 we get

**Proposition 3.3.** *Let* $(M, J)$ *be an almost complex manifold. Then the Nijenhuis tensors of the structure* $H$ *in* $TM$ *for the horizontal and vertical lifts have the form*

$$\begin{aligned} -N_1(X^H,Y^H) &= N_1(X^V,Y^V) = \{R(X,Y)u\}^V, \\ N_1(X^H,Y^V) &= N_1(X^V,Y^H) = -\{R(X,Y)u\}^H; \end{aligned} \tag{3.3}$$

(3.4)

$$N_2(X^H,Y^H) = \{J(\nabla_X J)(Y) - J(\nabla_Y J)(X)\}^H - \{R(X,Y)u\}^V,$$
$$N_2(X^V,Y^V) = -\{(\nabla_{JX} J)(Y) + (\nabla_{JY} J)(X)\}^H + \{R(JX,JY)u\}^V,$$
$$N_2(X^H,Y^V) = \{J(\nabla_X J)(Y) + (\nabla_{JY} J)(X)\}^V - \{JR(X,JY)u\}^H,$$
$$N_2(X^V,Y^H) = -\{(\nabla_{JX} J)(Y) + (J\nabla_Y J)(X)\}^V - \{JR(JX,Y)u\}^H;$$

(3.5)

$$\begin{aligned} N_3(X^H,Y^H) = \{&-R(X,Y)u + R(JX,JY)u \\ &+ JR(JX,Y)u + JR(X,JY)u\}^V, \end{aligned}$$
$$N_3(X^H,Y^V) = \{(\nabla_{JX} J)(Y) - (\nabla_X J)(JY)\}^V,$$
$$N_3(X^V,Y^H) = -\{(\nabla_Y J)(JX) - (\nabla_{JY} J)(X)\}^V, \ N_3(X^V,Y^V) = 0.$$

*for any* $X, Y \in \mathcal{T}_1^0(M)$ *at* $u \in T_pM$.

The last equalities for $N_\alpha$ imply the following necessary and sufficient conditions for integrability of $J_\alpha$ and $H$.

**Theorem 3.4.** *Let $TM$ be the tangent bundle manifold with an almost hypercomplex structure $H = (J_1, J_2, J_3)$ defined as in (3.1) and $M$ be its base manifold with an almost complex structure $J$. Then the following interconnections hold:*
*(1) $(TM, J_1)$ is complex iff $M$ is flat; (see also [10])*
*(2) $(TM, J_\beta)$ for $\beta = 2$ or $3$ is complex iff $M$ is flat and $J$ is parallel;*
*(3) $(TM, H)$ is hypercomplex iff $M$ is flat and $J$ is parallel.*

**Corollary 3.5.**
*(1) $(TM, J_2)$ is complex iff $(TM, J_3)$ is complex.*
*(2) If $(TM, J_2)$ or $(TM, J_3)$ is complex then $(TM, H)$ is hypercomplex.*

### 3.2. *The Sasaki Metric on the Tangent Bundle*

Let us introduce the *Sasaki metric* $\hat{g}$ on $TM$ as in [12] (i. e. $\hat{g}$ is the so-called *diagonal lift* of the base metric $g$) defined by

$$\hat{g}(X^H, Y^H) = \hat{g}(X^V, Y^V) = g(X, Y), \quad \hat{g}(X^H, Y^V) = 0. \tag{3.6}$$

Having in mind the definition (3.1) of the structure $H$, we verify immediately that the Sasaki metric satisfies the properties (1.1) and therefore it is valid the following

**Theorem 3.6.** *The tangent bundle $TM$ equipped with the almost hypercomplex structure $H$ and the Sasaki metric $\hat{g}$, defined by (3.1) and (3.6), respectively, is an almost hypercomplex pseudo-Hermitian manifold denoting by $(TM, H, \hat{G})$.*

Since $\hat{\nabla}$ is the Levi-Civita connection of $\hat{g}$ on $TM$ as $\nabla$ of $g$ on $M$, then using the property of type

$$\begin{aligned} 2g(\nabla_X Y, Z) = {} & Xg(Y, Z) + Yg(X, Z) - Zg(X, Y) \\ & + g([X, Y], Z) + g([Z, X], Y) + g([Z, Y], X) \end{aligned}$$

we obtain the covariant derivatives of the horizontal and vertical lifts of vector fields on $TM$ as follows

**Lemma 3.7.** *For any $X, Y \in \mathcal{T}_1^0(M)$ at $u \in T_pM$*

$$\hat{\nabla}_{X^H} Y^H = (\nabla_X Y)^H - \tfrac{1}{2}\{R(X, Y)u\}^V, \quad \hat{\nabla}_{X^V} Y^H = \tfrac{1}{2}\{R(u, X)Y\}^H,$$
$$\hat{\nabla}_{X^H} Y^V = \tfrac{1}{2}\{R(u, Y)X\}^H + (\nabla_X Y)^V, \quad \hat{\nabla}_{X^V} Y^V = 0.$$

After that we calculate the components of the curvature tensor $\hat{R}$ of $\hat{\nabla}$ with respect to the horizontal and vertical lifts of the vector fields on $M$ and we receive

**Proposition 3.8.** *The following interconnections between the curvature tensors $R$ and $\hat{R}$ corresponding to the metrics $g$ and $\hat{g}$ are valid*

$$
\begin{aligned}
&\hat{R}(X^H,Y^H,Z^H,W^H) = R(X,Y,Z,W) + \tfrac{1}{4}\{g\,(R(W,X)u, R(Y,Z)u)\\
&\qquad - g\,(R(W,Y)u, R(X,Z)u)\} - \tfrac{1}{2}g\,(R(X,Y)u, R(Z,W)u)\,,\\
&\hat{R}(X^H,Y^H,Z^H,W^V) = -\tfrac{1}{2}g\,((\nabla_X R)(Y,Z)u - (\nabla_Y R)(X,Z)u, W)\,,\\
&\hat{R}(X^H,Y^H,Z^V,W^V) = R(X,Y,Z,W)\\
&\qquad - \tfrac{1}{4}\{g\,(R(u,W)X, R(u,Z)Y) - g\,(R(u,W)Y, R(u,Z)X)\},\\
&\hat{R}(X^H,Y^V,Z^H,W^H) = \tfrac{1}{2}g\,((\nabla_X R)(u,Y)Z, W)\,,\\
&\hat{R}(X^H,Y^V,Z^H,W^V) = \tfrac{1}{2}R(X,Z,Y,W) - \tfrac{1}{4}g\,(R(u,Y)Z, R(u,W)X)\,,\\
&\hat{R}(X^V,Y^V,Z^H,W^H) = R(X,Y,Z,W)\\
&\qquad - \tfrac{1}{4}\{g\,(R(u,Y)Z, R(u,X)W) - g\,(R(u,X)Z, R(u,Y)W)\},\\
&\hat{R}(X^V,Y^V,Z^H,W^V) = \hat{R}(X^H,Y^V,Z^V,W^V) = \hat{R}(X^V,Y^V,Z^V,W^V) = 0.
\end{aligned}
$$

*for the lifts $(\cdot)^H, (\cdot)^V \in \mathcal{T}_1^0(TM)$ of any $X,Y,Z,W \in \mathcal{T}_1^0(M)$ at $u \in T_pM$.*

Hence we get

**Theorem 3.9.** *The tangent bundle $(TM, \hat{g})$ is flat if and only if its base manifold $(M, g)$ is flat.*

**Remark.** The results of the last three statements 3.7–3.9 are confirmed also by [7] (see also [6]), where $g$ is a Riemannian metric.

According to [7], the tangent bundle with Sasaki metric of a Riemannian manifold is never locally symmetric unless the base manifold is locally Euclidean. Having in mind the last reasonings, the proof of the corresponding theorem in [7] and that $g$ is a pseudo-Riemannian metric in our case, we obtain the following

**Conclusion.** If the tangent bundle $(TM, \hat{g})$ is locally symmetric then the curvature tensor $R$ of its base pseudo-Riemannian manifold $(M, g)$ is zero or isotropic, i.e. $R = 0$ or $g(R, R) = 0$, $R \neq 0$).

### 3.3. *The Tangent Bundles with Almost Hermitian, Almost Skew-Hermitian and Almost Hypercomplex Pseudo-Hermitian Structures*

Suppose that $(M, J)$ is an almost complex manifold with skew-Hermitian metrics $g$, $\tilde{g}$ and that $(TM, H)$ is its almost hypercomplex tangent bundle with the pseudo-Hermitian metric structure $\hat{G} = (\hat{g}, \hat{\Phi}, \hat{g}_2, \hat{g}_3)$ derived (as in (1.2)) from the Sasaki metric $\hat{g}$ on $TM$ - the diagonal lift of $g$. The generated $4n$-dimensional manifold we will denote by $(TM, H, \hat{G})$.

To characterize the structural tensors $F_\alpha(\cdot,\cdot,\cdot) = \hat{g}\left((\hat{\nabla}.J_\alpha)(\cdot),\cdot\right)$ at each $u \in T_pM$ on $(TM, H, \hat{G})$ we use Lemma 3.7 and the definitions (3.1) of $J_\alpha$, whence we obtain the following

**Proposition 3.10.** *The nonzero components of $F_\alpha$ with respect to the lifts of the vector fields depend on structural tensor $F$ and the curvature tensor $R$ on $(M, J, g, \tilde{g})$ by the following way:*

$$\begin{aligned} -F_1(X^H,Y^H,Z^H) &= F_1(X^H,Y^V,Z^V) = F_1(X^V,Y^H,Z^V) \\ &= F_1(X^V,Y^V,Z^H) = \tfrac{1}{2}R(Y,Z,X,u); \end{aligned} \tag{3.7}$$

$$\begin{aligned} F_2(X^H,Y^H,Z^H) &= -\tfrac{1}{2}R(X,Y,JZ,u) + \tfrac{1}{2}R(Z,X,JY,u), \\ F_2(X^H,Y^V,Z^V) &= \tfrac{1}{2}R(X,JY,Z,u) - \tfrac{1}{2}R(JZ,X,Y,u), \\ F_2(X^H,Y^H,Z^V) &= F_2(X^H,Y^V,Z^H) = F(X,Y,Z), \\ F_2(X^V,Y^H,Z^V) &= \tfrac{1}{2}R(Y,JZ,X,u), \\ F_2(X^V,Y^V,Z^H) &= -\tfrac{1}{2}R(JY,Z,X,u); \end{aligned} \tag{3.8}$$

$$\begin{aligned} F_3(X^H,Y^H,Z^H) &= -F_3(X^H,Y^V,Z^V) = -F(X,Y,Z), \\ F_3(X^H,Y^H,Z^V) &= -\tfrac{1}{2}R(X,JY,Z,u) - \tfrac{1}{2}R(X,Y,JZ,u), \\ F_3(X^H,Y^V,Z^H) &= \tfrac{1}{2}R(Z,X,JY,u) + \tfrac{1}{2}R(JZ,X,Y,u), \\ F_3(X^V,Y^H,Z^H) &= \tfrac{1}{2}R(JY,Z,X,u) - \tfrac{1}{2}R(Y,JZ,X,u). \end{aligned} \tag{3.9}$$

*where $(\cdot)^H, (\cdot)^V \in \mathcal{T}_1^0(TM)$ are the lifts of any $X, Y, Z \in \mathcal{T}_1^0(M)$ at $u \in T_pM$.*

Hence we compute the corresponding Lie forms with respect to the adapted frame.

Let $\{e_i^H, e_{\bar{i}}^H, e_i^V, e_{\bar{i}}^V\}$ be the adapted frame of type $(+,\dots,+,-,\dots,-,+,\dots,+,-,\dots,-)$ at each point of $TM$ derived by the orthonormal basis $\{e_{(i)}, e_{(\bar{i})}\}$ of signature $(n,n)$ at each point of $M$ denoting $e_{(\bar{i})} = Je_i$. The indices $i$ and $\bar{i}$ run over the ranges $\{1,\dots,n\}$ and $\{n+1,\dots,2n\}$, respectively. For example

$$\begin{aligned}\theta_3(Z^H) &= \textstyle\sum_{i=1}^n \{F_3(e_i^H, e_i^H, Z^H) - F_3(e_{\bar{i}}^H, e_{\bar{i}}^H, Z^H)\\ &\qquad + F_3(e_i^V, e_i^V, Z^H) - F_3(e_{\bar{i}}^V, e_{\bar{i}}^V, Z^H)\}\\ &= \textstyle\sum_{i=1}^n \{-F(e_i, e_i, Z) + F(e_{\bar{i}}, e_{\bar{i}}, Z)\} = -\theta(Z),\\ \theta_3(Z^V) &= \textstyle\sum_{i=1}^n \{F_3(e_i^H, e_i^H, Z^V) - F_3(e_{\bar{i}}^H, e_{\bar{i}}^H, Z^V)\\ &\qquad + F_3(e_i^V, e_i^V, Z^V) - F_3(e_{\bar{i}}^V, e_{\bar{i}}^V, Z^V)\}\\ &= \textstyle\sum_{i=1}^n \{-\frac{1}{2}R(e_i, e_{\bar{i}}, Z, u) - \frac{1}{2}R(e_{\bar{i}}, e_i, Z, u)\} = 0.\end{aligned}$$

Then we have

**Proposition 3.11.**
(1) $(TM, J_1, \hat{g})$ *has a zero Lie form* $\theta_1$;
(2) $(TM, J_2, \hat{g})$ *has a zero Lie form* $\theta_2$ *iff* $(M, J, g, \tilde{g})$ *has a zero Lie form* $\theta$ *and a zero associated Ricci tensor* $\tilde{\rho}$;
(3) $(TM, J_3, \hat{g})$ *has a zero Lie form* $\theta_3$ *iff* $(M, J, g, \tilde{g})$ *has a zero Lie form* $\theta$.

According (3.7) the following property is valid.

**Theorem 3.12.** $(TM, J_1, \hat{g})$ *is an almost Kähler manifold and it is a Kähler manifold if and only if* $(M, J, g, \tilde{g})$ *is flat.*

**Remark.** By comparison with the Riemannian case, in [10] it is shown that $(TM, J_1, \hat{g})$ is almost Kählerian (i.e. symplectic) for any Riemannian metric $g$ on the base manifold when the connection used to define the horizontal lifts is the Levi-Civita connection.

Having in mind (3.7)–(3.9) and Theorem 3.4 it is easy to conclude the following sequel of properties:

**Proposition 3.13.**
(1) $(TM, J_1, \hat{g})$ *is Kählerian iff* $M$ *is flat.*
(2) $(TM, J_\beta, \hat{g})$ *for* $\beta = 2$ *or* $3$ *is skew-Kählerian iff* $(M, J, g, \tilde{g})$ *is flat and skew-Kählerian.*
(3) $(TM, H, G)$ *is a pseudo-hyper-Kähler manifold iff* $(M, J, g, \tilde{g})$ *is flat and skew-Kählerian.*

**Corollary 3.14.**
(1) $(TM, J_2, \hat{g})$ *is skew-Kählerian iff* $(TM, J_3, \hat{g})$ *is skew-Kählerian.*
(2) *If* $(TM, J_2, \hat{g})$ *or* $(TM, J_3, \hat{g})$ *is skew-Kählerian then* $(TM, H, \hat{G})$ *is pseudo-hyper-Kählerian.*

**Corollary 3.15.**

(1) *The only complex manifolds* $(TM, J_\alpha, \hat{g})$ *for some* $\alpha = 1, 2, 3$ *are the Kähler manifolds with respect to that* $J_\alpha$.

(2) *The only hypercomplex manifolds* $(TM, H, \hat{G})$ *are the pseudo-hyper-Kähler manifolds.*

Let us recall that the class $\{\mathcal{W}_2 \oplus \mathcal{W}_3\}$ of the almost skew-Hermitian manifolds is determined by the condition for vanishing of the Lie form, the class $\mathcal{W}_3$ by $\sigma F = 0$ and the class $\mathcal{W}_0$ by $F = 0$ (i.e. the last is the class of the skew-Kähler manifolds). Then we obtain the following

**Theorem 3.16.** *Let* $M$ *be* $(M, J, g, \tilde{g})$ *and* $TM$ *be* $(TM, H, \hat{G})$.

(1) $TM \in \{\mathcal{W}_2 \oplus \mathcal{W}_3\}(J_2)$ *iff* $M \in \{\mathcal{W}_2 \oplus \mathcal{W}_3\}(J)$ *and* $\rho = \tilde{\rho} = 0$;

(2) $TM \in \mathcal{W}_3(J_2)$ *iff* $M \in \mathcal{W}_0(J)$ *and* $\rho = \tilde{\rho} = 0$;

(3) $TM \in \{\mathcal{W}_2 \oplus \mathcal{W}_3\}(J_3)$ *iff* $M \in \{\mathcal{W}_2 \oplus \mathcal{W}_3\}(J)$;

(4) $TM \in \mathcal{W}_3(J_3)$ *iff* $M \in \mathcal{W}_0(J)$,

*where* $\rho$ *and* $\tilde{\rho}$ *denote the Ricci tensor and the associated Ricci tensor on* $M$, *respectively.*

If we set some additional conditions for $(M, J, g, \tilde{g})$ then we receive the corresponding specialization of $(TM, H, \hat{G})$ given in the next properties:

**Proposition 3.17.** *Let* $M$ *be* $(M, J, g, \tilde{g})$ *and* $TM$ *be* $(TM, H, \hat{G})$.

(1) *If* $M \in \{\mathcal{W}_2 \oplus \mathcal{W}_3\}(J)$ *then* $TM \in \mathcal{AK}(J_1) \cap \{\mathcal{W}_1 \oplus \mathcal{W}_2 \oplus \mathcal{W}_3\}(J_2) \cap \{\mathcal{W}_2 \oplus \mathcal{W}_3\}(J_3)$.

(2) *If* $M \in \{\mathcal{W}_2 \oplus \mathcal{W}_3\}(J)$ *and* $\rho = \tilde{\rho} = 0$ *then* $TM \in \mathcal{AK}(J_1) \cap \{\mathcal{W}_2 \oplus \mathcal{W}_3\}(J_2) \cap \{\mathcal{W}_2 \oplus \mathcal{W}_3\}(J_3)$.

(3) *If* $M \in \{\mathcal{W}_2 \oplus \mathcal{W}_3\}(J)$ *and* $R = 0$ *then* $TM \in \mathcal{K}(J_1) \cap \{\mathcal{W}_2 \oplus \mathcal{W}_3\}(J_2) \cap \{\mathcal{W}_2 \oplus \mathcal{W}_3\}(J_3)$.

**Proposition 3.18.** *Let* $M$ *be* $(M, J, g, \tilde{g})$ *and* $TM$ *be* $(TM, H, \hat{G})$.

(1) *If* $M \in \mathcal{W}_0(J)$ *then* $TM \in \mathcal{AK}(J_1) \cap \{\mathcal{W}_1 \oplus \mathcal{W}_2 \oplus \mathcal{W}_3\}(J_2) \cap \mathcal{W}_3(J_3)$.

(2) *If* $M \in \mathcal{W}_0(J)$ *and* $\rho = \tilde{\rho} = 0$ *then* $TM \in \mathcal{AK}(J_1) \cap \mathcal{W}_3(J_2) \cap \mathcal{W}_3(J_3)$.

(3) *If* $M \in \mathcal{W}_0(J)$ *and* $R = 0$ *then* $TM \in \mathcal{K}(J_1) \cap \mathcal{W}_0(J_2) \cap \mathcal{W}_0(J_3)$.

In above properties $\mathcal{AK}(J_1)$ and $\mathcal{K}(J_1)$ denote the classes regarding $J_1$ of the almost Kähler manifolds and the Kähler manifolds determined by the conditions $\theta_1 = 0$ and $F_1 = 0$, respectively.

## 4. Eight-Dimensional Derived Examples

Let us recall the developed examples of 4-dimensional almost hypercomplex pseudo-Hermitian manifolds. In [4] and [8] we have constructed a lot of explicit examples of the investigated manifolds.

At first in [4], it is considered a pseudo-Riemannian spherical manifold $S_2^4$ in pseudo-Euclidean vector space $\mathbb{R}_2^5$ of signature $(2,3)$. It admits a hypercomplex pseudo-Hermitian structure, with respect to which it is of the class $\mathcal{W} = \bigcap_{\alpha=1}^{3} \mathcal{W}(J_\alpha)$.

Secondly, in the same paper it is considered the 4-dimensional compact homogenous space $L|_\Gamma$, where $L = H \times S^1$ is a connected Lie group, $H$ is the Heizenberg group, $S^1$ is the circle and $\Gamma$ is the discrete subgroup of $L$ consisting of all matrices of integer entries. The manifold $M$ generated by $L$ we equipped with an $(H,G)$-structure. Then it is a $\mathcal{W}(J_1)$-manifold but it does not belong to $\mathcal{W}$.

In [8] the sequel of examples begins at Examples 1 and 2 about two Engel manifolds with almost $(H,G)$-structure. The first one is equipped with double isotopic hyper-Kählerian structures which are neither hypercomplex nor symplectic. The second one – with double isotopic hyper-Kählerian structures which are non-integrable but symplectic.

The Example 3 is constructed as a real semi-space with almost $(H,G)$-structure of the class $\mathcal{W}$.

A real quarter-space with almost $(H,G)$-structure is shown in Example 4, which is a $\mathcal{K}(J_1)$-manifold and an isotropic hyper-Kähler manifold.

In Example 5, an almost $(H,G)$-structure is introduced on a real pseudo-hyper-cylinder in a pseudo-Euclidean real space $\mathbb{R}_2^5$. The received manifold is not integrable and the Lie forms are non-zero, regarding any $J_\alpha$.

The following three examples (Examples 6–8) concern several surfaces $S_\mathbb{C}^2$ in a 3-dimensional complex Euclidean space $\left(\mathbb{C}^3, \langle\cdot,\cdot\rangle\right)$. There is used that the natural decomplexification of an $n$-dimensional complex Euclidean space is the $2n$-dimensional real space with a complex skew-Hermitian structure. Namely, Example 6 shows a flat pseudo-hyper-Kähler manifold as a complex cylinder. Example 7 is a complex cone with almost $(H,G)$-structure and it is a flat hypercomplex manifold which is Kählerian with respect to $J_1$ but it does not belong to $\mathcal{W}(J_2)$ or $\mathcal{W}(J_3)$ and the Lie forms $\theta_2$ and $\theta_3$ are non-zero. In Example 8, a complex sphere with almost $(H,G)$-structure is given. There is shown that it is a $\mathcal{K}(J_2)$-manifold of pointwise constant totally real sectional curvatures, but with respect to $J_1$ and $J_3$, the corresponding Nijenhuis tensors and Lie forms are non-zero.

The last two examples in [8] are inspired from an example of a locally flat almost Hermitian surface constructed in [11] about a connected Lie subgroup of $\mathcal{GL}(4,\mathbb{R})$. There are introduced the appropriate metric in two variants and then the received $(H,G)$-manifold in Example 9 is complex with respect to $J_2$ but non-hypercomplex and the Lie forms do not vanish; and the constructed $(H, G)$-manifold in Example 10 is flat and it is Kählerian with respect to $J_1$ but with respect to $J_2$ and $J_3$ it is not complex.

We can use the given examples in [4] and [8] to construct 8-dimensional tangent bundle manifolds with almost hypercomplex pseudo-Hermitian structure which is almost Kählerian with respect to $J_1$. For this purpose we use $J_2$ or $J_3$ like a base almost complex structure $J$ of $(M, J, g, \tilde{g})$. Then applying the above mentioned results to the case $n = 2$, we receive the following

**Corollary 4.1.** *Let $M$ and $TM$ denote for short the base 4-dimensional manifold $(M, J, g, \tilde{g})$ with almost complex skew-Hermitian structure and its tangent bundle $(TM, H, \hat{G})$ with almost hypercomplex pseudo-Hermitian structure constructed as in § 3, respectively.*

(1) *If $M$ with $J := J_2$ (or $J_3$) is the complex cylinder introduced in Example 6 of [8], then $TM$ is a flat pseudo-hyper-Kählerian.*
(2) *If $M$ with $J := J_2$ (or $J_3$) is the complex cone introduced in Example 7 of [8] or the Lie group introduced in Example 10 of [8], then $TM$ is a flat almost hypercomplex pseudo-Hermitian and an $\mathcal{AK}(J_1)$-manifold.*
(3) *If $M$ with $J := J_2$ is the complex sphere introduced in Example 8 of [8], then $TM$ is non-flat and its $\theta_3$ is also zero as $\theta_1$.*
(4) *If $M$ with $J := J_2$ (or $J_3$), is some of the manifolds introduced in the other 8 known examples of [4,8], then $TM$ is a non-flat manifold which is almost Kählerian with respect to $J_1$, but with respect to $J_2$ and $J_3$ it is not complex and the Lie forms are not zero.*

### *Acknowledgement*

The results of this paper were obtained during the author's fellowship studies at Niigata University financed by Matsumae International Foundation, Japan. The author would like to express deep gratitude to his host scientist Kouei Sekigawa whose guidance and support were crucial for the successful completion of this project. The author dedicate this work to the 60th anniversary of Prof. Sekigawa.

### References

1. D.V. Alekseevsky and S. Marchiafava, *Quaternionic Structures on a Manifold and Subordinated Structures*, Ann. Mat. Pura Appl. (IV), CLXXI (1996), 205–273.

2. G. Ganchev and A. Borisov, *Note on the Almost Complex Manifolds with a Norden Metric*, Compt. rend. Acad. bulg. Sci., 39 (1986), no. 5, 31–34.
3. A. Gray and L.M. Hervella, *The Sixteen Classes of Almost Hermitian Manifolds and Their Linear Invariants*, Ann. Mat. Pura Appl. (IV), CXXIII (1980), 35–58.
4. K. Gribachev, M. Manev and S. Dimiev, *Almost Hypercomplex Pseudo-Hermitian Manifolds*, Boll. Unione Mat. Ital. Sez. A, (to appear).
5. K. Gribachev, M. Manev and S. Dimiev, *On Almost Hypercomplex Pseudo-Hermitian Manifolds*, Trends in Compl. Anal., Diff. Geom. & Math. Phys., World Sci. Publ., 2003, 51–62.
6. S. Gudmundsson, E. Kappos, *On the Geometry of Tangent Bundles*, Expo. Math., 20 (2002), 1–41.
7. O. Kowalski, *Curvature of the Induced Riemannian Metric on the Tangent Bundle of a Riemannian Manifold*, J. Reine Angew. Math., 250 (1971), 124–129.
8. M. Manev, *Some Four-Dimensional Almost Hypercomplex Pseudo-Hermitian Manifolds*, Contemporary Aspects in Compl. Anal., Diff. Geom. & Math. Phys., World Sci. Publ., 2005 (to appear).
9. A.P. Norden, *On the Structure of the Connections on the Manifolds of the Lines in an Eucleadian Space*, Russian Math. (Izv. VUZ), 17 (1960), 145–157 (in Russian).
10. Ph. Tondeur, *Structure Presque Kählerienne Naturelle sur le Fibre des Vecteurs Covariants d'une Variete Riemannienne*, C. R. Acad. Sci. Paris, 254 (1962), 407–408 (in French).
11. F. Tricerri and L. Vanhecke, *Flat almost Hermitian manifolds which are not Kähler manifolds*, Tensor (N.S.), 31 (1977), no. 3, 249–254.
12. S. Sasaki, *On the Differential Geometry of Tangent Bundles of Riemannian Manifolds*, Tôhoku Math. J., 10 (1958), 238–354.
13. A. Sommese, *Quaternionic Manifolds*, Math. Ann., 212 (1975), 191–214.
14. N. Steenrod, *The Topology of Fibre Bundles*, Princeton Univ. Press, Princeton, N. J., 1951.
15. K. Yano and S. Ishihara, *Tangent and Cotangent Bundles*, Marcel Dekker, New York, 1973.

# ON THE EXISTENCE OF 3-DIMENSIONAL INVARIANT SUBMANIFOLDS OF $S^6$

KATSUYA MASHIMO

*Department of Mathematics*
*Tokyo University of Agriculture and Technology*
*Koganei, Tokyo 184–8588, Japan*
*E-mail: mashimo@cc.tuat.ac.jp*

It is well-known that an almost complex structure $J$ is defined on the 6-dimensional unit sphere $S^6$ by using the multiplication of the Cayley algebra $\mathfrak{C}$ . The group $G_2$ of automorphisms of the Cayley algebra acts on $S^6$ transitively and we have a diffeomorphism $S^6 = G_2/SU(3)$. The action of $G_2$ on $S^6$ extends to the action on the Grassmann bundle $G_3^+(TS^6)$ consisting of oriented 3-dimensional tangent subspaces.

In this note we consider 3-dimensional submanifolds of $S^6$ such that the tangent space at any point is contanined in an orbit of the action of $G_2$ on $G_3^+(TS^6)$.

The set of orbits of the action of $G_2$ on $G_3^+(TS^6)$ is parametrized by a complex number $\lambda$ ($|\lambda| \leq 1$) (see Proposition 2). We denote by $\Sigma_\lambda$ the orbit which corresponds to $\lambda$ and call a 3-dimensional submanifold $N$ a $\Sigma_\lambda$-*submanifold* if and only if $T_xN \in \Sigma_\lambda$ for any $x \in N$.

The aim of this paper is to show the non-existence of compact $\Sigma_\lambda$-submanifold for $\lambda \notin \mathbf{R}$.

## 1. Preliminaries

Let $\mathfrak{C}$ be the Cayley algebra. Take an orthonormal basis $\{E_0 = 1, E_1, \ldots, E_7\}$ of the Cayley algebra $\mathfrak{C}$ such that

$$E_1E_2 = E_3, \qquad E_1E_4 = E_5, \qquad E_1E_7 = E_6, \qquad E_2E_5 = E_7,$$
$$E_2E_4 = E_6, \qquad E_3E_4 = E_7, \qquad E_3E_6 = E_5.$$

We put

$$\mathfrak{C}_0 = \{ u \in \mathfrak{C} \,:\, u + \overline{u} = 0 \}$$

where $^-$ denotes the conjugation in $\mathfrak{C}$.

The unit sphere $S^6 \subset \mathfrak{C}_0$ centered at the origin has an almost complex structure $J$ defined by

$$J_p(X) = p \cdot X, \quad p \in S^6, \ X \in T_pS^6.$$

**Lemma 1.** *For a pair of mutually orthogonal unit vectors $a_1$, $a_2$ in $\mathfrak{C}_0$ put $a_3 = a_1 \cdot a_2$. Take a unit vector $a_4 \in \mathfrak{C}_0$, which is perpendicular to $a_1$, $a_2$ and $a_3$. If we put $a_5 = a_1 \cdot a_4$, $a_6 = a_2 \cdot a_4$ and $a_7 = a_3 \cdot a_4$ then the matrix*

$$g = [\, a_1, a_2, a_3, a_4, a_5, a_6, a_7 \,] \in SO(7)$$

*is an element of $G_2$.*

We denote by $\omega_1, \ldots, \omega_7$ the dual 1-forms of the standard orthonormal base $E_1, \ldots, E_7$. The $SU(3)$-invariant 3-form defined by

$$\omega = (\omega_2 + \sqrt{-1}\,\omega_3) \wedge (\omega_4 + \sqrt{-1}\,\omega_5) \wedge (\omega_7 + \sqrt{-1}\,\omega_6)$$

is extended to a $G_2$ invariant 3-form on $S^6$, which we also denote by $\omega$.

**Proposition 2.** *If $\xi$ is an oriented 3-dimensional subspace of $T_xS^6$ with $\omega(\xi) = c + \sqrt{-1}\,b$ $(b, c \in \boldsymbol{R})$ there exists an oriented orthonormal base $u_1$, $u_2$, $u_3$ of $\xi$ and an element $g \in G_2$ such that*

$$x = g(E_1), \quad u_1 = g(E_2), \quad u_2 = g(E_4),$$
$$u_3 = \sqrt{1 - b^2 - c^2}\, g(E_5) + bg(E_6) + cg(E_7).$$

*Especially $\xi_1$, $\xi_2 \in G_3^+(TS^6)$ are congruent under the action of $G_2$ if and only if $\omega(\xi_1) = \omega(\xi_2)$.*

**Proof.** Since $J$ is skew-symmetric, 0 is an eigenvalue of the restriction of $J$ on $\xi$. Take a unit eigenvector $u_1$ of $J|_\xi$ corresponding to the eigenvalue 0 and extend it to an orthonormal base $u_1, u_2, u_3$ of $\xi$. By the choice of $u_1$, we have

$$\langle J(u_1), u_2 \rangle = \langle J(u_1), u_3 \rangle = \langle u_1, u_2 \rangle = \langle u_1, u_3 \rangle = \langle u_2, u_3 \rangle = 0 \qquad (1)$$

By Lemma 1, there exists an element $g \in G_2$ with

$$g(E_1) = x, \quad g(E_2) = u_1, \quad g(E_4) = u_2.$$

From (1), $u_3$ is of the following form

$$u_3 = ag(E_5) + bg(E_6) + cg(E_7).$$

We can take the base $\{u_1, u_2, u_3\}$ such that $a \geq 0$, for we can adopt $\{-u_1, u_2, -u_3\}$ as the oriented orthonormal base of $\xi$ instead of $\{u_1, u_2, u_3\}$. □

## 2. Existence of $\Sigma_\lambda$-submanifold

For a complex number $\lambda$ with $|\lambda| \le 1$ we put

$$\Sigma_\lambda = \{\, \xi \; : \; \omega(\xi) = \lambda \,\}.$$

An oriented 3-dimensional submanifold $N$ is said to be a $\Sigma_\lambda$-submanifold if and only if the tangent space is contained in $\Sigma_\lambda$ at any point $x \in N$. By definition a 3-dimensional CR-submanifold of $S^6$ is a $\Sigma_0$-submanifold.

A parallel 3-form defined by

$$\phi(x, y, z) = \langle xy, z \rangle\,, \quad x, y, z \in \mathfrak{C}_0,$$

is called the *associative 3-form* and its dual 4-form $*\phi$ is called the *coassociative 4-form* ([1]).

In [2], the author proved that a 3-dimensional submanifold $N$ of $S^6$ is a totally real submanifold if and only if the truncated cone $\tilde{M} = \{\, tx \; : \; t > 0,\; x \in N \,\}$ is a coassociative submanifold (namely $*\phi(T_p\tilde{M}) = 1$ holds at any point $p \in \tilde{M}$). If $M$ is a $\Sigma_\lambda$-submanifold then the truncated cone $\tilde{M}$ satisfies $*\phi(T_p(\tilde{M})) = \mathrm{Re}\lambda$. Thus we have

**Theorem 3.** *A 3-dimensional totally real submanifold of $S^6$ is a $\Sigma_{\pm 1}$-submanifold.*

**Theorem 4.** *If there exists a compact 3-dimensional $\Sigma_\lambda$-submanifold then $\lambda$ is a real number.*

**Proof.** Let $\lambda = c + \sqrt{-1}\, b$ be a complex number and put $a = \sqrt{1 - |\lambda|^2}$. Assume that $|\lambda| < 1$ and there exists an oriented 3-dimensional $\Sigma_\lambda$-submanifold $N \subset S^6$. As it was shown in the proof of Proposition 2, we can take a locally defined oriented orthonormal frame $u_1, u_2, u_3$ with

$$\langle J(u_1), u_2 \rangle = \langle J(u_1), u_3 \rangle = 0.$$

The vector $u_1$ is defined up to sign. Since an orientation on $\mathrm{span}\{u_2, u_3\}$ is defined by the sign of $\langle J(u_2), u_3 \rangle$, we can take a global vector field $U_1$ which satisfies $J(U_1) \perp \xi$.

We denote by $\nabla$ the Levi-Civita connection of $S^6$. Let $p$ be the unit normal vector field of $S^6$ and $\sigma$ be the second fundamental form of the submanifold $N \subset S^6$. By an elementary calculation, we have

$$\langle (\nabla_{U_1} J)(u_2), u_3 \rangle = \langle \sigma(U_1, u_2), J(u_3) \rangle - \langle \sigma(U_1, u_3), J(u_2) \rangle\,,$$
$$\langle (\nabla_{u_2} J)(u_3), U_1 \rangle = a \langle \nabla_{u_2} U_1, u_2 \rangle - \langle \sigma(U_1, u_2), J(u_3) \rangle + \langle \sigma(u_2, u_3), J(U_1) \rangle\,,$$
$$\langle (\nabla_{u_3} J)(U_1), u_2 \rangle = a \langle \nabla_{u_3} U_1, u_3 \rangle - \langle \sigma(u_3, u_2), J(U_1) \rangle + \langle \sigma(u_3, U_1), J(u_2) \rangle\,.$$

The left hand sides of the above three are equal, since we have

$$\langle(\nabla_X J)(Y),\ Z\rangle = \langle\nabla_X(p\cdot Y) - p\cdot(\nabla_X Y), Z\rangle = \phi(X,Y,Z).$$

Thus we obtain

$$3\,\langle(\nabla_{U_1}J)u_2, u_3\rangle = a\,\mathrm{div}(U_1) = -3b$$

Integrating both sides on $N$, we have

$$0 = -3b\,\mathrm{vol}(M).$$

Thus $\lambda = c + \sqrt{-1}\,b = c$ is a real number. □

## References

1. R. Harvey and H.B. Lawson, *Calibrated geometries*, Acta Math., 148(1982), 47–157.
2. K. Mashimo, *Homogeneous totally real submanifolds of* $S^6$, Tsukuba J. Math., 9(1985), 185–202.

# INFINITESIMAL FOURIER TRANSFORMATION FOR THE SPACE OF FUNCTIONALS

TAKASHI NITTA

*Department of Education*
*Mie University*
*Kamihama, Tsu, 514-8507, Japan*
*E-mail: nitta@edu.mie-u.ac.jp*

TOMOKO OKADA

*Graduate school of Mathematics*
*Nagoya University*
*Chikusa-ku, Nagoya, 464-8602, Japan*
*E-mail: m98122c@math.nagoya-u.ac.jp*

The purpose of the present paper is to formulate a Fourier transformation for the space of functionals, as an infinitesimal meaning. We extend the real numbers $\mathbf{R}$ to $^*(^*\mathbf{R})$ on the basis of nonstandard methods for the construction. The domain of a functional is the set of all internal functions from a $^*$-finite lattice to a $^*$-finite lattice with a double meaning. Considering a $^*$-finite lattice with a double meaning, we propose a new treatment of the domains for functionals in the theory of Fourier transformation, and exhibit two typical examples.

## 0. Introduction

Recently many kinds of geometric invariants are defined on manifolds and they are used for studying low dimensional manifolds, for example, Donaldson's invariant, Chern-Simon's invariant and so on. They are originally defined as Feynman path integrals in physics. The Feynman path integral is in a sense an integral of a functional on an infinite dimensional space of functions. We would like to study the Feynman path integral and the originally defined invariants. For the purpose, we would be sure that it is necessary to construct a theory of Fourier transformation on the space of functionals. For it, as the later argument, we would need many stages of infinitesimals and infinites, that is, we need to put a concept of stage on the field of real numbers. We use nonstandard analysis to develop a theory of Fourier transformation on the space of functionals.

Historically, for the theories of Fourier transformations in nonstandard analysis, in 1972, Luxemburg([5]) developed a theory of Fourier series with *-finite summation on the basis of nonstandard analysis. The basic idea of his approach is to replace the usual $\infty$ of the summation to an infinite natural number $N$. He approximated the Fourier transformation on the unit circle by the Fourier transformation on the group of $N$th roots of unity.

In 1988, Kinoshita([3]) defined a discrete Fourier transformation for each even *-finite number $H(\in {}^*\mathbf{N})$: $(F\varphi)(p) = \sum_{-\frac{H^2}{2}\leq z<\frac{H^2}{2}} \frac{1}{H}\exp(-2\pi ip\frac{1}{H}z)$ $\varphi(\frac{1}{H}z)$, called "infinitesimal Fourier transformation". He developed a theory for the infinitesimal Fourier transformation and studied the distribution space deeply, and proved the same properties hold as usual Fourier transformation of $L^2(\mathbf{R})$.

In 1989, Gordon([2]) independently defined a generic, discrete Fourier transformation for each infinitesimal $\Delta$ and *-finite number $M$, defined by $(F_{\Delta,M}\,\varphi)(p) = \sum_{-M\leq z\leq M}\Delta\exp(-2\pi ip\Delta z)\varphi(\Delta z)$. He studied under which condition the discrete Fourier transformation $F_{\Delta,M}$ approximates the usual Fourier transformation $\mathcal{F}$ for $L^2(\mathbf{R})$. His proposed condition is $(A')$ of his notation: let $\Delta$ be an infinitely small and $M$ an infinitely large natural number such that $M\cdot\Delta$ is infinitely large. He showed that under the condition $(A')$ the standard part of $F_{\Delta,M}\,\varphi$ approximates the usual $\mathcal{F}\varphi$ for $\varphi\in L^2(\mathbf{R})$. One of the different points between Kinoshita's and Gordon's is that there is the term $\Delta\exp(-2\pi ip\Delta M)\varphi(\Delta M)$ in the summation of their two definitions or not. We mention that both definitions are same for the standard part of the discrete Fourier transformation for $\varphi\in L^2(\mathbf{R})$ and Kinoshita's definition satisfies the condition $(A')$ for an even infinite number $H$ if $\Delta=\frac{1}{H}$, $M=\frac{H^2}{2}$.

We shall extend their theory of Fourier transformation for the space of functions to a theory of Fourier transformation for the space of functionals. For the purpose of this, we shall represent a space of functions from $\mathbf{R}$ to $\mathbf{R}$ as a space of functions from a set of lattices in an infinite interval $\left[-\frac{H}{2},\frac{H}{2}\right)$ to a set of lattices in an infinite interval $\left[-\frac{H'}{2},\frac{H'}{2}\right)$. We consider what $H'$ is to treat any function from $\mathbf{R}$ to $\mathbf{R}$. If we put a function $a(x)=x^n (n\in\mathbf{Z}^+)$, we need that $\frac{H'}{2}$ is greater than $\left(\frac{H}{2}\right)^n$, and if we choose a function $a(x)=e^x$, we need that $\frac{H'}{2}$ is greater than $e^{\frac{H}{2}}$. If we choose any infinite number, there exists a function whose image is beyond the infinite number. Since we treat all functions from $\mathbf{R}$ to $\mathbf{R}$, we need to put $\frac{H'}{2}$ as an infinite number greater than any infinite number of ${}^*\mathbf{R}$. Hence we make $\left[-\frac{H'}{2},\frac{H'}{2}\right)$ not in ${}^*\mathbf{R}$

but in $^*(^*\mathbf{R})$, where $^*(^*\mathbf{R})$ is a double extension of $\mathbf{R}$, that is, $H'$ is an infinite number in $^*(^*\mathbf{R})$. First we shall develop an infinitesimal Fourier transformation theory for the space of functionals, and secondly we calculate fundamental two examples for our infinitesimal Fourier transformation. These are a functional $f$ and an infinite-dimensional Gaussian distribution $g$ where $\mathbf{st}(f(\alpha)) = \exp\bigl(\pi i \int_{-\infty}^{\infty} \alpha^2(t)dt\bigr)$, $\mathbf{st}(g(\alpha)) = \exp\bigl(-\pi \int_{-\infty}^{\infty} \alpha^2(t)dt\bigr)$ for $\alpha \in L^2(\mathbf{R})$. We obtain the following results of standard meanings: $(Ff)(b) = \overline{f(b)}$ or $-\overline{f(b)}$ and $(Fg)(b) = C_2(b)g(b)$, $\mathbf{st}(C_2(b)) = 1$ if $b$ is finite valued. Our infinitesimal Fourier transformation of $g$ is also $g$ when the domain of $g$ is standard.

## 1. Formulation

To explain our infinitesimal Fourier transformation for the space of functionals, we introduce Kinoshita's infinitesimal Fourier transformation for the space of functions. We fix an infinite set $\Lambda$ and an ultrafilter $F$ of $\Lambda$ so that $F$ includes the Fréchet filter $F_0(\Lambda)$. We remark that the set of natural numbers is naturally imbedded in $\Lambda$. Let $H$ be an even infinite number where the definition being even is the following: if $H$ is written as $[(H_\lambda, \lambda \in \Lambda)]$ then $\{\lambda \in \Lambda \mid H_\lambda \text{ is even}\} \in F$, where $[\ ]$ denotes the equivalence class with respect to the ultrafilter $F$. Let $\varepsilon$ be $\frac{1}{H}$, that is, if $\varepsilon$ is $[(\varepsilon_\lambda, \lambda \in \Lambda)]$ then $\varepsilon_\lambda$ is $\frac{1}{H_\lambda}$. Then we shall define a lattice space $\mathbf{L}$, a sublattice space $L$ and a space of functions $R(L)$:

$$\begin{aligned}
\mathbf{L} &:= \varepsilon^*\mathbf{Z} = \{\varepsilon z \mid z \in {}^*\mathbf{Z}\},\\
L &:= \{\varepsilon z \mid z \in {}^*\mathbf{Z},\ -\tfrac{H}{2} \le \varepsilon z < \tfrac{H}{2}\}\\
&= \{[(\varepsilon_\lambda z_\lambda), \lambda \in \Lambda] \mid \varepsilon_\lambda z_\lambda \in L_\lambda\}\ (\subset \mathbf{L}),\\
R(L) &:= \{\varphi \mid \varphi \text{ is an internal function from } L \text{ to } {}^*\mathbf{C}\}\\
&= \{[(\varphi_\lambda, \lambda \in \Lambda)] \mid \varphi_\lambda \text{ is a function from } L_\lambda \text{ to } \mathbf{C}\},
\end{aligned}$$

where $L_\lambda := \{\varepsilon_\lambda z_\lambda \mid z_\lambda \in \mathbf{Z},\ -\frac{H_\lambda}{2} \le \varepsilon_\lambda z_\lambda < \frac{H_\lambda}{2}\}$.

Kinoshita([3]) introduced an infinitesimal delta function $\delta(x)(\in R(L))$ and an infinitesimal Fourier transformation on $R(L)$. From now on, functions in $R(L)$ are extended to periodic functions on $\mathbf{L}$ with the period $H$ and we denote them by the same notations. For $\varphi(\in R(L))$, the infinitesimal Fourier transformation $F\varphi$, the inverse infinitesimal Fourier transformation $\overline{F}\varphi$, and the convolution of $\varphi$, $\psi(\in R(L))$ are defined as follows:

$$\delta(x) := \begin{cases} H & (x = 0), \\ 0 & (x \neq 0), \end{cases}$$

$$(F\varphi)(p) := \sum_{x \in L} \varepsilon \exp(-2\pi ipx)\, \varphi(x), \quad (\overline{F}\varphi)(p) := \sum_{x \in L} \varepsilon \exp(2\pi ipx)\, \varphi(x),$$

$$(\varphi * \psi)(x) := \sum_{y \in L} \varepsilon \varphi(x - y)\psi(y).$$

He obtained the following equalities as same as the usual Fourier analysis:

$$\delta = F1 = \overline{F}1, \quad F \text{ is unitary}, \ F^4 = 1, \quad \overline{F}F = F\overline{F} = 1,$$
$$\varphi * \delta = \delta * \varphi = \varphi, \quad \varphi * \psi = \psi * \varphi,$$
$$F(\varphi * \psi) = (F\varphi)(F\psi), \quad F(\varphi\psi) = (F\varphi) * (F\psi),$$
$$\overline{F}(\varphi * \psi) = (\overline{F}\varphi)(\overline{F}\psi), \quad \overline{F}(\varphi\psi) = (\overline{F}\varphi) * (\overline{F}\psi).$$

On the other hand, we obtain the following theorem from his result and an elementary calculation:

**Theorem 1.1.** *For an internal function with two variables $f : L \times L \to {}^*\mathbf{C}$ and $g(\in R(L))$,*

$$F_x\Big(\sum_{y \in L} \varepsilon f(x - y, y)g(y)\Big)(p) = \big\{F_y(F_u(f(u, y))(p)) * F_y(g(y))\big\}(p),$$

*where $F_x$, $F_y$, $F_u$ are Fourier transformations for $x$, $y$, $u$, and $*$ is the convolution for the variable paired with $y$ by the Fourier transformation.*

**Proof.** By the above Kinoshita's result, $F(\varphi\psi) = (F\varphi) * (F\psi)$. We use it and obtain the following:

$$\begin{aligned}
&F_x\Big(\sum_{y \in L} \varepsilon f(x - y, y)g(y)\Big)(p) \\
&= \sum_{x, y \in L} \varepsilon \exp(-2\pi ipx)\varepsilon f(x - y, y)g(y) \\
&= \sum_{y, u \in L} \varepsilon^2 \exp(-2\pi ip(y + u))f(u, y)g(y) \ (u := x - y) \\
&= \sum_{y \in L} \Big(\varepsilon \exp(-2\pi ipy)\Big(\sum_{u \in L} \varepsilon \exp(-2\pi pu)f(u, y)\Big)g(y)\Big) \\
&= F_y(F_u(f(u, y))(p) \cdot g(y))(p) \\
&= \{F_y(F_u(f(u, y))(p)) * F_y(g(y))\}(p).
\end{aligned}$$

□

To treat a *-unbounded functional $f$ in the nonstandard analysis, we need a second nonstandardization. Let $F_2 := F$ be a nonprincipal ultrafilter on an infinite set $\Lambda_2 := \Lambda$ as above. Denote the ultraproduct of a set $S$ with respect to $F_2$ by ${}^*S$ as above. Let $F_1$ be another nonprincipal ultrafilter on an infinite set $\Lambda_1$. Take the *-ultrafilter ${}^*F_1$ on ${}^*\Lambda_1$. For an internal set $S$ in the sense of *-nonstandardization, let ${}^\star S$ be the *-ultraproduct of $S$ with respect to ${}^*F_1$. Thus, we define a double ultraproduct ${}^\star({}^*\mathbf{R})$, ${}^\star({}^*\mathbf{Z})$, etc for the set $\mathbf{R}$, $\mathbf{Z}$, etc. It is shown easily that

$$^\star({}^*\mathbf{S}) = S^{\Lambda_1 \times \Lambda_2}/F_1^{F_2},$$

where $F_1^{F_2}$ denotes the ultrafilter on $\Lambda_1 \times \Lambda_2$ such that for any $A \subset \Lambda_1 \times \Lambda_2$, $A \in F_1^{F_2}$ if and only if

$$\{\, \lambda \in \Lambda_1 \mid \{\, \mu \in \Lambda_2 \mid (\lambda, \mu) \in A \,\} \in F_2 \,\} \in F_1.$$

We always work with this double nonstandardization. The natural imbedding ${}^\star S$ of an internal element $S$ which is not considered as a set in *-nonstandardization is often denoted simply by $S$.

**Definition 1.2.** Let $H(\in {}^*\mathbf{N})$, $H'(\in {}^\star({}^*\mathbf{N}))$ be even positive numbers such that $H'$ is larger than any element in ${}^*\mathbf{Z}$, and let $\varepsilon(\in {}^*\mathbf{R})$, $\varepsilon'(\in {}^\star({}^*\mathbf{R}))$ be infinitesimals satisfying $\varepsilon H = 1$, $\varepsilon' H' = 1$. We define as follows:

$$\mathbf{L} := \varepsilon^*\mathbf{Z} = \{\, \varepsilon z \mid z \in {}^*\mathbf{Z} \,\}, \quad \mathbf{L}' := \varepsilon'^\star({}^*\mathbf{Z}) = \{\, \varepsilon' z' \mid z' \in {}^\star({}^*\mathbf{Z}) \,\},$$
$$L := \{\, \varepsilon z \mid z \in {}^*\mathbf{Z},\ -\tfrac{H}{2} \le \varepsilon z < \tfrac{H}{2} \,\} \ (\subset \mathbf{L}),$$
$$L' := \{\, \varepsilon' z' \mid z' \in {}^\star({}^*\mathbf{Z}),\ -\tfrac{H'}{2} \le \varepsilon' z' < \tfrac{H'}{2} \,\} \ (\subset \mathbf{L}').$$

Here $L$ is an ultraproduct of lattices

$$L_\mu := \{\, \varepsilon_\mu z_\mu \mid z_\mu \in \mathbf{Z},\ -\tfrac{H_\mu}{2} \le \varepsilon_\mu z_\mu < \tfrac{H_\mu}{2} \,\} \ (\mu \in \Lambda_2)$$

in $\mathbf{R}$, and $L'$ is also an ultraproduct of lattices

$$L'_\lambda := \{\, \varepsilon'_\lambda z'_\lambda \mid z'_\lambda \in {}^*\mathbf{Z},\ -\tfrac{H'_\lambda}{2} \le \varepsilon'_\lambda z'_\lambda < \tfrac{H'_\lambda}{2} \,\} \ (\lambda \in \Lambda_1)$$

in ${}^*\mathbf{R}$ that is an ultraproduct of

$$L'_{\lambda\mu} := \{\, \varepsilon'_{\lambda\mu} z'_{\lambda\mu} \mid z'_{\lambda\mu} \in \mathbf{Z},\ -\tfrac{H'_{\lambda\mu}}{2} \le \varepsilon'_{\lambda\mu} z'_{\lambda\mu} < \tfrac{H'_{\lambda\mu}}{2} \,\} \ (\mu \in \Lambda_2).$$

We define a latticed space of functions $X$ as follows,

$$\begin{aligned} X := \{\, a \mid\ & a \text{ is an internal function with double meanings,} \\ & \qquad\qquad\qquad\qquad \text{from } {}^\star L \text{ to } L' \,\} \\ = \{\, [(a_\lambda), &\lambda \in \Lambda_1] \mid a_\lambda \text{ is an internal function from } L \text{ to } L'_\lambda \,\}, \end{aligned}$$

where $a_\lambda : L \to L'_\lambda$ is $a_\lambda = [(a_{\lambda\mu}), \mu \in \Lambda_2]$, $a_{\lambda\mu} : L_\mu \to L'_{\lambda\mu}$. We define three equivalence relations $\sim_H$, $\sim_{\star(H)}$ and $\sim_{H'}$ on $\mathbf{L}$, ${}^\star\mathbf{L}$ and $\mathbf{L}'$:

$$x \sim_H y \Longleftrightarrow x - y \in H^*\mathbf{Z}, \quad x \sim_{\star(H)} y \Longleftrightarrow x - y \in \star(H)^\star({}^*\mathbf{Z}),$$
$$x \sim_{H'} y \Longleftrightarrow x - y \in H'^\star({}^*\mathbf{Z}).$$

Then we identify $\mathbf{L}/{\sim_H}$, ${}^\star\mathbf{L}/{\sim_{\star(H)}}$ and $\mathbf{L}'/{\sim_{H'}}$ as $L$, ${}^\star L$ and $L'$. Since ${}^\star L$ is identified with $L$, the set ${}^\star\mathbf{L}/{\sim_{\star(H)}}$ is identified with $\mathbf{L}/{\sim_H}$. Furthermore we represent $X$ as the following internal set: $\{\, a \mid a \text{ is an internal function with double meanigs, from } {}^\star\mathbf{L}/{\sim_{\star(H)}} \text{ to } \mathbf{L}'/{\sim_{H'}} \,\}$. We use the same notation as a function from ${}^\star L$ to $L'$ to represent a function in the above internal set. We define the space $A$ of functionals as follows:

$$A := \{\, f \mid f \text{ is an internal function with double meanigs, from } X \text{ to } {}^\star({}^*\mathbf{C}) \,\}.$$

Then $f$ is written as $f = [(f_\lambda), \lambda \in \Lambda_1]$, $f_\lambda$ is an internal function from the set $\{\, a_\lambda \mid a_\lambda \text{ is an internal function from } L \text{ to } L'_\lambda \,\}$ to ${}^*\mathbf{C}$, and $f_\lambda$ is written as $f_\lambda = [(f_{\lambda\mu}), \mu \in \Lambda_2]$, $f_{\lambda\mu} : \{a_{\lambda\mu} : L_\mu \to L'_{\lambda\mu}\} \to \mathbf{C}$.

We define an infinitesimal delta function $\delta(a)(\in A)$, an infinitesimal Fourier transformation of $f(\in A)$, an inverse infinitesimal Fourier transformation of $f$ and a convolution of $f$, $g(\in A)$, by the following:

**Definition 1.3.**

$$\delta(a) := \begin{cases} (H')^{({}^\star H)^2} & (a = 0), \\ 0 & (a \neq 0), \end{cases}$$
$$\varepsilon_0 := (H')^{-({}^\star H)^2} \in {}^\star({}^*\mathbf{R}),$$
$$(Ff)(b) := \sum_{a \in X} \varepsilon_0 \exp\left(-2\pi i \sum_{k \in L} a(k)b(k)\right) f(a),$$
$$(\overline{F}f)(b) := \sum_{a \in X} \varepsilon_0 \exp\left(2\pi i \sum_{k \in L} a(k)b(k)\right) f(a),$$
$$(f * g)(a) := \sum_{a' \in X} \varepsilon_0 f(a - a')g(a').$$

We define an inner product on $A$ : $(f, g) := \sum_{b \in X} \varepsilon_0 \overline{f(b)} g(b)$, where $\overline{f(b)}$ is the complex conjugate of $f(b)$. Then we obtain the following theorem:

**Theorem 1.4.**
(1) $\delta = F1 = \overline{F}1$, (2) $F$ *is unitary*, $F^4 = 1$, $\overline{F}F = F\overline{F} = 1$,

(3) $f * \delta = \delta * f = f$, (4) $f * g = g * f$,
(5) $F(f * g) = (Ff)(Fg)$, (6) $\overline{F}(f * g) = (\overline{F}f)(\overline{F}g)$,
(7) $F(fg) = (Ff) * (Fg)$, (8) $\overline{F}(fg) = (\overline{F}f) * (\overline{F}g)$.

We define two types of infinitesimal divided differences. Let $f$ and $a$ be elements of $A$ and $X$ respectively and let $b(\in X)$ be an internal function whose image is in ${}^\star({}^*\mathbf{Z}) \cap L'$. We remark that $\varepsilon' b$ is an element of $X$.

**Definition 1.5.** $(D_{+,b}f)(a) := \frac{f(a+\varepsilon' b)-f(a)}{\varepsilon'}$, $(D_{-,b}f)(a) := \frac{f(a)-f(a-\varepsilon' b)}{\varepsilon'}$.

For $k, l \in {}^*L$, let $\alpha_k$, $b_l(\in X)$ be defined by $\alpha_k(a) := a(k)$ and $b_l = 1$ or $0$ if $k = l$ or otherwise. Then $((\alpha_k(-i\varepsilon' D_{+,b_l}) - (-i\varepsilon' D_{+,b_l})\alpha_k)f)(a) = i\varepsilon'\delta_{k,l}f(a+\varepsilon' b_l)$. Let $\lambda_b(a) := \frac{\exp(2\pi i\varepsilon' ab)-1}{\varepsilon'}$, $\overline{\lambda}_b(a) := \frac{\exp(-2\pi i\varepsilon' ab)-1}{\varepsilon'}$. Then we obtain the following theorem corresponding to Kinoshita's result for the relationship between the infinitesimal Fourier transformation and the infinitesimal divided differences:

**Theorem 1.6.**
(1) $(F(D_{+,b}f))(a) = \lambda_b(a)(Ff)(a)$, (2) $(F(D_{-,b}f))(a) = -\overline{\lambda}_b(a)(Ff)(a)$,
(3) $(F(\lambda_b f))(a) = -(D_{-,b}(Ff))(a)$, (4) $(F(\overline{\lambda_b} f))(a) = (D_{+,b}(Ff))(a)$,
(5) $(D_{+,b}(\overline{F}f))(a) = (\overline{F}(\lambda_b f))(a)$, (6) $(D_{-,b}(\overline{F}f))(a) = -(\overline{F}(\overline{\lambda_b} f))(a)$,
(7) $\lambda_b(a) = 2\pi i(\frac{\sin(\pi\varepsilon' ab)}{\pi\varepsilon'})\exp(\pi i\varepsilon' ab)$.

Replacing the definitions of $L'$, $\delta$, $\varepsilon_0$, $F$, $\overline{F}$ in Definition 1.2 and Definition 1.3 by the following, we shall define another type of infinitesimal Fourier transformation. The different point is only the definition of an inner product of the space of functions $X$. In Definition 1.3, the inner product of $a, b$ $(\in X)$ is $\sum_{k\in L} a(k)b(k)$, and in the following definition, it is ${}^\star\varepsilon \sum_{k\in L} a(k)b(k)$.

**Definition 1.7.**

$$L' := \{\, \varepsilon' z' \mid z' \in {}^\star({}^*\mathbf{Z}),\ -{}^\star H\tfrac{H'}{2} \le \varepsilon' z' < {}^\star H \tfrac{H'}{2} \,\},$$

$$\delta(a) := \begin{cases} ({}^\star H)^{\frac{({}^\star H)^2}{2}} H'^{({}^\star H)^2} & (a = 0), \\ 0 & (a \neq 0), \end{cases}$$

$$\varepsilon_0 := ({}^\star H)^{-\frac{({}^\star H)^2}{2}} H'^{-({}^\star H)^2}$$

$$(Ff)(b) := \sum_{a\in X} \varepsilon_0 \exp\Big(-2\pi i {}^\star\varepsilon \sum_{k\in L} a(k)b(k)\Big) f(a),$$

$$(\overline{F}f)(b) := \sum_{a\in X} \varepsilon_0 \exp\Big(2\pi i {}^\star\varepsilon \sum_{k\in L} a(k)b(k)\Big) f(a).$$

In this case, we obtain the same theorems as Theorem 1.4 and Theorem 1.6, and the following theorem corresponding to Theorem 1.1:

**Theorem 1.8.** *For an internal function with two variables* $f : X \times X \to {}^{\star}({}^{*}\mathbf{C})$ *and* $g(\in A)$,

$$F_a\Big(\sum_{b\in X} \varepsilon_0 f(a-b,b)g(b)\Big)(d) = \big\{F_b(F_c(f(c,b))(d)) * F_b(g(b))\big\}(d),$$

*where* $F_a$, $F_b$, $F_c$ *are Fourier transformations for* $a$, $b$, $c$, *and* $*$ *is the convolution for the variable pairing with* $b$ *by the Fourier transformation.*

## 2. Proofs of Theorems

Proof of Theorem 1.4.

(1) $(F1)(0) = \sum_{a\in X} \varepsilon_0 = \varepsilon_0 (H'^2)^{({}^*H)^2} = (H')^{({}^*H)^2}$. If $b \neq 0$, then

$$\begin{aligned}(F1)(b) &= \textstyle\sum_{a\in X} \varepsilon_0 \exp\left(-2\pi i \sum_{k\in L} a(k)b(k)\right)\\ &= \varepsilon_0 \textstyle\prod_{k\in L} \sum_{a(k)\in L'} \exp(-2\pi i a(k)b(k))\\ &= \textstyle\prod_{k\in L, b(k)\neq 0} \varepsilon_0 \frac{\exp(-2\pi i \varepsilon'(-\frac{H'^2}{2})b(k))(1-\exp(-2\pi i \varepsilon' H'^2 b(k)))}{1-\exp(-2\pi i \varepsilon' b(k))}\\ &\quad \cdot \textstyle\prod_{k\in L, b(k)=0} \sum_{a(k)\in L'} \exp(-2\pi i a(k)b(k)) = 0.\end{aligned}$$

Hence $F1 = \delta$. The same argument implies that $\overline{F}1 = \delta$.

(2)
$$\begin{aligned}(Ff, Fg) &= \textstyle\sum_{b\in X} \varepsilon_0 \overline{(Ff)(b)}(Fg)(b)\\ &= \textstyle\sum_{a\in X} \sum_{c\in X} \varepsilon_0^2 \overline{f(a)} g(c)\delta(c-a)\\ &= \textstyle\sum_{a\in X} \varepsilon_0 \overline{f(a)} g(a) = (f,g).\end{aligned}$$

Hence $F$ is unitary. Since $(F^2 f)(c) = (F(Ff))(c) = f(-c)$, $F^4 = 1$. Thus the eigenvalues of $F$ are $1$, $-1$, $-i$, $i$. Furthermore, $(\overline{F}(Ff))(c) = \sum_{a\in X} \varepsilon_0 \delta(a-c) f(a) = f(c)$. The same argument implies $(F(\overline{F}f))(c) = f(c)$.

(3)
$$\begin{aligned}(f*\delta)(a) &= \textstyle\sum_{b\in X} \varepsilon_0 f(a-b)\delta(b) = f(a),\\ (\delta * f)(a) &= \textstyle\sum_{b\in X} \varepsilon_0 \delta(a-b) f(b) = f(a).\end{aligned}$$

(4)
$$\begin{aligned}(f*g)(a) &= \textstyle\sum_{b\in X} \varepsilon_0 f(a-b)g(b)\\ &= \textstyle\sum_{(a-b)\in X} \varepsilon_0 f(a-b) g(a-(a-b))\\ &= (g*f)(a).\end{aligned}$$

(5) $(F(f*g))(c) = \sum_{a\in X} \varepsilon_0 \exp\left(-2\pi i \sum_{k\in L} c(k)a(k)\right)$
$\sum_{a\in X} \varepsilon_0 f(a-b)g(b)$
$= \sum_{a\in X} \varepsilon_0 \exp\left(-2\pi i \sum_{k\in L} c(k)(b(k)+d(k))\right)$
$\sum_{b\in X} \varepsilon_0 f(a-b)g(b), \quad \text{where } d(k) := a(k)-b(k),$
$= \sum_{b\in X} \varepsilon_0 \exp\left(-2\pi i \sum_{k\in L} c(k)b(k)\right) g(b)$
$\sum_{d\in X} \varepsilon_0 \exp\left(-2\pi i \sum_{k\in L} c(k)d(k)\right) f(d)$
$= (Fg)(c)(Ff)(c) = (Ff)(c)(Fg)(c).$

(6) Similarly, $\overline{F}(f'*g') = (\overline{F}f')(\overline{F}g')$.
(7) The above (6) implies $f'*g' = F((\overline{F}f')(\overline{F}g'))$. We put $f' = Ff$, $g' = Fg$. Then we obtain $(Ff)*(Fg) = F(fg)$.
(8) Similarly, $(\overline{F}f)*(\overline{F}g) = \overline{F}(fg)$.

Proof of Theorem 1.6.

$$(1) \quad (F(D_{+,b}f))(a) = \sum_{c\in X} \varepsilon_0 \exp(-2\pi iac)\tfrac{1}{\varepsilon'}(f(c+\varepsilon' b)-f(c)) = \tfrac{1}{\varepsilon'}(\exp(2\pi i\varepsilon' ab)-1)(Ff)(a) = \lambda_b(a)Ff(a),$$

$$(2) \quad (F(D_{-,b}f))(a) = \sum_{c\in X} \varepsilon_0 \exp(-2\pi iac)\tfrac{1}{\varepsilon'}(f(c)-f(c-\varepsilon b')) = \sum_{c\in X} \varepsilon_0 (\tfrac{1}{\varepsilon'}(\exp(-2\pi iac)f(c) - \exp(-2\pi i\varepsilon' ab)\exp(-2\pi ia(c-\varepsilon' b))f(c-\varepsilon' b))) = \tfrac{1}{\varepsilon'}(1-\exp(-2\pi i\varepsilon' ab))(Ff)(a) = -\overline{\lambda}_b(a)Ff(a),$$

$$(3) \quad (F(\lambda_b f))(a) = \sum_{c\in X} \varepsilon_0 \exp(-2\pi iac)(\lambda_b f)(c) = \sum_{c\in X} \varepsilon_0 \frac{\exp(-2\pi i(a-b\varepsilon')c)-\exp(-2\pi iac)}{\varepsilon'} f(c) = -D_{-,b}(Ff)(a),$$

$$(4) \quad (F(\overline{\lambda}_b f))(a) = \sum_{c\in X} \varepsilon_0 \exp(-2\pi iac)(\overline{\lambda}_b f)(c) = \sum_{c\in X} \varepsilon_0 \frac{\exp(-2\pi i(a+b\varepsilon')c)-\exp(-2\pi iac)}{\varepsilon'} f(c) = D_{+,b}(Ff)(a).$$

(1), (2) imply (5), (6).

## 3. Examples

We calculate two examples of the infinitesimal Fourier transformation for the space $A$ of functionals. Let $\star \circ * : \mathbf{R} \to {}^\star({}^*\mathbf{R})$ be the natural elementary imbedding and let $\mathbf{st}(c)$ for $c \in {}^\star({}^*\mathbf{R})$ be the standard part of $c$ with respect to the natural elementary imbedding $\star \circ *$. The first is for

$\exp\left(i\pi^\star\varepsilon\sum_{k\in L}a^2(k)\right)$ and the second is for $\exp\left(-\pi^\star\varepsilon\sum_{k\in L}a^2(k)\right)$. We denote the two functionals by $f(a), g(a)$. If there is an $L^2$-function $\alpha(t)$ on $\mathbf{R}$ for $a(k)$ so that $a(k) = \star((^*\alpha)(k))$, then $\mathbf{st}(f(a)) = \exp\left(i\pi\int_{-\infty}^{\infty}\alpha^2(t)dt\right)$, and $\mathbf{st}(g(a)) = \exp\left(-\pi\int_{-\infty}^{\infty}\alpha^2(t)dt\right)$. Then we obtain the following results:

**Example 1.** $(Ff)(b) = C_1\overline{f(b)}$, where $C_1 = \sum_{a\in X}\varepsilon_0\exp\left(i\pi^\star\varepsilon\sum_{k\in L}a^2(k)\right)$, that is just a standard number $(-1)^{\frac{H}{2}}$.

**Example 2.** $(Fg)(b) = C_2(b)g(b)$, where $C_2(b) = \sum_{a\in X}\varepsilon_0\exp\left(-\pi^\star\varepsilon\sum_{k\in L}(a(k)+ib(k))^2\right)$. If $b$ is a finite valued function then it satisfies that $\mathbf{st}(C_2(b)) = 1$.

For the calculation, we use Kinishita's infinitesimal Fourier transformation $\varphi_1(x) = \exp(i\pi x^2)$, $\varphi_2(x) = \exp(-\pi x^2)$ for the space $R(L)$ of functions. We obtain:

$$(F\varphi_1)(p) = \exp\left(i\tfrac{\pi}{4}\right)\overline{\varphi_1(p)}, \quad (F\varphi_2)(p) = c(p)\varphi_2(p),$$

where $\mathrm{st}\,(c(p)) = \int_{-\infty}^{\infty}\exp(-\pi t^2)dt$, in the case of finite $p$.

We denote the following:

$$R(\mathbf{L}) := \left\{\varphi' \mid \varphi' \text{ is an internal function from } \mathbf{L} \text{ to } {}^*\mathbf{C}\right\},$$
$$R_H(\mathbf{L}) := \left\{\varphi' \in R(\mathbf{L}) \mid \varphi'(x+H) = \varphi'(x)\right\}.$$

Let $e$ be a mapping from $R(L)$ to $R_H(\mathbf{L})$ defined by $(e(\varphi))(x) = \varphi(\hat{x})$, where $\hat{x}$ is an element of $L$ satisfying $x \sim_H \hat{x}$. Now $\exp(i\pi x^2)$ is an element of $R_H(\mathbf{L})$, in fact, putting $x \in \mathbf{L}$:

$$\begin{aligned}\exp(i\pi(x+H)^2) &= \exp(i\pi(x^2+2xH+H^2))\\ &= \exp(i\pi x^2)\exp(2\pi i\varepsilon zH)\exp(i\pi H^2) \quad (x = \varepsilon z(z\in {}^*\mathbf{Z}))\\ &= \exp(i\pi x^2), \text{ as } \varepsilon H = 1 \text{ and } H \text{ is even.}\end{aligned}$$

Hence $e(\exp(i\pi x^2)) = \exp(i\pi x^2)$, that is, $e(\varphi_1(x)) = \varphi_1(x)$. We do the infinitesimal Fourier transformation of $\varphi_1(x)$,

$$\begin{aligned}(F\varphi_1)(p) &= \textstyle\sum_{x\in L}\varepsilon\exp(i\pi x^2)\exp(-2\pi ixp)\\ &= \textstyle\sum_{x\in L}\varepsilon\exp(i\pi(x-p)^2)\exp(-i\pi p^2)\\ &= \textstyle\sum_{x-y\in L}\varepsilon\exp(i\pi(x-p)^2)\exp(-i\pi p^2)\\ &= \left(\textstyle\sum_{x\in L}\varepsilon\exp(i\pi x^2)\right)\overline{\varphi_1(p)}.\end{aligned}$$

By Gauss sums (cf. [6], p. 409): $\sum_{n=0}^{N-1}\exp\left(\frac{2\pi i}{N}n^2\right) = \frac{1+(-i)^N}{1-i}\sqrt{N}$, when $N = 4m$ $(m \in \mathbf{N})$, $\sum_{n=0}^{N-1}\exp\left(\frac{2\pi i}{N}n^2\right) = (1+i)\sqrt{N}$. Using it, we obtain the following:

$$\sum_{x\in L}\varepsilon\exp(i\pi x^2) = \sum_{z=-\frac{H^2}{2}}^{\frac{H^2}{2}-1}\varepsilon\exp(i\pi(\varepsilon z)^2) = exp\left(i\tfrac{\pi}{4}\right).$$

Nextly we calculate the infinitesimal Fourier transformation of $e(\varphi_2(x))$.

$$(F(e(\varphi_2)))(p) = \left(\sum_{x\in L}\varepsilon\exp(-\pi(x+ip)^2)\right)\varphi_2(p).$$

We assume that $p(\in L)$ is finite. Since $\exp(-\pi(x+ip)^2)$ is proved to be an $S$-integrable function directly, the term $\sum_{x\in L}\varepsilon\exp(-\pi(x+ip)^2)$ satisfies the following (cf. [4], [1]):

$$\text{st}\left(\sum_{x\in L}\varepsilon\exp(-\pi(x+ip)^2)\right) = \int_{-\infty}^{\infty}\exp(-\pi(t+i\,{}^{\circ}p)^2)dt,$$

where $p \in L$, ${}^{\circ}p = \text{st}(p)(\in \mathbf{R})$. We remark that the integral $\int_{-\infty}^{\infty}\exp(-\pi(t+i\,{}^{\circ}p)^2)dt$ is equal to $\int_{-\infty}^{\infty}\exp(-\pi t^2)dt$.

We define an equivalent relation $\sim_{{}^{\star}HH'}$ in $\mathbf{L}'$ by $x \sim_{{}^{\star}HH'} y \Leftrightarrow x - y \in {}^{\star}HH'{}^{\star}({}^{*}\mathbf{Z})$. We identify $\mathbf{L}'/\sim_{{}^{\star}HH'}$ with $L'$. Let

$$X_{H,{}^{\star}HH'} := \left\{ a' \mid a' \text{ is an internal function with double meanings, from } {}^{\star}\mathbf{L}/\sim_{\star(H)} \text{ to } \mathbf{L}'/\sim_{{}^{\star}HH'} \right\},$$

and let $\mathbf{e}$ is a mapping from $X$ to $X_{H,{}^{\star}HH'}$, defined by $(\mathbf{e}(a))([k]) = [a(\hat{k})]$, where $[\ ]$ in left hand side represents the equivalent class for the equivalent relation $\sim_{{}^{\star}H}$ in ${}^{\star}\mathbf{L}$, $\hat{k}$ is a representative in ${}^{\star}L$ satisfying $k \sim_{{}^{\star}H} \hat{k}$, and $[\ ]$ in right hand side represents the equivalent class for the equivalent relation $\sim_{{}^{\star}HH'}$ in $\mathbf{L}'$. We consider an example $f(a) = \exp\left(i\pi^{\star}\varepsilon\sum_{k\in L}a^2(k)\right)$ in the space $A$ of functionals, for $a \in X$.

$$\exp\left(i\pi^{\star}\varepsilon\sum_{k\in L}(a(k) + {}^{\star}HH')^2\right) = \exp\left(i\pi^{\star}\varepsilon\sum_{k\in L}a^2(k)\right).$$

Hence if $\mathbf{e}^{\sharp}(f)(a)$ is defined by $f(\mathbf{e}(a))$, then $\mathbf{e}^{\sharp}(f) = f$. We do the infinitesimal Fourier transformation of $f(a)$.

$$\begin{aligned}(Ff)(b) &= \left(F\left(\exp\left(i\pi^{\star}\varepsilon\sum_{k\in L}a^2(k)\right)\right)\right)(b)\\ &= \left(\sum_{a\in X}\varepsilon_0\exp\left(i\pi^{\star}\varepsilon\sum_{k\in L}(a(k)-b(k))^2\right)\right)\exp\left(-i\pi^{\star}\varepsilon\sum_{k\in L}b^2(k)\right).\end{aligned}$$

The term $\sum_{a\in X}\varepsilon_0\exp\left(i\pi^{\star}\varepsilon\sum_{k\in L}(a(k)-b(k))^2\right)$ is represented as

$$\begin{aligned}&\sum_{a_{\lambda\mu}\in X_{\lambda\mu}}(\varepsilon_0)_{\lambda\mu}\exp(i\pi\varepsilon_\mu\sum_{k_\mu\in L_\mu}((a(k))_{\lambda\mu} - (b(k))_{\lambda\mu})^2)\\ &= \prod_{k_\mu\in L_\mu}(\sum_{((a(k))_{\lambda\mu}\in L'_{\lambda\mu}}\sqrt{\varepsilon_\mu}\varepsilon'_{\lambda\mu}\exp(i\pi\varepsilon_\mu((a(k))_{\lambda\mu} - (b(k))_{\lambda\mu})^2))\end{aligned}$$

for $\lambda\mu$ component, as $\varepsilon_0 = ({}^\star\varepsilon)^{\frac{({}^\star H)^2}{2}}\varepsilon'^{({}^\star H)^2}$. In the above, $X_{\lambda\mu}$ is the set of functions from $L_\mu$ to $L'_{\lambda\mu}$. If we put $(a(k))_{\lambda\mu} = \varepsilon'_{\lambda\mu}z^a_{\lambda\mu}$ ($z^a_{\lambda\mu} \in \mathbf{Z}$) and $(b(k))_{\lambda\mu} = \varepsilon'_{\lambda\mu}z^b_{\lambda\mu}$ ($z^b_{\lambda\mu} \in \mathbf{Z}$), we remark that $z^a_{\lambda\mu}$ and $z^b_{\lambda\mu}$ depend on $k_\mu$. The above is

$$\begin{aligned}&\textstyle\prod_{k_\mu\in L_\mu}(\sum_{\varepsilon'_{\lambda\mu}z^a_{\lambda\mu}\in L'_{\lambda\mu}} \sqrt{\varepsilon_\mu}\varepsilon'_{\lambda\mu}\exp(i\pi\varepsilon_\mu(\varepsilon'_{\lambda\mu}z^a_{\lambda\mu} - \varepsilon'_{\lambda\mu}z^b_{\lambda\mu})^2))\\ &\textstyle= \prod_{k_\mu\in L_\mu}(\sum_{\varepsilon'_{\lambda\mu}z^a_{\lambda\mu}\in L'_{\lambda\mu}} \sqrt{\varepsilon_\mu}\varepsilon'_{\lambda\mu}\exp(i\pi(\sqrt{\varepsilon_\mu}\varepsilon'_{\lambda\mu}z^a_{\lambda\mu} - \sqrt{\varepsilon_\mu}\varepsilon'_{\lambda\mu}z^b_{\lambda\mu})^2)).\end{aligned}$$

Since $\exp(i\pi\varepsilon x^2)$ is a periodic function with period ${}^\star HH'$ on $\mathbf{L}'$, we obtain

$$\begin{aligned}&\textstyle\sum_{\varepsilon'_{\lambda\mu}z^a_{\lambda\mu}\in L'_{\lambda\mu}} \sqrt{\varepsilon_\mu}\varepsilon'_{\lambda\mu}\exp\left(i\pi(\sqrt{\varepsilon_\mu}\varepsilon'_{\lambda\mu}z^a_{\lambda\mu} - \sqrt{\varepsilon_\mu}\varepsilon'_{\lambda\mu}z^b_{\lambda\mu})^2\right)\\ &\textstyle= \sum_{\varepsilon'_{\lambda\mu}z^a_{\lambda\mu}\in L'_{\lambda\mu}} \sqrt{\varepsilon_\mu}\varepsilon'_{\lambda\mu}\exp\left(i\pi(\sqrt{\varepsilon_\mu}\varepsilon'_{\lambda\mu}z^a_{\lambda\mu})^2\right).\end{aligned}$$

Hence

$$\begin{aligned}&\textstyle\sum_{a\in X}\varepsilon_0\exp\left(i\pi{}^\star\varepsilon\sum_{k\in L}(a(k)-b(k))^2\right)\\ &\qquad\textstyle= \sum_{a\in X}\varepsilon_0\exp\left(i\pi{}^\star\varepsilon\sum_{k\in L}(a(k))^2\right) = C_1.\end{aligned}$$

We calculate $C_1$ as follows :

$$\begin{aligned}&\textstyle\sum_{a\in X}\varepsilon_0\exp\left(i\pi{}^\star\varepsilon\sum_{k\in L}(a(k))^2\right)\\ &\textstyle= \varepsilon_0\prod_{x\in L}\sum_{a(k)\in L'}\exp\left(i\pi{}^\star\varepsilon(a(k))^2\right)\\ &\textstyle= \prod_{x\in L}(\sqrt{{}^\star H}H')^{-1}\frac{1}{2}\sum_{z'=0}^{2{}^\star HH'^2-1}\exp\left(2\pi i\frac{z^2}{2{}^\star HH'^2}\right)\\ &\quad\text{(Gauss sums (cf. [6], p. 409)}\\ &\quad\textstyle\sum_{n=0}^{N-1}\exp\left(\frac{2\pi i}{N}n^2\right) = (1+i)\sqrt{N}(N = 4m\ (m\in\mathbf{N}))\\ &\textstyle= \prod_{x\in L}(\sqrt{{}^\star H}H')^{-1}\frac{1}{2}(1+i)\sqrt{2{}^\star HH'^2} = \prod_{x\in L}\frac{1+i}{\sqrt{2}}\\ &= \left(\frac{1+i}{\sqrt{2}}\right)^{H^2} = (-1)^{\frac{H}{2}}.\end{aligned}$$

We do the infinitesimal Fourier transformation of $g(a)$.

$$\begin{aligned}(Fg)(b) = &\textstyle\left(\sum_{a\in X}\varepsilon_0\exp\left(-\pi{}^\star\varepsilon\sum_{k\in L}(a(k)+ib(k))^2\right)\right)\\ &\textstyle\exp\left(-\pi{}^\star\varepsilon\sum_{k\in L}b^2(k)\right).\end{aligned}$$

We consider the term $\sum_{a\in X}\varepsilon_0\exp\left(-{}^\star\varepsilon\pi\sum_{k\in L}(a(k)+ib(k))^2\right)$. We write $(a(k))_{\lambda\mu} = \varepsilon'_{\lambda\mu}z^a_{\lambda\mu}(k_\mu)$ ($z^a_{\lambda\mu}(k_\mu) \in \mathbf{Z}$), $(b(k))_{\lambda\mu} = \varepsilon'_{\lambda\mu}z^b_{\lambda\mu}(k_\mu)$ ($z^b_{\lambda\mu}(k_\mu) \in \mathbf{Z}$). From now on, we denote $z^a_{\lambda\mu}(k_\mu)$, $z^b_{\lambda\mu}(k_\mu)$ by $z^a_{\lambda\mu}$, $z^b_{\lambda\mu}$ for simplicity. Then the $\lambda\mu$-component of $\sum_{a\in X}\varepsilon_0\exp\left(-\pi{}^\star\varepsilon\sum_{k\in L}(a(k)+ib(k))^2\right)$

is equal to

$$\sum_{a_{\lambda\mu}\in X_{\lambda\mu}}(\varepsilon_0)_{\lambda\mu}\exp\left(-\pi\varepsilon_\mu\sum_{k_\mu\in L_\mu}((a(k))_{\lambda\mu}+i(b(k))_{\lambda\mu})^2\right)$$
$$=\prod_{k_\mu\in L_\mu}\left(\sum_{\varepsilon'_{\lambda\mu}z^a_{\lambda\mu}\in L'_{\lambda\mu}}\sqrt{\varepsilon_\mu}\varepsilon'{}_{\lambda\mu}\exp\left(-\pi(\sqrt{\varepsilon_\mu}\varepsilon'_{\lambda\mu}z^a_{\lambda\mu}+i\sqrt{\varepsilon_\mu}\varepsilon'_{\lambda\mu}z^b_{\lambda\mu})^2\right)\right).$$

We assume that $b$ ($\in X$) is finitely valued, that is, $\exists b_0\in\mathbf{R}$ s.t. $k\in L\Rightarrow |b(k)|\leq\star(*(b_0))$. The $\lambda\mu$-component of

$$\frac{\sum_{a\in X}\varepsilon_0\exp\left(-\pi^\star\varepsilon\sum_{k\in L}(a(k)+ib(k))^2\right)}{\star(*(\int_{-\infty}^{\infty}\exp(-\pi x^2)dx)^{H^2})}$$

is equal to

$$\prod_{k_\mu\in L_\mu}\frac{\sum_{\varepsilon'_{\lambda\mu}z^a_{\lambda\mu}\in L'_{\lambda\mu}}\sqrt{\varepsilon_\mu}\varepsilon'{}_{\lambda\mu}\exp(-\pi(\sqrt{\varepsilon_\mu}\varepsilon'_{\lambda\mu}z^a_{\lambda\mu}+i\sqrt{\varepsilon_\mu}\varepsilon'_{\lambda\mu}z^b_{\lambda\mu})^2)}{\int_{-\infty}^{\infty}\exp(-\pi x^2)dx}.$$

We write

$$B_{\lambda\mu}(k_\mu):=\sum_{\varepsilon'_{\lambda\mu}z^a_{\lambda\mu}\in L'_{\lambda\mu}}\sqrt{\varepsilon_\mu}\varepsilon'{}_{\lambda\mu}\exp\left(-\pi(\sqrt{\varepsilon_\mu}\varepsilon'_{\lambda\mu}z^a_{\lambda\mu}+i\sqrt{\varepsilon_\mu}\varepsilon'_{\lambda\mu}z^b_{\lambda\mu})^2\right)$$
$$-\int_{-\infty}^{\infty}\exp(-\pi(x+i\sqrt{\varepsilon_\mu}\varepsilon'_{\lambda\mu}z^b_{\lambda\mu})^2)dx.$$

It is equal to

$$-2\int_{-\infty}^{-\sqrt{\varepsilon_\mu}\frac{H'_{\lambda\mu}}{2}}\exp(-\pi(x+ib_{\lambda\mu})^2)dx-\int_{-\sqrt{\varepsilon_\mu}\frac{H'_{\lambda\mu}}{2}}^{\sqrt{\varepsilon_\mu}\frac{H'_{\lambda\mu}}{2}}\exp(-\pi(x+ib_{\lambda\mu})^2)dx$$
$$+\sum_{\varepsilon'_{\lambda\mu}z^a_{\lambda\mu}\in L'_{\lambda\mu}}\sqrt{\varepsilon_\mu}\varepsilon'{}_{\lambda\mu}\exp\left(-\pi(\sqrt{\varepsilon_\mu}\varepsilon'_{\lambda\mu}z^a_{\lambda\mu}+i\sqrt{\varepsilon_\mu}\varepsilon'_{\lambda\mu}z^b_{\lambda\mu})^2\right).\qquad\cdots(*_1)$$

Then the above is equal to

$$\prod_{k_\mu\in L_\mu}\frac{\sum_{\varepsilon'_{\lambda\mu}z^a_{\lambda\mu}\in L'_{\lambda\mu}}\sqrt{\varepsilon_\mu}\varepsilon'{}_{\lambda\mu}\exp\left(-\pi(\sqrt{\varepsilon_\mu}\varepsilon'_{\lambda\mu}z^a_{\lambda\mu}+i\sqrt{\varepsilon_\mu}\varepsilon'_{\lambda\mu}z^b_{\lambda\mu})^2\right)}{\int_{-\infty}^{\infty}\exp(-\pi x^2)dx}$$
$$\left(\left(1+\frac{1}{\int_{-\infty}^{\infty}\exp(-\pi x^2)dx\cdot\frac{1}{B_{\lambda\mu}(k_\mu)}}\right)^{\frac{1}{B_{\lambda\mu}(k_\mu)}}\right)^{B_{\lambda\mu}(k_\mu)H^2_\mu}.\qquad\cdots(*_2)$$

We show that $[(B_{\lambda\mu}(k_\mu))]$ is infinitesimal in $^\star(^*\mathbf{C})$ with respect to $\mathbf{C}$. It implies that $[(\frac{1}{B_{\lambda\mu}(k_\mu)})]$ is infinite in $^\star(^*\mathbf{C})$. For a sequence $a_n$, we remark that $\lim_{n\to\infty}a_n=a\iff\forall N$ : infinite with respect to $\mathbf{C}$ ($\in{}^\star(^*\mathbf{C})$), $\mathbf{st}(^\star(^*a_N))=a$. Hence

$$\mathbf{st}\left(\left[\left(1+\frac{1}{\int_{-\infty}^{\infty}\exp(-\pi x^2)dx\cdot\frac{1}{B_{\lambda\mu}(k_\mu)}}\right)^{\frac{1}{B_{\lambda\mu}(k_\mu)}}\right]\right)=\exp\left(-\int_{-\infty}^{\infty}\exp(-\pi x^2)dx\right)$$

and $\mathbf{st}([(*_1))])=1$.

Since $b_{\lambda\mu}$ is finite and $\left[\left(\sqrt{\varepsilon_\mu}\frac{H'_{\lambda\mu}}{2}\right)\right]$ is infinitesimal in ${}^\star({}^*\mathbf{R})$ with respect to $\mathbf{R}$, the first term of $(*_1)$ is infinitesimal in ${}^\star({}^*\mathbf{C})$ with respect to $\mathbf{C}$. In order to show that $[(B_{\lambda\mu}(k_\mu))]$ is infinitesimal in ${}^\star({}^*\mathbf{C})$, we consider the second and third terms in $(*_1)$, and we prove that it represents an infinitesimal number. First we calculate

$$\exp(-\pi(x+i\sqrt{\varepsilon_\mu}\varepsilon'_{\lambda\mu}z^b_{\lambda\mu})^2)-\exp(-\pi(\sqrt{\varepsilon_\mu}\varepsilon'_{\lambda\mu}z^a_{\lambda\mu}+i\sqrt{\varepsilon_\mu}\varepsilon'_{\lambda\mu}z^b_{\lambda\mu})^2).$$

For simplicity we write $\sqrt{\varepsilon_\mu}\varepsilon'_{\lambda\mu}z^a_{\lambda\mu}$, $\sqrt{\varepsilon_\mu}\varepsilon'_{\lambda\mu}z^b_{\lambda\mu}$ as $a_{\lambda\mu}$, $b_{\lambda\mu}$. It is

$$\begin{aligned}&\exp(-\pi(x+ib_{\lambda\mu})^2)-\exp(-\pi(a_{\lambda\mu}+ib_{\lambda\mu})^2)\\&=\{\exp(-\pi(x^2-b^2_{\lambda\mu}))\cos(2\pi b_{\lambda\mu}x)-\exp(-\pi(a^2_{\lambda\mu}-b^2_{\lambda\mu})\}\cos(2\pi b_{\lambda\mu}a_{\lambda\mu})\}\\&-i\{\exp(-\pi(x^2-b^2_{\lambda\mu}))\sin(2\pi b_{\lambda\mu}x)-\exp(-\pi(a^2_{\lambda\mu}-b^2_{\lambda\mu}))\sin(2\pi b_{\lambda\mu}a_{\lambda\mu})\}.\\&\quad\cdots(*_3)\end{aligned}$$

We consider the first term of $(*_3)$. Then

$$\exp(-\pi(x^2-b^2_{\lambda\mu}))\cos(2\pi b_{\lambda\mu}x)=\exp(\pi b^2_{\lambda\mu})\exp(-\pi x^2)\cos(2\pi b_{\lambda\mu}x).$$

We put $f(x)=\exp(-\pi x^2)\cos(2\pi b_{\lambda\mu}x)$. We assume that $0\le b_{\lambda\mu}$.

$$\begin{aligned}f'(x)&=-2\pi x\exp(-\pi x^2)\cos(2\pi b_{\lambda\mu}x)-\exp(-\pi x^2)2\pi b_{\lambda\mu}\sin(2\pi b_{\lambda\mu}x)\\&=-2\pi\sqrt{x^2+b^2_{\lambda\mu}}\exp(-\pi x^2)\cos(2\pi b_{\lambda\mu}x+\alpha_x),\end{aligned}$$

where $\cos\alpha_x=\frac{x}{\sqrt{x^2+b^2_{\lambda\mu}}}$, $-\sin\alpha_x=\frac{b_{\lambda\mu}}{\sqrt{x^2+b^2_{\lambda\mu}}}$. Since $-\sin\alpha_x=\frac{b_{\lambda\mu}}{\sqrt{x^2+b^2_{\lambda\mu}}}$, $\alpha_x$ is negative. There is a unique maximum of $|f(x)|$ in $\left\{x\in\mathbf{R}\;\middle|\;\frac{\pi}{2}(2m-1)\le 2\pi b_{\lambda\mu}x<\frac{\pi}{2}(2m+1)\right\}$ for each $m\in\mathbf{Z}$, that is, $x$ satisfies

$$\begin{aligned}&f'(x)=0,\ \frac{\pi}{2}(2m-1)\le 2\pi b_{\lambda\mu}x<\frac{\pi}{2}(2m+1)\\&\qquad\iff 2\pi b_{\lambda\mu}x+\alpha_x=\frac{\pi}{2}(2m-1).\qquad\cdots(*_4)\end{aligned}$$

We write the value of $x$ having the maximum of $|f(x)|$ in the interval as $x_m$. On the other hand, we denote the value $\alpha_x$ at $x=A_{2m}$ by $\alpha_{A_{2m}}$. Then $A_{2m}=\frac{m-1/2-\alpha_{A_{2m}}/\pi}{2b_{\lambda\mu}}$. The maximum of $f(x)$ is $f(A_{2m})=\exp\left(-\pi x^2_m\right)\cos\left(m\pi-\frac{\pi}{2}-\alpha_{A_{2m}}\right)$. Hence $|f(A_{2m})|\le\exp\Big(-\pi\left(\frac{m-1/2-\alpha_{A_{2m}}/\pi}{2b_{\lambda\mu}}\right)^2\Big)\le\exp\Big(-\pi\left(\frac{m-1/2}{2b_{\lambda\mu}}\right)^2\Big)$. Then

$$\Big|\sum_{\varepsilon'_{\lambda\mu} z^a_{\lambda\mu}\in L'_{\lambda\mu}} \sqrt{\varepsilon_\mu}\varepsilon'{}_{\lambda\mu} \exp(-\pi((\sqrt{\varepsilon_\mu}\varepsilon'_{\lambda\mu} z^a_{\lambda\mu})^2 - (\varepsilon'_{\lambda\mu} z^b_{\lambda\mu})^2))$$

$$\cos(2\pi\varepsilon'_{\lambda\mu} z^b_{\lambda\mu}\sqrt{\varepsilon_\mu}\varepsilon'_{\lambda\mu} z^a_{\lambda\mu}) - \int_{-\sqrt{\varepsilon_\mu}\frac{H'_{\lambda\mu}}{2}}^{\sqrt{\varepsilon_\mu}\frac{H'_{\lambda\mu}}{2}} \exp(-\pi(x^2 - b^2_{\lambda\mu}))\cos(2\pi b_{\lambda\mu} x)dx\Big|$$

$$= \exp(\pi b^2_{\lambda\mu})\Big|\sum_{\varepsilon'_{\lambda\mu} z^a_{\lambda\mu}\in L'_{\lambda\mu}} \sqrt{\varepsilon_\mu}\varepsilon'{}_{\lambda\mu} f((\sqrt{\varepsilon_\mu}\varepsilon'_{\lambda\mu} z^a_{\lambda\mu})^2) - \int_{-\sqrt{\varepsilon_\mu}\frac{H'_{\lambda\mu}}{2}}^{\sqrt{\varepsilon_\mu}\frac{H'_{\lambda\mu}}{2}} f(x)dx\Big|.$$

We denote $A_{2m+1} = \frac{1}{4b_{\lambda\mu}}(2m+1)$. Since $[(b(k)_{\lambda\mu})]$ is finite, there exists a positive real number $c$ so that $\star(*(c)) \le \left[\left(\left|\frac{1}{4b_{\lambda\mu}}\right|\right)\right]$, that is, $\left\{\lambda \mid \left\{\mu \mid c \le \left|\frac{1}{4b_{\lambda\mu}}\right|\right\} \in F_2\right\} \in F_1$. Furthermore since $\sqrt{{}^*\varepsilon}\,\varepsilon'$ is infinitesimal in ${}^\star({}^*\mathbf{R})$, $k \in L \Rightarrow \sqrt{{}^*\varepsilon}\,\varepsilon' < \left|\frac{1}{4b(k)}\right|$, that is, $\left\{\lambda \mid \left\{\mu \mid \sqrt{\varepsilon_\mu}\varepsilon'_{\lambda\mu} < \left|\frac{1}{4b_{\lambda\mu}}(k_\mu)\right|\right\} \in F_2\right\} \in F_1$. We assume $\lambda\mu$ satisfies the above condition. We denote $\sqrt{\varepsilon_\mu}\varepsilon'_{\lambda\mu}$ by $\Delta$. We shall show that the following term is infinitesimal in ${}^\star({}^*\mathbf{R})$: $\Big|\sum_{\varepsilon'_{\lambda\mu} z^a_{\lambda\mu}\in L'_{\lambda\mu}} \sqrt{\varepsilon_\mu}\varepsilon'{}_{\lambda\mu} f((\sqrt{\varepsilon_\mu}\varepsilon'_{\lambda\mu} z^a_{\lambda\mu})^2) - \int_{-\sqrt{\varepsilon_\mu}\frac{H'_{\lambda\mu}}{2}}^{\sqrt{\varepsilon_\mu}\frac{H'_{\lambda\mu}}{2}} f(x)dx\Big|$. We write the maximum of $j$ as l so that $A_j \in \left[0, \sqrt{\varepsilon_\mu}\frac{H'_{\lambda\mu}}{2}\right]$. Let $x_k = \Delta k$, $x_{i_j} \le A_j < x_{i_j+1}$ $(1 \le j \le l)$. Then we divide the interval $\left[0, \sqrt{\varepsilon_\mu}\frac{H'_{\lambda\mu}}{2}\right]$ into suitable intervals and prove it.

Case 1.

$$\Big|\int_{-A_1}^{A_1} f(x)dx - \Big(\sum_{i=-i_1}^{i_1} f(x_i)\Big)\Delta\Big|$$
$$= \Big|\int_{-A_1}^{-x_{i_1}} f(x)dx + \int_{-x_{i_1}}^{-x_{i_1-1}}(f(x) - f(-x_{i_1}))dx$$
$$+ \cdots + \int_{-x_1}^{x_0}(f(x) - f(-x_1))dx$$
$$+ \int_{x_0}^{x_1}(f(x) - f(x_0))dx + \cdots + \int_{x_{i_1-1}}^{x_{i_1}}(f(x) - f(x_{i_1-1}))dx$$
$$+ \int_{x_{i_1}}^{A_1}(f(x) - f(x_{i_1}))dx - f(x_{i_1})(x_{i_1+1} - A_1)\Big|, \text{ where } x_0 = 0,$$
$$\le \int_{-A_1}^{-x_{i_1}} f(x)dx + \int_{-x_{i_1}}^{-x_{i_1-1}}(f(x) - f(-x_{i_1}))dx$$
$$+ \cdots + \int_{-x_1}^{x_0}(f(x) - f(-x_1))dx$$
$$+ \int_{x_0}^{x_1}(f(x_0) - f(x))dx + \cdots + \int_{x_{i_1-1}}^{x_{i_1}}(f(x_{i_1-1}) - f(x))dx$$
$$+ \int_{x_{i_1}}^{A_1}(f(x_{i_1}) - f(x))dx + f(x_{i_1})(x_{i_1+1} - A_1),$$

since $f$ is an even function, $f(-x) = f(x)$,

$$= (f(x_0) - f(x_1) + f(x_1) - f(x_2) + \cdots + f(x_{i_1-1}) - f(x_{i_1}) + f(x_{i_1}))\,\Delta$$

$= f(x_0)\Delta = f(0)\Delta.$

Case 2.

$$\left|\int_{-x_{i_2}}^{-A_1} f(x)dx + \int_{A_1}^{x_{i_2+1}} f(x)dx - \left(\sum_{i=-i_2}^{-(i_1+1)} f(x_i) + \sum_{i=i_1+1}^{i_2} f(x_i)\right)\Delta\right|$$
$$\leq \int_{-x_{i_2}}^{-x_{i_2-1}} (f(x) - f(-x_{i_2}))dx + \cdots + \int_{-x_{i_1+2}}^{-x_{i_1+1}} (f(x) - f(-x_{i_1+2}))dx$$
$$+ \int_{-x_{i_1+1}}^{-A_1} (f(x) - f(-x_{i_1+1}))dx - f(-x_{i_1+1})(A_1 - x_{i_1})$$
$$+ \int_{A_1}^{x_{i_1+1}} (-f(x))dx + \int_{x_{i_1+1}}^{x_{i_1+2}} (f(x_{i_1+1}) - f(x))dx$$
$$+ \cdots + \int_{x_{i_2}}^{x_{i_2+1}} (f(x_{i_2}) - f(x))dx,$$

since $f$ is an even function, $f(-x) = f(x)$,

$$= \int_{x_{i_2-1}}^{x_{i_2}} (f(x) - f(x_{i_2}))dx + \cdots + \int_{x_{i_1+1}}^{x_{i_1+2}} (f(x) - f(x_{i_1+2}))dx$$
$$+ \int_{A_1}^{x_{i_1+1}} (f(x) - f(x_{i_1+1}))dx - f(-x_{i_1+1})(A_1 - x_{i_1})$$
$$+ \int_{A_1}^{x_{i_1+1}} (-f(x))dx + \int_{x_{i_1+1}}^{x_{i_1+2}} (f(x_{i_1+1}) - f(x))dx$$
$$+ \cdots + \int_{x_{i_2}}^{x_{i_2+1}} (f(x_{i_2}) - f(x))dx,$$

since $\int_{x_{i_2}}^{x_{i_2+1}} (f(x_{i_2}) - f(x))dx \leq \int_{x_{i_2}}^{x_{i_2+1}} (f(x_{i_2}) - f(A_2))dx,$

$$\leq (-f(x_{i_1+1}) + f(x_{i_1+1}) - f(x_{i_1+2})$$
$$+ \cdots + f(x_{i_2-1}) - f(x_{i_2}) + f(x_{i_2}) - f(A_2))\Delta$$
$$= (-f(A_2))\Delta.$$

Since the next steps are just same, we obtain the following:

$$\left|\sum_{\varepsilon'_{\lambda\mu} z^a_{\lambda\mu} \in L'_{\lambda\mu}} \sqrt{\varepsilon_\mu}\varepsilon'{}_{\lambda\mu} f((\sqrt{\varepsilon_\mu}\varepsilon'_{\lambda\mu} z^a_{\lambda\mu})^2) - \int_{-\sqrt{\varepsilon_\mu}\frac{H'_{\lambda\mu}}{2}}^{\sqrt{\varepsilon_\mu}\frac{H'_{\lambda\mu}}{2}} f(x)dx\right|$$
$$\leq 2\sqrt{\varepsilon_\mu}\varepsilon'{}_{\lambda\mu} \sum_{m=0}^{l} \exp\left(-\pi\left(\frac{m-\frac{1}{2}}{2b_{\lambda\mu}}\right)^2\right).$$

Hence if $0 \leq b_{\lambda\mu}$,

$$\left|\sum_{\varepsilon'_{\lambda\mu} z^a_{\lambda\mu}, \varepsilon'_{\lambda\mu} z^b_{\lambda\mu} \in L'_{\lambda\mu}} \sqrt{\varepsilon_\mu}\varepsilon'{}_{\lambda\mu} \exp(-\pi((\sqrt{\varepsilon_\mu}\varepsilon'_{\lambda\mu} z^a_{\lambda\mu})^2 - (\varepsilon'_{\lambda\mu} z^b_{\lambda\mu})^2))\right.$$
$$\left.\cos(2\pi\varepsilon'_{\lambda\mu} z^b_{\lambda\mu}\sqrt{\varepsilon_\mu}\varepsilon'_{\lambda\mu} z^a_{\lambda\mu}) - \int_{-\sqrt{\varepsilon_\mu}\frac{H'_{\lambda\mu}}{2}}^{\sqrt{\varepsilon_\mu}\frac{H'_{\lambda\mu}}{2}} \exp(-\pi(x^2 - b^2_{\lambda\mu}))\cos(2\pi b_{\lambda\mu} x)dx\right|$$
$$\leq \exp(\pi b^2_{\lambda\mu}) \cdot 2\sqrt{\varepsilon_\mu}\varepsilon'{}_{\lambda\mu} \sum_{m=0}^{l} \exp\left(-\pi\left(\frac{m-\frac{1}{2}}{2b_{\lambda\mu}}\right)^2\right).$$

Since $\left[\left(\sum_{m=0}^{l}\exp\left(-\pi\left(\frac{m-\frac{1}{2}}{2b_{\lambda\mu}}\right)^2\right)\right)\right] \leq \sum_{m=0}^{l}\exp\left(-\pi\left(\frac{m-\frac{1}{2}}{2b_0}\right)^2\right)$, it is finite. Hence $\left[\left(\exp(\pi b_{\lambda\mu}^2)\cdot 2\sqrt{\varepsilon_\mu}\varepsilon'_{\lambda\mu}\sum_{m=0}^{l}\exp\left(-\pi\left(\frac{m-\frac{1}{2}}{2b_{\lambda\mu}}\right)^2\right)\right)\right]$ is infinitesimal in ${}^\star({}^*\mathbf{R})$. If $b_{\lambda\mu} < 0$, the argument is parallel, and also, for the term of sin in $(*_3)$, though sin is not an even function, the same argument holds. Hence $[(B_{\lambda\mu}(k_\mu))]$ is infinitesimal in ${}^\star({}^*\mathbf{C})$ with respect to $\mathbf{C}$. Put $N_{\lambda\mu} = \frac{1}{B_{\lambda\mu}(k_\mu)}$. Then it is infinite, and since $\lim_{n\to\infty}\left(1+\frac{1}{an}\right)^n = e^{-a}$,

$$\mathbf{st}\left(\left[\left(1+\frac{1}{\int_{-\infty}^{\infty}\exp(-\pi x^2)dx\cdot\frac{1}{B_{\lambda\mu}(k_\mu)}}\right)^{\frac{1}{B_{\lambda\mu}(k_\mu)}}\right]\right) = \exp\left(-\int_{-\infty}^{\infty}\exp(-\pi x^2)dx\right).$$

There is an infinitesimal $C$ $(= [(C_{\lambda\mu})])$ in ${}^\star({}^*\mathbf{C})$ so that

$$\left(1+\frac{1}{\int_{-\infty}^{\infty}\exp(-\pi x^2)dx\cdot N_{\lambda\mu}}\right)^{N_{\lambda\mu}} = \exp\left(-\int_{-\infty}^{\infty}\exp(-\pi x^2)dx\right) + C_{\lambda\mu}.$$

Then $[((*_1))]$ is equal to

$$\left[\left(\left(1+\frac{1}{\int_{-\infty}^{\infty}\exp(-\pi x^2)dx\cdot N_{\lambda\mu}}\right)^{N_{\lambda\mu}}\right)^{[(B_{\lambda\mu}(k_\mu)H_\mu^2)]}\right] = \left(\star\left(*\left(\exp\left(-\int_{-\infty}^{\infty}\exp(-\pi x^2)dx\right)\right)\right)+C\right)^{[(B_{\lambda\mu}(k_\mu)H_\mu^2)]}.$$

Since $[(B_{\lambda\mu}(k_\mu)H_\mu^2)]$ is infinitesimal,

$$\mathbf{st}\left(\frac{\sum_{a\in X}\varepsilon_0\exp\left(-\pi^\star\varepsilon\sum_{k\in L}(a(k)+ib(k))^2\right)}{\star\left(*\left(\int_{-\infty}^{\infty}\exp(-\pi x^2)dx\right)^{H^2}\right)}\right) = 1.$$

Since $\int_{-\infty}^{\infty}\exp(-\pi x^2)dx = 1$, then $\mathbf{st}(C_2(b)) = 1$.

**Acknowledgement.** We would like to thank Prof. T. Kamae for a useful suggestion about Gauss sums.

## References

1. R.M. Anderson: *A non-standard representation for Brownian motion and Itô integration*, Israel J. Math. **25** (1976), 15–46.
2. E.I. Gordon: *On the Fourier transformation in nonstandard analysis*, Izv. Vyssh. Uchebn. Zaved. Mat. 1989, no. **2**, 17–25: English transl. in Soviet Math. (Iz. VUZ) **33** (1989).
3. M. Kinoshita: *Nonstandard representation of distribution I*, Osaka J. Math. **25** (1988), 805–824.

4. P.A. Loeb: *Conversion from nonstandard to standard measure spaces and application in probability theory*, Trans. Amer. Math. Soc. **211** (1975), 113–122.
5. W.A. Luxemburg: *A Nonstandard approach to Fourier analysis*, Contributions to Nonstandard Analysis, North-Holland, Amsterdam, pp.16–39, 1972.
6. R. Remmert: Theory of complex functions, Graduate Texts in Mathematics **122**, Springer, Berlin-Heidelberg-New York, 1992.

# A LIOUVILLE THEOREM FOR HARMONIC MAPS WITH FINITE TOTAL $N$-ENERGY


TAKASHI OKAYASU

*Faculty of Education,*
*Ibaraki University*
*E-mail: okayasu@mx.ibaraki.ac.jp*



*Keywords*: Harmonic map, Liouville theorem


## 1. Introduction

The classical theorem of Liouville says that a bounded harmonic function on $\mathbf{R}^n$ is necessarily a constant. Let $u : (M, g) \to (N, h)$ be a harmonic map. There are many results giving conditions for $u$ to be a constant map. Those results are usually referred to as Liouville theorems for harmonic maps (see [3], [5], [10] and the references therein).

Here we consider Liouville theorem concerning the growth rate of the energy of $u$. In [11] Sealey proved that if $u : \mathbf{R}^n \to N$ $(n \geq 3)$ is a harmonic map of finite energy $\int_{\mathbf{R}^n} |du|^2 < \infty$, then $u$ is a constant map. By using the monotonicity formula (see [9]) , we can easily extend this result to $p$-energy. If $u : \mathbf{R}^n \to N$ $(n \geq 3)$ is a harmonic map with $\int_{\mathbf{R}^n} |du|^p < \infty$ for some $2 \leq p < n$, then $u$ is a constant map.

To extend this result to $p = n$, we need another technique because $p = n$ is a critical exponent. In [6] and [7] this difficulty was overcome.

**Theorem 1.1.** ([6], [7]) *Let $N$ be a Riemannian manifold with sectional curvature bounded from above by a positive constant. If $u : \mathbf{R}^n \to N$ $(n \geq 3)$ is a harmonic map with $\int_{\mathbf{R}^n} |du|^n < \infty$, then $u$ is a constant map.*

The purpose of this note is to extend this result to conformally flat Riemannian manifolds with nonnegative Ricci curvature. Here is our main theorem.

---


2000 *Mathematics Subject Classification.* Primary 53C43; Secondary 58E20.
Partially supported by Grant-in-Aid for Scientific Research (No.15540057 ), Ministry of Education, Science and Culture, Japan

**Theorem 1.2.** *Let* $(M, g) = (\mathbf{R}^n,\ f^2 g_0)$ *be a complete conformally flat Riemannian manifold with nonnegative Ricci curvature. Suppose that* $\liminf_{R\to\infty} \operatorname{vol}(B(o;R))/R^n > 0$ *for some point* $o \in M$ $(n \geq 3)$, *where* $B(o;R)$ *is the geodesic ball of* $M$ *centered at* $o$ *with radius* $R$. *Suppose, moreover, that* $f$ *satisfies*

$$1 + r\frac{\partial}{\partial r}\log f \geq 0, \tag{1.1}$$

*where* $r$ *is the Euclidean distance from* $o$. *Let* $N$ *be a Riemannian manifold with sectional curvature bounded from above by a positive constant. If* $u : M^n \to N$ *is a harmonic map with* $\int_{M^n} |du|^n < \infty$, *then* $u$ *is a constant map.*

For the proof we use several apriori estimates derived by Bando, Kasue and Nakajima in [1].

## 2. Proof

First we prove the following theorem.

**Theorem 2.1.** *Let* $M^n$ *be a noncompact complete* $n$*-dimensional Riemannian manifold with nonnegative Ricci curvature and* $\liminf_{R\to\infty} \operatorname{vol}(B(o;R))/R^n > 0$ *for some point* $o \in M$ $(n \geq 3)$. *Let* $N$ *be a Riemannian manifold with sectional curvature bounded from above by a positive constant* $K$. *If* $u : M^n \to N$ *is a harmonic map with* $\int_{M^n} |du|^n < \infty$, *then* $|du| = O(R^{-\alpha})$ *for any* $\alpha < n - 1$.

**Proof.** We first remark that the assumptions of the section 4 (page. 331) of [1] are all satisfied. From $\operatorname{Ric}_M \geq 0$, by the Bishop-Gromov volume comparison theorem we have

$$\operatorname{vol}(B(o;t)) \leq V t^n \quad \text{for all} \quad t > 0. \tag{2.1}$$

From $\liminf_{R\to\infty} \operatorname{vol}(B(o;R))/R^n > 0$ and $\operatorname{Ric}_M \geq 0$, the Sobolev inequality

$$\left\{\int_M v^{2\gamma}\right\}^{1/\gamma} \leq S \int_M |\nabla v|^2 \quad \text{for all} \quad v \in C_0^1(M) \tag{2.2}$$

holds for $C^1$ functions with compact support on $M$, where $\gamma = \frac{n}{n-2}$ ([12]). So we can use the apriori estimates of [1].

Let's recall the Weitzenböck formula ([4]).

$$\frac{1}{2}\Delta|du|^2 = |\nabla du|^2 + \sum_{i=1}^{n} \mathrm{Ric}^M(du(e_i), du(e_i)) - \sum_{i,j=1}^{n} h\left(R^N(du(e_i), du(e_j))du(e_j), du(e_i)\right), \tag{2.3}$$

where $\Delta$ is the Laplace-Beltrami operator of $M$, $\{e_i\}$ is a local orthonormal frame of $M$, $\mathrm{Ric}^M$ is the Ricci tensor of $M$ and $R^N$ is the curvature tensor of $N$, respectively. From the assumption on curvatures, we get

$$\frac{1}{2}\Delta|du|^2 \geq -K|du|^4 + |\nabla du|^2. \tag{2.4}$$

On the other hand we have the following improved Kato's inequality ([8]).

$$\left(1 + \frac{1}{n-1}\right) |\nabla|du||^2 \leq |\nabla du|^2. \tag{2.5}$$

Applying this to (2.4), we obtain

$$|du|\Delta|du| \geq -K|du|^4 + \frac{1}{n-1}|\nabla|du||^2. \tag{2.6}$$

From (2.4) we also get

$$\frac{1}{2}\Delta|du|^2 \geq -K|du|^4. \tag{2.7}$$

Since $\int_{M^n} |du|^n < \infty$, we can apply Proposition (4.8) 2) in [1] to get

$$\int_{D(R)} |du|^{\gamma n} = O(R^{-2\gamma}), \tag{2.8}$$

where $R$ is the geodesic distance from $o \in M$ and $D(R) = M \setminus B(o;R)$. From (2.6) we obtain

$$\Delta|du|^{1-\frac{1}{n-1}} \geq -\frac{n-2}{n-1}K|du|^{3-\frac{1}{n-1}}. \tag{2.9}$$

Since we have (2.8), we can apply Proposition (4.8) 1) of [1] to this equation. We get

$$|du|^{1-\frac{1}{n-1}} = O(R^{-\alpha}) \quad \text{for any} \quad \alpha < n-2. \tag{2.10}$$

That is

$$|du| = O(R^{-\alpha}) \quad \text{for any } \alpha < n-1. \tag{2.11}$$

Thus we have proved the theorem. □

**Proof of Theorem 1.2.** Suppose $u$ is not a constant map. By a result of Sealey [11], we have $\int_M |du|^2 = \infty$. On the other hand, from theorem 2.1 and $\mathrm{vol}\,(B(o;R)) \leq VR^n$, we have

$$\int_{B(o;R)} |du|^2 \leq c_1 - c_2 R^{n-2\alpha} \tag{2.12}$$

for some positive constants $c_1$, $c_2$. Since $n \geq 3$ we can choose $\alpha$ so that $n - 2\alpha < 0$. Thus letting $R \to \infty$ we get a contradiction. □

**Remark 2.2.** We give examples of $(M, g) = (\mathbf{R}^n, f^2 g_0)$ which satisfy all the conditions of theorem 1.2. Let $f = (1 + r^2)^{-\beta}$. It is easy to see that if $0 \leq \beta < \frac{1}{2}$, then the metric is complete ant the Ricci curvature is nonnegative (see, for example, [13]). Moreover, $f$ satisfies (1.1). We can easily deform this function $f$ keeping the conditions of theorem 1.2. So we can produce many examples of theorem 1.2.

It is believed that there exist many complete manifolds with nonnegative Ricci curvature and with maximal volume growth which has infinite topological type (see [14], p. 249). At the moment we cannot prove that our theorem holds for all noncompact complete manifolds with nonnegative Ricci curvature and maximal volume growth without any further assumptions. But we can give a Liouville theorem if the $n$-energy is sufficiently small. Using the same technique as in [2] for (2.6), we get the following gap theorem. The argument is a standard one. For the completeness we give its proof.

**Theorem 2.3.** *Let $M^n$ be a noncompact complete $n$-dimensional Riemannian manifold with nonnegative Ricci curvature and $\liminf_{R\to\infty}$ $\mathrm{vol}\,(B(o;R))/R^n > 0$ for some point $o \in M$ $(n \geq 3)$. Let $N$ be a Riemannian manifold with sectional curvature bounded from above by a positive constant $K$. Then there exists a constant $C = C(n, S) > 0$, depending only on $n$ and the Sobolev constant $S$, such that if $u : M^n \to N$ is a harmonic map with $\int_{M^n} |du|^n < C/K^{n/2}$, then $u$ is a constant map.*

**Proof.** From (2.6) we have

$$|du|\Delta|du| \geq -K|du|^4 + \frac{1}{n-1}|\nabla|du||^2. \tag{2.13}$$

For simplicity we set $a = |df|$ and rewrite the above inequality as

$$\frac{1}{n-1}|\nabla a|^2 \leq a\Delta a + Ka^4. \tag{2.14}$$

Let $\lambda \in C_0^1(M)$. Multiplying $\lambda^2$ and integrating over $M$, we get

$$\frac{1}{n-1}\int \lambda^2|\nabla a|^2 \leq \int \lambda^2 a\Delta a + K\int \lambda^2 a^4. \tag{2.15}$$

Since

$$\int 2\lambda a\nabla\lambda\cdot\nabla a + \int \lambda^2|\nabla a|^2 + \int \lambda^2 a\Delta a = 0,$$

we get from (2.15)

$$\frac{n}{n-1}\int \lambda^2|\nabla a|^2 + 2\int \lambda a\nabla\lambda\cdot\nabla a \leq K\int \lambda^2 a^4. \tag{2.16}$$

We have

$$2\int |\lambda a\nabla\lambda\cdot\nabla a| \leq \epsilon\int \lambda^2|\nabla a|^2 + \frac{1}{\epsilon}\int a^2|\nabla\lambda|^2$$

for any $\epsilon > 0$. Plugging this into (2.16), we get

$$\Big(\frac{n}{n-1} - \epsilon\Big)\int \lambda^2|\nabla a|^2 \leq K\int \lambda^2 a^4 + \frac{1}{\epsilon}\int a^2|\nabla\lambda|^2. \tag{2.17}$$

On the other hand we have Sobolev's inequality

$$\Big(\int |v|^{2\gamma}\Big)^{\frac{1}{\gamma}} \leq S\int |\nabla v|^2 \quad \text{for} \quad \forall v \in C_0^1(M).$$

We apply this inequality for $\lambda a$ and get

$$\Big(\int \lambda^{2\gamma}a^{2\gamma}\Big)^{\frac{1}{\gamma}} \leq 2S\Big\{\int \lambda^2|\nabla a|^2 + \int a^2|\nabla\lambda|^2\Big\}. \tag{2.18}$$

From (2.17) and (2.18), we get

(2.19)

$$\Big(\int \lambda^{2\gamma}a^{2\gamma}\Big)^{\frac{1}{\gamma}} \leq \frac{2S}{\frac{n}{n-1} - \epsilon}\Big\{K\int \lambda^2 a^4 + \frac{1}{\epsilon}\int a^2|\nabla\lambda|^2\Big\} + 2S\int a^2|\nabla\lambda|^2.$$

Applying Hölder's inequality

$$\int \lambda^2 a^4 \leq \Big(\int \lambda^{2\gamma}a^{2\gamma}\Big)^{\frac{1}{\gamma}}\Big(\int a^n\Big)^{\frac{2}{n}}$$

to (2.19), we get

$$\Big\{1 - \frac{2SK}{\frac{n}{n-1} - \epsilon}\Big(\int a^n\Big)^{\frac{2}{n}}\Big\}\Big(\int \lambda^{2\gamma}a^{2\gamma}\Big)^{\frac{1}{\gamma}} \leq \Big\{\frac{2SK}{\epsilon(\frac{n}{n-1} - \epsilon)} + 2S\Big\}\int a^2|\nabla\lambda|^2.$$

If we have

$$1 - \frac{2SK}{\frac{n}{n-1} - \epsilon}\Big(\int a^n\Big)^{\frac{2}{n}} > 0, \tag{2.20}$$

then

$$\left(\int \lambda^{2\gamma} a^{2\gamma}\right)^{\frac{1}{\gamma}} \leq C \int a^2 |\nabla \lambda|^2. \tag{2.21}$$

For any $R > 0$ we can choose the cut-off function $\lambda$ so that

$$\lambda(x) = \begin{cases} 1, & \text{when } x \in B(o;R) \\ 0, & \text{when } x \in M \setminus B(o;2R) \end{cases} \tag{2.22}$$

and $|\nabla \lambda| \leq \frac{1}{R}$. Then from (2.21), theorem 2.1 and the volume estimate, we get

$$\begin{aligned} \left(\int_{B(0;R)} a^{2\gamma}\right)^{\frac{1}{\gamma}} &\leq \frac{C}{R^2} \int_{B(o;2R)} a^2 \\ &\leq \frac{C}{R^{2+2\alpha}} \mathrm{vol}\,(B(o;2R)) \leq \frac{C}{R^{2+2\alpha-n}} \end{aligned} \tag{2.23}$$

for any $\alpha < n-1$. Since $n \geq 3$ we can choose $\alpha$ so that $2 + 2\alpha - n > 0$. Letting $R \to \infty$ we obtain $a = |df| \equiv 0$. Since we can choose $\epsilon$ arbitrarily small, (2.20) is satisfied if

$$\int_M |df|^n < \left\{\frac{n}{2(n-1)SK}\right\}^{n/2}. \qquad \square$$

## References


1. S. Bando, A. Kasue and H. Nakajima, *On a construction of coordinates at infinity on manifolds with fast curvature decay and maximal volume growth*, Inv. Math. 97 (1989), 313–349.
2. P. Bérard, *Remarques sur l'equation de J. Simons*, Differential Geometry, Longman, Harlow, 1991, 47–57.
3. S.Y .Cheng, *Liouville theorem for harmonic maps*, Proc. Sympos. Pure Math., XXXVI. (1980), 147–151.
4. J. Eells and L. Lemaire, *A report on harmonic maps*, Bull. London Math. Soc. 10 (1978), 1–68
5. H. Karcher, and J. C. Wood, *Nonexistence results and growth properties for harmonic mapsand forms*, J. Reine Angew. Math. 353 (1984), 165–180.
6. N. Nakauchi, S. Takakuwa, *A remark on p-harmonic maps*, Nonlinear Anal. 25 (1995), 169–185.
7. T. Okayasu, *Harmonic maps from noncompact manifolds with nonnegative Ricci curvature and maximal volume growth*, unpublished preprint in 1993.
8. T. Okayasu, *Regularity of minimizing harmonic maps into $S^4, S^5$ and symmetric spaces*, Math. Ann. 298 (1994),193–205.
9. P. Price, *A monotonicity formula for Yang-Mills fields*, Manuscripta Math. 43 (1983), 131–166.
10. M. Rigoli and A.G. Setti, *Energy estimates and Liouville theorems for harmonic maps*, Internat. J. Math. 11 (2000), 413–448.

11. H.C. Sealey, *Some conditions ensuring the vanishing of harmonic differential forms with applications to harmonic maps and Yang-Mills theory*, Math. Proc. Camb. Phil. Soc. 91 (1982), 441–452.
12. S.T. Yau, *Survey on partial differential equations in differential geometry*, Seminar on Differential Geometry, Ann. of Math. Stud. 102 (1982), 3–71.
13. S. Zhu, *The classification of complete locally conformally flat manifolds of nonnegative Ricci curvature*, Pacific J. Math. 163 (1994), 189–199.
14. S. Zhu, *The comparison geometry of Ricci curvature*, Comparison geometry (Berkeley, CA, 1993–94), Math. Sci. Res. Inst. Publ. 30. (1997), 221–262.

# ON A FAMILY OF ALMOST KÄHLER STRUCTURES ON THE TANGENT BUNDLES OVER SOME STATISTICAL MODELS

TAKUJI SATO

*Graduate School of Natural Science and Technology*
*Kanazawa University*
*Kanazawa 920-1192, Japan*
*E-mail: tsato@t.kanazawa-u.ac.jp*

We investigate the $\alpha$-family of almost Kähler structures on the tangent bundle $TS$ over a statistical model $S$. We show that $TS$ becomes a Kähler manifold of constant holomorphic sectional curvature in particular cases.

*Keywords*: almost Kähler structure, tangent bundle, statistical model, holomorphic sectional curvature

## 1. Introduction

It is well known that an almost Kähler structure can be defined on the tangent bundle over a Riemannian manifold $(M, g)$ [T-O,Y-I]. More generally, from an affine connection on a manifold $M$, we can define an almost complex structure on the tangent bundle $TM$, and if this affine connection and its dual connection with respect to a metric $g$ are both torsion free, then we can introduce on $TM$ an almost Kähler structure.

In this paper, we shall consider a statistical model $S$ which can be regarded as a $C^\infty$-manifold in information geometry, and discuss a family of almost Kähler structures on the tangent bundle $TS$ induced from the Fisher metric and the $\alpha$-connection on $S$. For the $\alpha$-family of almost Kähler structures on $TS$, where $S$ is the manifold of normal distributions (resp. of discrete distributions on a set of 3-points), we show that $TS$ is a Kähler

1991 *Mathematics Subject Classification.* 53A15, 53B35, 53C15, 53C55

manifold of constant holomorphic sectional curvature provided that $\alpha = -1$ (resp. $\alpha = 1$).

The author wishes to express his deepest gratitude to Professor Kouei Sekigawa for his helpful guidance and constant encouragement.

## 2. Statistical models

In this section, we shall summarize some basic facts on statistical models in information geometry for the later use. For the details, we refer to [Am,A-N].

Let $S$ be a set of probability distributions on a set $\mathcal{X}$. Suppose each element of $S$ may be parametrized by $n$ real variables $(\theta^1, \dots, \theta^n)$ so that

$$S = \left\{ p_\theta = p(x, \theta) \mid \theta = (\theta^1, \dots, \theta^n) \in \Theta \right\},$$

where $\Theta$ is an open subset of $\mathbb{R}^n$ and the mapping $\theta \mapsto p_\theta$ is injective. We call such $S$ an $n$-dimensional *statistical model*. Under some regularity conditions, $S$ may be considered as a $C^\infty$-manifold with a coordinate system $\varphi(p_\theta) = \theta \in \Theta \subset \mathbb{R}^n$. With respect to a coordinate system $(\theta^1, \dots, \theta^n)$, the *Fisher metric* $h = (h_{ij})$ of $S$ is defined by

$$h_{ij} := E_\theta[\partial_i l_\theta \partial_j l_\theta] = \int \partial_i l(x, \theta) \partial_j l(x, \theta) p(x, \theta)\, dx,$$

where $\partial_i = \frac{\partial}{\partial \theta^i}$, $l_\theta(x) = l(x, \theta) = \log p(x, \theta)$ and $E_\theta$ denotes the expectation with respect to $p_\theta$. The Fisher metric is a 'natural' Riemannian metric in information geometry.

Besides the Levi-Civita connection of $h$, a family of torsion free affine connections $\nabla^{(\alpha)}$, so called $\alpha$-connection, plays an important role in information geometry. The *$\alpha$-connection* is given by

$$\nabla^{(\alpha)} := \nabla - \frac{\alpha}{2} T,$$

where $\nabla = \nabla^{(0)}$ is the Levi-Civita connection, and $T = (T_{ij}^k)$ is the (1,2)-tensor given by $T_{ijk} = h_{kl} T_{ij}^l = E_\theta[\partial_i l \partial_j l \partial_k l]$. $\nabla^{(\alpha)}$ and $\nabla^{(-\alpha)}$ are mutually *dual* with respect to $h$, i.e.,

$$Xh(Y, Z) = h(\nabla_X^{(\alpha)} Y, Z) + h(Y, \nabla_X^{(-\alpha)} Z), \tag{2.1}$$

holds for any vector fields $X, Y, Z$ on $S$.

We take up two typical examples of statistical models in the following.

### 2.1. *Normal distributions*

Let $\mathcal{X} = \mathbb{R}$, $n = 2$, $\theta = (\theta^1, \theta^2) = (\mu, \sigma)$ and $\Theta = \mathbb{R} \times \mathbb{R}_+$. Then the statistical model $S$ of normal distributions is defined by $S = \{\, p(x,\theta) \mid \theta \in \Theta \,\}$, where

$$p(x,\theta) = \frac{1}{\sqrt{2\pi}\sigma} e^{-\frac{(x-\mu)^2}{2\sigma^2}}.$$

The Fisher metric of $S$ is given by

$$h = \begin{pmatrix} \frac{1}{\sigma^2} & 0 \\ 0 & \frac{2}{\sigma^2} \end{pmatrix},$$

and $(S,h)$ has constant Gauss curvature $-1/2$. The components of $\alpha$-connection $\nabla^{(\alpha)}$ are given by, with respect to a coordinate system $\theta = (\theta^1, \theta^2) = (\mu, \sigma)$,

$$(2.2)\qquad \begin{aligned} &{\Gamma^{(\alpha)}}_{11}^{\ \ 1} = 0, \quad {\Gamma^{(\alpha)}}_{11}^{\ \ 2} = \frac{1-\alpha}{2\sigma}, \quad {\Gamma^{(\alpha)}}_{12}^{\ \ 1} = -\frac{1+\alpha}{\sigma}, \\ &{\Gamma^{(\alpha)}}_{12}^{\ \ 2} = 0, \quad {\Gamma^{(\alpha)}}_{22}^{\ \ 1} = 0, \qquad\quad {\Gamma^{(\alpha)}}_{22}^{\ \ 2} = -\frac{1+2\alpha}{\sigma}. \end{aligned}$$

Note that this connection is flat when $\alpha = \pm 1$.

### 2.2. *Discrete distributions on finite $\mathcal{X}$*

Let $\mathcal{X} = \{z_0, z_1, \dots, z_n\}$ and $\Theta = \{\, (\theta^1, \dots, \theta^n) \mid \theta^i > 0, \sum_{i=1}^n \theta^i < 1 \,\}$. The $n$-dimensional statistical model $S$ of discrete distributions is defined by $S = \{\, p(z,\theta) \mid \theta \in \Theta \,\}$, where

$$p(z_i,\theta) = \begin{cases} \theta^i & (1 \le i \le n) \\ \theta^0 = 1 - \sum_{i=1}^n \theta^i & (i = 0). \end{cases}$$

The Fisher metric $h$ of $S$ is given by

$$h_{ij} = \begin{cases} \frac{1}{\theta^0} + \frac{1}{\theta^i} & (i = j) \\ \frac{1}{\theta^0} & (i \ne j), \end{cases}$$

and $(S,h)$ has constant curvature $1/4$. The components of $\alpha$-connection are given by

$$(2.3)\qquad \Gamma^{(\alpha)}_{ijk} = \begin{cases} \frac{1+\alpha}{2}\left(\frac{1}{(\theta^0)^2} - \frac{1}{(\theta^i)^2}\right) & (i = j = k) \\ \frac{1+\alpha}{2}\frac{1}{(\theta^0)^2} & (\text{otherwise}). \end{cases}$$

This connection is $(\pm 1)$-flat as well.

## 3. Almost Kähler structures on the tangent bundle

Let $(M,h)$ be a Riemannian manifold with a torsion free affine connection $\nabla = (\Gamma^i_{jk})$, and $TM$ be its tangent bundle. As usually, we adopt $(x^i, y^i)$ as a coordinate of $u = \sum_{i=1}^n y^i \frac{\partial}{\partial x^i} \in T_pM$, where $(x^i)$ denotes a coodinate system around $p$ in $M$. Then $X = \xi^i \frac{\partial}{\partial x^i} + \eta^i \frac{\partial}{\partial y^i} \in T_u(TM)$ can be decomposed into horizontal and vertical parts $X^h, X^v$ with respect to $\nabla$:

$$X^h = \xi^i \frac{\partial}{\partial x^i} - \Gamma^i_{jk} y^j \xi^k \frac{\partial}{\partial y^i}, \qquad X^v = (\eta^i + \Gamma^i_{jk} y^j \xi^k) \frac{\partial}{\partial y^i}.$$

By identification $X^h \leftrightarrow X_h = \xi^i \frac{\partial}{\partial x^i}$, $X^v \leftrightarrow X_v = (\eta^i + \Gamma^i_{jk} y^j \xi^k) \frac{\partial}{\partial x^i}$, we have

$$T_u(TM) = H_u \oplus V_u \cong T_pM \oplus T_pM.$$

By using this identification, an almost complex structure $J$ and a Riemannian metric $g$ on $TM$ are defined by

$$JX = J(X_h, X_v) := (-X_v, X_h), \tag{3.1}$$

$$g := h \oplus h. \tag{3.2}$$

It is obvious that $(J,g)$ is almost Hermitian, and its Kähler form is given by $\Omega = -h_{ik}\Gamma^k_{rj} y^r dx^i \wedge dx^j - h_{ij} dx^i \wedge dy^j$. If the dual connction $\nabla^* = (\Gamma^{*\,i}_{jk})$ of $\nabla$ is also torsion free, then we have

$$\begin{aligned}
\Omega =& h_{il}\Gamma^l_{kj} y^j dx^k \wedge dx^i - h_{ij} dx^i \wedge dy^j \\
=& (\Gamma^{*\,l}_{ki} h_{lj} + \Gamma^l_{kj} h_{il}) y^j dx^k \wedge dx^i - h_{ij} dx^i \wedge dy^j \\
=& \frac{\partial h_{ij}}{\partial x^k} y^j dx^k \wedge dx^i + h_{ij} dy^j \wedge dx^i \\
=& d(h_{ij} y^j dx^i).
\end{aligned}$$

So we have the following

**Proposition 3.1.** *Let $(M,h)$ be a Riemannian manifold with a torsion free affine connection $\nabla$. If its dual connection $\nabla^*$ with respect to $h$ is torsion free, then the almost Hermitian structure $(J,g)$ on $TM$ induced by $\nabla$ is almost Kählerian.*

According to [Ku], the above Proposition can be restated as

**Proposition 3.2.** *Let $(M,\nabla,h)$ be a statistical manifold. Then the almost Hermitian structure on $TM$ induced by $\nabla$ is almost Kählerian.*

## 4. $\alpha$-family of almost Kähler structures on $TS$

In this section, we shall investigate the $\alpha$-family of almost Kähler structures on the tangent bundles over the standard statistical models in § 2.1 and 2.2. Let $(S,h)$ be a manifold of statistical model with $\alpha$-connection $\nabla^{(\alpha)} = (\Gamma_{ij}^{(\alpha)k})$. Then, by definition, the Riemannian metric $g^{(\alpha)} = (g_{ij})$ and the almost complex structure $J^{(\alpha)} = (J_i^k)$ on $TS$ are expressed locally by

$$
\begin{aligned}
g_{ij} &= h_{ij} + h_{kl}\Gamma_i^{(\alpha)k}\Gamma_j^{(\alpha)l},\\
g_{i\bar{j}} &= h_{kj}\Gamma_i^{(\alpha)k},\\
g_{\bar{i}\bar{j}} &= h_{ij},
\end{aligned}
\tag{4.1}
$$

$$
\begin{aligned}
J_i^k &= -\Gamma_i^{(\alpha)k}, & J_{\bar{i}}^k &= -\delta_i^k,\\
J_i^{\bar{k}} &= \delta_i^k + \Gamma_j^{(\alpha)k}\Gamma_i^{(\alpha)j}, & J_{\bar{i}}^{\bar{k}} &= \Gamma_i^{(\alpha)k},
\end{aligned}
\tag{4.2}
$$

where $\Gamma_i^{(\alpha)k} = \Gamma_{ij}^{(\alpha)k}y^j$ and we put $(x^i, x^{\bar{i}}) = (x^i, y^i)$, so the indices $\bar{i} = n+i, \bar{j} = n+j, \cdots$ run over the range $\{n+1,\ldots,2n\}$. From now on, we shall use the simple notation $g$ and $J$ instead of $g^{(\alpha)}$ and $J^{(\alpha)}$, respectively.

Let $\{f_1,\ldots,f_n\}$ be a local orthonormal frame on $S$. Then the local vector fields $e_{2i} (i=1,\ldots,n)$ on $TS$ such that $(e_{2i})_h = 0$, $(e_{2i})_v = f_i$ are also orthnomal by (4.1). Setting $e_{2i-1} := -Je_{2i}$, we obtain a local unitary frame $\{e_1,\ldots,e_{2n}\}$ on $TS$ satisfying $Je_{2i-1} = e_{2i}$. With respect to this unitary frame, we shall compute explicitly some geometric quantities on $TS$ for each statistical model.

### 4.1. *Tangent budle of the manifold of normal distributions*

Let $(S,h)$ be the manifold of normal distributions with $\alpha$-connection $\nabla^{(\alpha)}$. Set $x_1 = \mu$, $x_2 = \sigma$, $x_3 = y^1$, $x_4 = y^2$ for a coordinate system on $TS$. Taking account of (2.2), we obtain an unitary frame

$$
\begin{aligned}
e_1 &= x_2\frac{\partial}{\partial x_1} + (1+\alpha)x_4\frac{\partial}{\partial x_3} - \frac{(1-\alpha)x_3}{2}\frac{\partial}{\partial x_4}, & e_2 &= x_2\frac{\partial}{\partial x_3},\\
e_3 &= \frac{1}{\sqrt{2}}\left\{x_2\frac{\partial}{\partial x_2} + (1+\alpha)x_3\frac{\partial}{\partial x_3} + (1+2\alpha)x_4\frac{\partial}{\partial x_4}\right\}, & e_4 &= \frac{1}{\sqrt{2}}x_2\frac{\partial}{\partial x_4},
\end{aligned}
$$

and

$$
\begin{aligned}
&[e_1,e_2] = \frac{1-\alpha}{\sqrt{2}}e_4, \quad [e_1,e_3] = -\frac{1}{\sqrt{2}}e_1 + \frac{(1-\alpha^2)x_4}{\sqrt{2}x_2}e_2 - \frac{(1-\alpha^2)x_3}{2x_2}e_4,\\
&[e_1,e_4] = -\frac{1+\alpha}{\sqrt{2}}e_2, \quad [e_2,e_3] = \frac{\alpha}{\sqrt{2}}e_2,
\end{aligned}
$$

$$[e_2, e_4] = 0, \qquad [e_3, e_4] = -\sqrt{2}\alpha e_4.$$

For the Levi-Civita connection $\nabla$ with respect to $g$ on $TS$, we have

$$(4.3) \qquad \begin{aligned}
(\nabla_{e_1} J)e_1 &= -(\nabla_{e_2} J)e_2 = -\frac{(1-\alpha^2)x_4}{2\sqrt{2}x_2}\, e_3, \\
(\nabla_{e_1} J)e_2 &= (\nabla_{e_2} J)e_1 = \frac{(1-\alpha^2)x_4}{2\sqrt{2}x_2}\, e_4, \\
(\nabla_{e_1} J)e_3 &= -(\nabla_{e_2} J)e_4 = \frac{(1-\alpha^2)x_4}{2\sqrt{2}x_2}\, e_1, \\
(\nabla_{e_1} J)e_4 &= (\nabla_{e_2} J)e_3 = -\frac{(1-\alpha^2)x_4}{2\sqrt{2}x_2}\, e_2, \\
(\nabla_{e_3} J)e_1 &= -(\nabla_{e_4} J)e_2 = \frac{(1-\alpha^2)x_3}{4x_2}\, e_3, \\
(\nabla_{e_3} J)e_2 &= (\nabla_{e_4} J)e_1 = -\frac{(1-\alpha^2)x_3}{4x_2}\, e_4, \\
(\nabla_{e_3} J)e_3 &= -(\nabla_{e_4} J)e_4 = -\frac{(1-\alpha^2)x_3}{4x_2}\, e_1, \\
(\nabla_{e_3} J)e_4 &= (\nabla_{e_4} J)e_3 = \frac{(1-\alpha^2)x_3}{4x_2}\, e_2.
\end{aligned}$$

In view of (4.3), we can reconfirm that $(TS, J^{(\alpha)}, g^{(\alpha)})$ is almost Kähler, and see that $(TS, J^{(\alpha)}, g^{(\alpha)})$ is Kähler if and only if $\alpha = \pm 1$. This fact coincides with the notice by [Ku].

The components of the Riemannian curvature tensor with respect to $\{e_i\}$ are as follows:

(4.4)

$$R_{1212} = -\Big(\frac{\alpha(3-\alpha)}{2} + \frac{(1-\alpha^2)^2 x_4^2}{8x_2^2}\Big), \qquad R_{1213} = \frac{\alpha(1-\alpha^2)x_3}{4\sqrt{2}x_2},$$

$$R_{1214} = \frac{(1-\alpha^2)^2 x_3 x_4}{8\sqrt{2}x_2^2}, \qquad R_{1223} = 0,$$

$$R_{1224} = \frac{\alpha(1-\alpha^2)x_3}{4\sqrt{2}x_2}, \qquad R_{1234} = \frac{1+3\alpha^2}{4},$$

$$R_{1313} = \frac{1}{2} + \frac{3(1-\alpha^2)^2(x_3^2 + 2x_4^2)}{16x_2^2}, \qquad R_{1314} = -\frac{\alpha(1-\alpha^2)x_4}{4x_2},$$

$$R_{1323} = \frac{\alpha(1-\alpha^2)x_4}{x_2}, \qquad R_{1324} = \frac{1}{2},$$

$$R_{1334} = \frac{5\alpha(1-\alpha^2)x_3}{4\sqrt{2}x_2}, \qquad R_{1414} = \frac{\alpha^2}{2} - \frac{(1-\alpha^2)^2 x_3^2}{16x_2^2},$$

$$R_{1423} = -\Big(\frac{\alpha^2}{2} - \frac{1-\alpha^2}{4}\Big), \qquad R_{1424} = \frac{\alpha(1-\alpha^2)x_4}{2x_2},$$

$$R_{1434} = 0, \qquad R_{2323} = \frac{\alpha^2}{2} - \frac{(1-\alpha^2)^2 x_4^2}{8x_2^2},$$

$$R_{2324} = \frac{\alpha(1-\alpha^2)x_4}{4x_2}, \qquad R_{2334} = -\frac{(1-\alpha^2)^2 x_3 x_4}{8\sqrt{2}x_2^2},$$

$$R_{2424} = \frac{\alpha^2}{2}, \qquad R_{2434} = -\frac{\alpha(1-\alpha^2)x_3}{4\sqrt{2}x_2},$$

$$R_{3434} = 2\alpha^2 - \frac{(1-\alpha^2)^2 x_3^2}{16x_2^2}.$$

From (4.4), the Ricci tensor $\rho$, the Ricci $*$-tensor $\rho^*$, the scalar curvature $\tau$ and the $*$-scalar curvature $\tau^*$ are given by respectively

$$\begin{aligned}
\rho_{11} &= -\frac{(1-\alpha)(1-2\alpha)}{2} - \frac{(1-\alpha^2)^2(x_3^2+2x_4^2)}{8x_2^2},\\
\rho_{22} &= \frac{3\alpha(1-\alpha)}{2} + \frac{(1-\alpha^2)^2 x_4^2}{4x_2^2},\\
\rho_{33} &= -\frac{1+5\alpha^2}{2} - \frac{(1-\alpha^2)^2(x_3^2+2x_4^2)}{8x_2^2},\\
\rho_{44} &= -3\alpha^2 + \frac{(1-\alpha^2)^2 x_3^2}{8x_2^2},\\
\rho_{12} &= -\frac{3\alpha(1-\alpha^2)x_4}{2x_2}, \qquad \rho_{14} = \frac{3\alpha(1-\alpha^2)x_3}{2\sqrt{2}x_2},\\
\rho_{24} &= -\frac{(1-\alpha^2)^2 x_3 x_4}{4\sqrt{2}x_2^2}, \qquad \rho_{13} = \rho_{23} = \rho_{34} = 0,
\end{aligned} \tag{4.5}$$

$$\begin{aligned}
\rho^*_{11} = \rho^*_{22} &= -\frac{(1-\alpha)(1-5\alpha)}{4} + \frac{(1-\alpha^2)^2 x_4^2}{8x_2^2},\\
\rho^*_{33} = \rho^*_{44} &= -\frac{1+11\alpha^2}{4} + \frac{(1-\alpha^2)^2 x_3^2}{16x_2^2},\\
\rho^*_{13} = \rho^*_{31} = \rho^*_{24} = \rho^*_{42} &= -\frac{(1-\alpha^2)^2 x_3 x_4}{8\sqrt{2}x_2^2},\\
\rho^*_{14} = -\rho^*_{32} &= \frac{3\alpha(1-\alpha^2)x_3}{2\sqrt{2}x_2},\\
\rho^*_{12} = \rho^*_{21} = \rho^*_{34} = \rho^*_{43} = \rho^*_{23} &= -\rho^*_{41} = 0,
\end{aligned} \tag{4.6}$$

$$\tau = -(1 - 3\alpha + 8\alpha^2) - \frac{(1-\alpha^2)^2(x_3^2 + 2x_4^2)}{8x_2^2}, \tag{4.7}$$

$$\tau^* = -(1 - 3\alpha + 8\alpha^2) + \frac{(1-\alpha^2)^2(x_3^2 + 2x_4^2)}{8x_2^2}. \tag{4.8}$$

Especially, in the case of $\alpha = -1$, we obtain the following

$$\begin{aligned} &R_{1212} = R_{3434} = 2, \\ &R_{1313} = R_{1414} = R_{2323} = R_{2424} = \frac{1}{2}, \\ &R_{1234} = 1, \\ &R_{1324} = -R_{1423} = \frac{1}{2}, \\ &R_{1213} = R_{1214} = R_{1223} = R_{1224} = 0, \\ &R_{1314} = R_{1323} = R_{1334} = R_{1424} = 0, \\ &R_{1434} = R_{2324} = R_{2334} = R_{2434} = 0, \end{aligned} \tag{4.9}$$

$$\rho = -3g, \tag{4.10}$$

$$\tau = -12. \tag{4.11}$$

(4.9), (4.10) and (4.11) show that $(TS, J^{(-1)}, g^{(-1)})$ is Kähler, Einstein, and moreover, it has constant holomorphic sectional curvature $-2$.

Note that, when $\alpha = 1$, $(TS, J^{(1)}, g^{(1)})$ is not Einstein, but has constant scalar curvature $\tau = -6$.

Summarizing the above results, we get the following

**Theorem 4.1.** *Let $(S, h)$ be the statistical model of normal distributions. The almost Kähler structure on $TS$ induced by the $\alpha$-connection $\nabla^{(\alpha)}$ is Kählerian if and only if $\alpha = \pm 1$, and if $\alpha = -1$, then $TS$ has constant holomorphic sectional curvature $-2$.*

### 4.2. *Tangent budle of the statistical model of discrete distributions*

Let $(S, h)$ be the 2-dimensional statistical model of discrete distributions on $\mathcal{X} = \{z_0, z_1, z_2\}$ with $\alpha$-connection $\nabla^{(\alpha)}$. For the sake of simplicity, we shall deal with only 2-dimensional case. Put $x_1 = \theta^1$, $x_2 = \theta^2$, $x_3 = y^1$, $x_4 = y^2$ for a coordinate system on $TS$ and $x_0 = 1 - x_1 - x_2$. We obtain

$$e_1 = \sqrt{\frac{x_0 x_1}{x_0 + x_1}} \left\{ \frac{\partial}{\partial x_1} - \Gamma_1^1 \frac{\partial}{\partial x_3} - \Gamma_1^2 \frac{\partial}{\partial x_4} \right\},$$

$$e_2 = \sqrt{\frac{x_0 x_1}{x_0 + x_1}} \frac{\partial}{\partial x_3},$$

$$e_3 = \sqrt{x_2(x_0+x_1)}\Big\{ -\frac{x_1}{x_0+x_1}\frac{\partial}{\partial x_1} + \frac{\partial}{\partial x_2} - (\Gamma_2^1 - \frac{x_1}{x_0+x_1}\Gamma_1^1)\frac{\partial}{\partial x_3} - (\Gamma_2^2 - \frac{x_1}{x_0+x_1}\Gamma_1^2)\frac{\partial}{\partial x_4}\Big\},$$

$$e_4 = \sqrt{x_2(x_0+x_1)}\Big\{ -\frac{x_1}{x_0+x_1}\frac{\partial}{\partial x_3} + \frac{\partial}{\partial x_4}\Big\},$$

where $\Gamma_i^k = \Gamma_i^{(\alpha)k}$, and

$$[e_1, e_2] = -\frac{\alpha}{2}\frac{x_0 - x_1}{x_0 + x_1}\sqrt{\frac{x_0+x_1}{x_0 x_1}}\, e_2 + \frac{1+\alpha}{2}\sqrt{\frac{x_2}{x_0+x_1}}\, e_4,$$

$$[e_1, e_3] = -\frac{1}{2}\sqrt{\frac{x_2}{x_0+x_1}}\, e_1 - \frac{1-\alpha^2}{4}\frac{x_4}{\sqrt{x_2(x_0+x_1)}}\, e_2 + \frac{1-\alpha^2}{4}\Big(\sqrt{\frac{x_0+x_1}{x_0x_1}}\, x_3 + \sqrt{\frac{x_1}{x_0(x_0+x_1)}}\, x_4\Big)\, e_4,$$

$$[e_1, e_4] = -\frac{1-\alpha}{2}\sqrt{\frac{x_2}{x_0+x_1}}\, e_2,$$

$$[e_2, e_3] = -\frac{\alpha}{2}\sqrt{\frac{x_2}{x_0+x_1}}\, e_2,$$

$$[e_2, e_4] = 0,$$

$$[e_3, e_4] = -\frac{\alpha}{2}\frac{1-2x_2}{\sqrt{x_2(x_0+x_1)}}\, e_4.$$

Straightforward computations show that

$$(4.12)\qquad (\nabla_{e_1}J)e_1 = -(\nabla_{e_2}J)e_2 = \frac{(1-\alpha^2)x_4}{8\sqrt{x_2(x_0+x_1)}}\, e_3,$$

$$(\nabla_{e_1}J)e_2 = (\nabla_{e_2}J)e_1 = -\frac{(1-\alpha^2)x_4}{8\sqrt{x_2(x_0+x_1)}}\, e_4,$$

$$(\nabla_{e_1}J)e_3 = -(\nabla_{e_2}J)e_4 = -\frac{(1-\alpha^2)x_4}{8\sqrt{x_2(x_0+x_1)}}\, e_1,$$

$$(\nabla_{e_1}J)e_4 = (\nabla_{e_2}J)e_3 = \frac{(1-\alpha^2)x_4}{8\sqrt{x_2(x_0+x_1)}}\, e_2,$$

$$(\nabla_{e_3}J)e_1 = -(\nabla_{e_4}J)e_2 = -\frac{1-\alpha^2}{8}\Big(\sqrt{\frac{x_0+x_1}{x_0x_1}}\, x_3 + \sqrt{\frac{x_1}{x_0(x_0+x_1)}}\, x_4\Big)\, e_3,$$

$$
\begin{aligned}
(\nabla_{e_3}J)e_2 &= (\nabla_{e_4}J)e_1 \\
&= \frac{1-\alpha^2}{8}\Big(\sqrt{\frac{x_0+x_1}{x_0x_1}}\,x_3 + \sqrt{\frac{x_1}{x_0(x_0+x_1)}}\,x_4\Big)\ e_4, \\
(\nabla_{e_3}J)e_3 &= -(\nabla_{e_4}J)e_4 \\
&= \frac{1-\alpha^2}{8}\Big(\sqrt{\frac{x_0+x_1}{x_0x_1}}\,x_3 + \sqrt{\frac{x_1}{x_0(x_0+x_1)}}\,x_4\Big)\ e_1, \\
(\nabla_{e_3}J)e_4 &= (\nabla_{e_4}J)e_3 \\
&= -\frac{1-\alpha^2}{8}\Big(\sqrt{\frac{x_0+x_1}{x_0x_1}}\,x_3 + \sqrt{\frac{x_1}{x_0(x_0+x_1)}}\,x_4\Big)\ e_2,
\end{aligned}
$$

where $\nabla$ is the Levi-Civita connection with respect to $g$ on $TS$.

In view of (4.12), we see again that $(TS, J^{(\alpha)}, g^{(\alpha)})$ is Kähler if and only if $\alpha = \pm 1$ (cf. [Ku]).

We can compute the components of the Riemannian curvature tensor, Ricci tensor, Ricci $*$-tensor and so on, as the prevoious example. However we shall skip these because it is somewhat long and tedious. The case of $\alpha = 1$ is remarkable; in this case, we have the following

$$
\begin{aligned}
&R_{1212} = R_{3434} = -1, \\
&R_{1313} = R_{1414} = R_{2323} = R_{2424} = -\frac{1}{4}, \\
&R_{1234} = -\frac{1}{2}, \\
&R_{1324} = -R_{1423} = -\frac{1}{4}, \\
&R_{1213} = R_{1214} = R_{1223} = R_{1224} = 0, \\
&R_{1314} = R_{1323} = R_{1334} = R_{1424} = 0, \\
&R_{1434} = R_{2324} = R_{2334} = R_{2434} = 0.
\end{aligned} \tag{4.13}
$$

(4.13) shows that $(TS, J^{(1)}, g^{(1)})$ is Kähler manifold of constant holomorphic sectional curvature 1. Of course, it is Einstein $\rho = \frac{3}{2}g$, and has constant scalar curvature $\tau = 6$.

On the contrary, when $\alpha = -1$, we see that $(TS, J^{(-1)}, g^{(-1)})$ is Kähler manifold with non-constant scalar curvature $\tau = 3 - (\frac{1}{x_1} + \frac{1}{x_2} + \frac{1}{1-x_1-x_2})$.

Consequently, we obtain the following

**Theorem 4.2.** *Let $(S, h)$ be the 2-dimensional statistical model of discrete distributions on a 3-points set $\mathcal{X}$. The almost Kähler structure on $TS$ induced by the $\alpha$-connection $\nabla^{(\alpha)}$ is Kählerian if and only if $\alpha = \pm 1$, and if $\alpha = 1$, then $TS$ has constant holomorphic sectional curvature 1.*

## References


Am. S. Amari, *Differntial-Geometrical Methods in Statistics*, Lecture Notes in Statistics 28. Springer-Verlag, 1985.

A-N. S. Amari and H. Nagaoka, *Methods of Information Geometry*, American Mathematical Society, Oxford University Press, 2000.

Ku. T. Kurose, Geometry of statistical manifolds, Mathematics in the 21st century, 2004, 34–43 (in Japanese).

T-O. S. Tachibana and M. Okumura, On the almost complex structure of tangent bundles of Riemannian spaces, Tôhoku Math. J. 14(1962), 156–161.

Y-I. K. Yano and S. Ishihara, *Tangent and Cotangent bundles*, Marcel Dekker, INC. New York, 1973.

# RECENT PROGRESS ON REAL HYPERSURFACES IN COMPLEX TWO-PLANE GRASSMANNIANS

YOUNG JIN SUH

*Kyungpook National University,*
*Department of Mathematics,*
*Taegu 702-701, Korea*
*E-mail: yjsuh@wmail.knu.ac.kr*

In this survey we consider the notion of $\xi$-invariant or $\mathfrak{D}^{\perp}$-invariant real hypersurfaces in complex two-plane Grassmannians $G_2(\mathbb{C}^{m+2})$ and give a complete classification for real hypersurfaces satisfying these both two conditions. On the other hand, we prove that there do not exist any real hypersurfaces in $G_2(\mathbb{C}^{m+2})$ with parallel second fundamental tensor and give some generalizations of such situations for real hypersurfaces in $G_2(\mathbb{C}^{m+2})$ with parallel second fundamental tensor on a distribution $\mathfrak{F}$ defined by $\mathfrak{F} = \xi \cup \mathfrak{D}^{\perp}$, where $\mathfrak{D}^{\perp} = \text{Span }\{\xi_1, \xi_2, \xi_3\}$. Related to this topics we introduce some works due to Berndt and Suh [4], Suh [14] for the characterizations of type $A$ with isometric Reeb flow or vanishing Lie derivative. Finally we introduce a new characterization of type $B$ in the paper due to Suh [15].

## 1. Riemannian geometry of $G_2(\mathbb{C}^{m+2})$

Now let us denote by $G_2(\mathbb{C}^{m+2})$ the set of all complex 2-dimensional linear subspaces in $\mathbb{C}^{m+2}$. The special unitary group $G = SU(m+2)$ acts transitively on $G_2(\mathbb{C}^{m+2})$ with stabilizer isomorphic to $K = S(U(2) \times U(m)) \subset G$. Then $G_2(\mathbb{C}^{m+2})$ can be identified with the homogeneous space $G/K$, which we equip with the unique analytic structure for which the natural action of $G$ on $G_2(\mathbb{C}^{m+2})$ becomes analytic. Denote by $\mathfrak{g}$ and $\mathfrak{k}$ the Lie algebra of $G$ and $K$, respectively, and by $\mathfrak{m}$ the orthogonal complement of $\mathfrak{k}$ in $\mathfrak{g}$ with respect to the Cartan-Killing form $B$ of $\mathfrak{g}$. Then $\mathfrak{g} = \mathfrak{k} \oplus \mathfrak{m}$ is an $Ad(K)$-invariant reductive decomposition of $\mathfrak{g}$. We put $o = eK$ and identify $T_oG_2(\mathbb{C}^{m+2})$ with $\mathfrak{m}$ in the usual manner. Since $B$ is negative definite on $\mathfrak{g}$, its negative restricted to $\mathfrak{m} \times \mathfrak{m}$ yields a positive definite inner product

2000 *Mathematics Subject Classification.* Primary 53C40; Secondary 53C15.
This work was supported by grant Proj. No. R14-2002-003-01001-0 from the Korea Science & Engineering Foundation.

on $\mathfrak{m}$. By $Ad(K)$-invariance of $B$ this inner product can be extended to a $G$-invariant Riemannian metric $g$ on $G_2(\mathbb{C}^{m+2})$. In this way $G_2(\mathbb{C}^{m+2})$ becomes a Riemannian homogeneous space, even a Riemannian symmetric space. For computational reasons we normalize $g$ such that the maximal sectional curvature of $(G_2(\mathbb{C}^{m+2}), g)$ is eight.

The Lie algebra $\mathfrak{k}$ has the direct sum decomposition $\mathfrak{k} = \mathfrak{su}(m) \oplus \mathfrak{su}(2) \oplus \mathfrak{R}$, where $\mathfrak{R}$ is the center of $\mathfrak{k}$. Viewing $\mathfrak{k}$ as the holonomy algebra of $G_2(\mathbb{C}^{m+2})$, the center $\mathfrak{R}$ induces a Kähler structure $J$ and the $\mathfrak{su}(2)$-part a quaternionic Kähler structure $\mathfrak{J}$ on $G_2(\mathbb{C}^{m+2})$. If $J_1$ is any almost Hermitian structure in $\mathfrak{J}$, then $JJ_1 = J_1J$, and $JJ_1$ is a symmetric endomorphism with $(JJ_1)^2 = I$ and $\mathrm{Tr}\ (JJ_1) = 0$.

A canonical local basis $J_1, J_2, J_3$ of $\mathfrak{J}$ consists of three local almost Hermitian structures $J_\nu$ in $\mathfrak{J}$ such that $J_\nu J_{\nu+1} = J_{\nu+2} = -J_{\nu+1}J_\nu$, where the index is taken modulo three. Since $\mathfrak{J}$ is parallel with respect to the Riemannian connection $\bar{\nabla}$ of $(G_2(\mathbb{C}^{m+2}), g)$, there exist for any canonical local basis $J_1, J_2, J_3$ of $\mathfrak{J}$ three local one-forms $q_1, q_2, q_3$ such that

$$\bar{\nabla}_X J_\nu = q_{\nu+2}(X)J_{\nu+1} - q_{\nu+1}(X)J_{\nu+2} \tag{1.1}$$

for all vector fields $X$ on $G_2(\mathbb{C}^{m+2})$.

The Riemannian curvature tensor $\bar{R}$ of $G_2(\mathbb{C}^{m+2})$ is locally given by

$$\begin{aligned}\bar{R}(X,Y)Z &= g(Y,Z)X - g(X,Z)Y + g(JY,Z)JX \\ &\quad - g(JX,Z)JY - 2g(JX,Y)JZ \\ &\quad + \sum_{\nu=1}^{3}\{g(J_\nu Y,Z)J_\nu X - g(J_\nu X,Z)J_\nu Y \\ &\quad - 2g(J_\nu X,Y)J_\nu Z\} \\ &\quad + \sum_{\nu=1}^{3}\{g(J_\nu JY,Z)J_\nu JX - g(J_\nu JX,Z)J_\nu JY\},\end{aligned} \tag{1.2}$$

where $J_1, J_2, J_3$ is any canonical local basis of $\mathfrak{J}$.

## 2. Some fundamental formulas for real hypersurfaces in $G_2(\mathbb{C}^{m+2})$

In this section we derive some fundamental formulas which will be used in the proof of our theorems (See [2],[3],[4],[11],[12],[13] and [14]).

Let $M$ be a real hypersurface in $G_2(\mathbb{C}^{m+2})$, that is, a hypersurface in $G_2(\mathbb{C}^{m+2})$ with real codimension one. The induced Riemannian metric on $M$ will also be denoted by $g$, and $\nabla$ denotes the Riemannian connection of

$(M,g)$. Let $N$ be a local unit normal field of $M$ and $A$ the shape operator of $M$ with respect to $N$.

The Kähler structure $J$ of $G_2(\mathbb{C}^{m+2})$ induces on $M$ an almost contact metric structure $(\phi,\xi,\eta,g)$. Furthermore, let $J_1,J_2,J_3$ be a canonical local basis of $\mathfrak{J}$. Then each $J_\nu$ induces an almost contact metric structure $(\phi_\nu,\xi_\nu,\eta_\nu,g)$ on $M$. Using the above expression (1.2) for the curvature tensor $\bar{R}$, the Codazzi equation is respectively given by

$$\begin{aligned}(\nabla_X A)Y-(\nabla_Y A)X &= \eta(X)\phi Y-\eta(Y)\phi X-2g(\phi X,Y)\xi\\ &\quad+\sum_{\nu=1}^{3}\{\eta_\nu(X)\phi_\nu Y-\eta_\nu(Y)\phi_\nu X-2g(\phi_\nu X,Y)\xi_\nu\}\\ &\quad+\sum_{\nu=1}^{3}\{\eta_\nu(\phi X)\phi_\nu\phi Y-\eta_\nu(\phi Y)\phi_\nu\phi X\}\\ &\quad+\sum_{\nu=1}^{3}\{\eta(X)\eta_\nu(\phi Y)-\eta(Y)\eta_\nu(\phi X)\}\xi_\nu\,.\end{aligned}$$

The following identities can be proved in a straightforward method and will be used frequently in subsequent calculations:

$$\begin{aligned}&\phi_{\nu+1}\xi_\nu=-\xi_{\nu+2},\quad \phi_\nu\xi_{\nu+1}=\xi_{\nu+2},\\ &\phi\xi_\nu=\phi_\nu\xi,\quad \eta_\nu(\phi X)=\eta(\phi_\nu X),\\ &\phi_\nu\phi_{\nu+1}X=\phi_{\nu+2}X+\eta_{\nu+1}(X)\xi_\nu,\\ &\phi_{\nu+1}\phi_\nu X=-\phi_{\nu+2}X+\eta_\nu(X)\xi_{\nu+1}.\end{aligned}\tag{2.1}$$

Now let us put

$$JX=\phi X+\eta(X)N,\quad J_\nu X=\phi_\nu X+\eta_\nu(X)N$$

for any tangent vector $X$ of a real hypersurface $M$ in $G_2(\mathbb{C}^{m+2})$, where $N$ denotes a normal vector of $M$ in $G_2(\mathbb{C}^{m+2})$. Then from this and the formulas in section 1 we have that

$$(\nabla_X\phi)Y=\eta(Y)AX-g(AX,Y)\xi,\quad \nabla_X\xi=\phi AX,\tag{2.2}$$

$$\nabla_X\xi_\nu=q_{\nu+2}(X)\xi_{\nu+1}-q_{\nu+1}(X)\xi_{\nu+2}+\phi_\nu AX,\tag{2.3}$$

$$\begin{aligned}(\nabla_X\phi_\nu)Y&=-q_{\nu+1}(X)\phi_{\nu+2}Y+q_{\nu+2}(X)\phi_{\nu+1}Y+\eta_\nu(Y)AX\\ &\quad-g(AX,Y)\xi_\nu.\end{aligned}\tag{2.4}$$

Summing up these formulas, we know the following

$$\begin{aligned}(2.5)\qquad \nabla_X(\phi_\nu\xi) &= (\nabla_X\phi_\nu)\xi + \phi_\nu(\nabla_X\xi)\\ &= -q_{\nu+1}(X)\phi_{\nu+2}\xi + q_{\nu+2}(X)\phi_{\nu+1}\xi + \eta_\nu(\xi)AX\\ &\quad - g(AX,\xi)\xi_\nu + \phi_\nu\phi AX.\end{aligned}$$

Moreover, from $JJ_\nu = J_\nu J$, $\nu = 1, 2, 3$, it follows that

$$(2.6)\qquad \phi\phi_\nu X = \phi_\nu\phi X + \eta_\nu(X)\xi - \eta(X)\xi_\nu.$$

## 3. Recent results for real hypersurfaces in $G_2(\mathbb{C}^{m+2})$

In Riemannian geometry, the general problem in submanifold theory is to determine or describe the submanifolds of a given Riemannian manifold which satisfy certain geometrical data. Submanifold theory in Euclidean spaces is classically well-established in spaces of constant curvature . In more general ambient spaces one encounters two basic problems. On the one hand, the fundamental equations for submanifolds become rather complicated and are difficult to handle. On the other hand, it is not easy to decide what data are natural to consider, they should somehow be related to the geometrical structure of the ambient space.

To give an example, consider real hypersurfaces $M$ in a complex projective space $\mathbb{C}P^m$. The Codazzi equation is rather complicated in the generic case, but simplifies considerably when requiring that the one-dimensional distribution $J(\perp M)$ on $M$ obtained by applying the Kähler structure $J$ of $\mathbb{C}P^m$ to the normal bundle $\perp M$ of $M$ is invariant under the shape operator $A$ of $M$. The condition $AJ(\perp M) \subset J(\perp M)$ appears to be rather natural, and in fact there is a well-established theory for such hypersurfaces. Any tube around a complex submanifold in $\mathbb{C}P^m$ satisfies this geometrical condition. Cecil and Ryan proved in [5] that these tubes are essentially characterized by this feature. Here the word essentially refers to some additional condition on the focal map. So, roughly speaking, the theory of real hypersurfaces in $\mathbb{C}P^m$ with $AJ(\perp M) \subset J(\perp M)$ is the theory of tubes around complex submanifolds in $\mathbb{C}P^m$, where $J(\perp M)$ is also denoted by $[\xi] = \text{Span}\{\xi\}$, $\xi = -JN$.

The analogous question in quaternionic projective space $\mathbb{H}P^m$ leads to a surprise. The corresponding geometrical feature is that the three dimensional distribution $\mathfrak{J}(\perp M)$ on $M$, which is obtained by applying the almost Hermitian structures in the quaternionic Kähler structure $\mathfrak{J}$ of $\mathbb{H}P^m$ to $\perp M$, is invariant under $A$. In fact, every tube around a quaternionic submanifold of $\mathbb{H}P^m$ has this geometrical feature. (Note that by a result of Alekseevskii

[1] such a quaternionic submanifold is necessarily totally geodesic.) But the converse is not true. Berndt proved in [2] that also every tube around a totally geodesic $\mathbb{C}P^m$ in $\mathbb{H}P^m$ satisfies $A\mathfrak{J}(\perp M) \subset \mathfrak{J}(\perp M)$, and that there are no other ones. So the real hypersurfaces in $\mathbb{H}P^m$ with $A\mathfrak{J}(\perp M) \subset \mathfrak{J}(\perp M)$ are precisely the tubes around totally geodesic $\mathbb{H}P^k$, $k \in \{0, \dots, m-1\}$, and $\mathbb{C}P^m$, where $\mathfrak{J}(\perp M)$ is also denoted by $\mathfrak{D}^\perp = \mathrm{Span}\{\xi_1, \xi_2, \xi_3\}$, $\xi_\nu = -J_\nu N$, $\nu = 1, 2, 3$.

In this survey we study the analogous question in the complex Grassmann manifold $G_2(\mathbb{C}^{m+2})$ of all two-dimensional linear subspaces in $\mathbb{C}^{m+2}$. This Riemannian symmetric space has a remarkable geometrical structure. It is the unique compact irreducible Riemannian manifold being equipped with both a Kähler structure $J$ and a quaternionic Kähler structure $\mathfrak{J}$ not containing $J$. In other words, $G_2(\mathbb{C}^{m+2})$ is the unique compact, irreducible, Kähler, quaternionic Kähler manifold which is not a hyperkähler manifold. So, in $G_2(\mathbb{C}^{m+2})$ we have the two natural geometrical conditions for real hypersurfaces that $J(\perp M) = [\xi]$ and $\mathfrak{J}(\perp M) = \mathfrak{D}^\perp$ are invariant under the shape operator. The main result of this survey is the classification of all real hypersurfaces in $G_2(\mathbb{C}^{m+2})$ satisfying both conditions. From such a view point Berndt and the present author [3] have proved the following:

**Theorem 1.** *Let $M$ be a connected real hypersurface in $G_2(\mathbb{C}^{m+2})$, $m \geq 3$. Then both $[\xi]$ and $\mathfrak{D}^\perp$ are invariant under the shape operator of $M$ if and only if*

(1) *$M$ is an open part of a tube around a totally geodesic $G_2(\mathbb{C}^{m+1})$ in $G_2(\mathbb{C}^{m+2})$, or*
(2) *$m$ is even, say $m = 2n$, and $M$ is an open part of a tube around a totally geodesic $\mathbb{H}P^n$ in $G_2(\mathbb{C}^{m+2})$.*

Next motivation for this survey is to answer the problem: *For a given almost Hermitian manifold $\bar{M}$, classify all orientable real hypersurfaces $M$ in $\bar{M}$ for which the Reeb flow is isometric.* The almost Hermitian structure on $\bar{M}$ induces an almost contact metric structure on $M$. The corresponding unit tangent vector field on $M$ is the Reeb vector field, and its flow is the Reeb flow on $M$.

A classical example is the unit sphere $S^{2m-1}$ in $\mathbb{C}^m$, where the orbits of the Reeb flow induce the Hopf foliation on $S^{2m-1}$ by great circles. It is easy to prove that every connected real hypersurface in $\mathbb{C}^m$ with isometric Reeb flow is, up to isometric congruence, an open part of a hypersphere in $\mathbb{C}^m$, or of a tube around a complex subspace $\mathbb{C}^k \subset \mathbb{C}^m$ $(1 \leq k \leq m-1)$, or of the linear subspace $\mathbb{R}^{2m-1} \subset \mathbb{R}^{2m} \simeq \mathbb{C}^m$. The classification of all real hypersurfaces in complex projective space $\mathbb{C}P^m$ with isometric Reeb flow has been obtained by Okumura [8], and the corresponding classification in complex hyperbolic space $\mathbb{C}H^m$ is due to Montiel and Romero [7]. Then in this survey we will show that the classification of all real hypersurfaces with isometric Reeb flow in complex two-plane Grassmannians $G_2(\mathbb{C}^{m+2})$ due to Berndt and the present author [4] as follows:

**Theorem 2.** *Let $M$ be a connected orientable real hypersurface in $G_2(\mathbb{C}^{m+2})$, $m \geq 3$. Then the Reeb flow on $M$ is isometric if and only if $M$ is an open part of a tube around some totally geodesic $G_2(\mathbb{C}^{m+1})$ in $G_2(\mathbb{C}^{m+2})$.*

When the Reeb flow on $M$ in $G_2(\mathbb{C}^{m+2})$ is *isometric* in Theorem 2, we say that the Reeb vector field $\xi$ on $M$ is Killing. This means that the metric tensor $g$ is invariant under the Reeb flow of $\xi$ on $M$, that is, $\mathcal{L}_\xi g = 0$, where $\mathcal{L}_\xi$ denotes the Lie derivative along the direction of the structure vector field $\xi$. Moreover, this is equivalent to the fact that the structure tensor $\phi$ commutes with the shape operator $A$ of $M$ in $G_2(\mathbb{C}^{m+2})$.

In this survey, specifically we also give another characterization of real hypersurfaces of type $A$ in Theorem 1 by geometric Lie invariant, that is, the shape operator $A$ of $M$ in $G_2(\mathbb{C}^{m+2})$ is invariant under the Reeb flow on $M$. That is, we give a result due to Suh [14] as follows:

**Theorem 3.** *Let $M$ be a connected orientable real hypersurface in $G_2(\mathbb{C}^{m+2})$, $m \geq 3$. Then the Reeb flow on $M$ satisfies $\mathcal{L}_\xi A = 0$ if and only if $M$ is an open part of a tube around some totally geodesic $G_2(\mathbb{C}^{m+1})$ in $G_2(\mathbb{C}^{m+2})$.*

In the geometry of real hypersurfaces in complex space forms $M_n(c)$ or in quaternionic space forms it can be easily checked that there do not exist any real hypersurfaces with parallel shape operator $A$ by virtue of the equation of Codazzi.

From this point of a view many differential geometers have considered much more weaker notion than such a parallel second fundamental form, that is, $\nabla A = 0$. In particular, Kimura and Maeda [6] have proved that a real hypersurface $M$ in a complex projective space $\mathbb{C}P^m$ satisfying $\nabla_\xi A = 0$ is locally congruent to a real hypersurface of type $A_1, A_2$, that is, a tube over a totally geodesic complex submanifold $\mathbb{C}P^k$ with radius $0 < r < \frac{\pi}{2}$. The structure vector field $\xi$ mentioned above is defined by $\xi = -JN$, where $J$ denotes a Kähler structure of $\mathbb{C}P^m$ and $N$ a local unit normal field of $M$ in $\mathbb{C}P^m$. A real hypersurface $M$ in $\mathbb{C}P^m$ is said to be *Hopf* if the structure vector field $\xi$ is principal. Moreover, in a class of Hopf hypersurfaces it was proved that there do not exist any real hypersurfaces with parallel Ricci tensor, that is $\nabla S = 0$, where $S$ denotes the Ricci tensor of a real hypersurface $M$ in $\mathbb{C}P^m$.

On the other hand, in a quaternionic projective space $\mathbb{H}P^m$ Pérez [9] has considered the notion of $\nabla_{\xi_i} A = 0$, $i = 1, 2, 3$, for real hypersurfaces in $\mathbb{H}P^m$ and classified that $M$ is locally congruent to of $A_1, A_2$-type, that is, a tube over $\mathbb{H}P^k$ with radius $0 < r < \pi/4$. The almost contact structure vector fields $\{\xi_1, \xi_2, \xi_3\}$ are defined by $\xi_i = -J_i N$, $i = 1, 2, 3$, where $J_i$ denotes a quaternionic Kähler structure of $\mathbb{H}P^m$ and $N$ a unit normal field of $M$ in $\mathbb{H}P^m$. Moreover, Pérez and the present author [10] have considered the notion of $\nabla_{\xi_i} R = 0$, $i = 1, 2, 3$, where $R$ denotes the curvature tensor of a real hypersurface $M$ in $\mathbb{H}P^m$, and proved that $M$ is locally congruent to a tube of radius $\pi/4$ over $\mathbb{H}P^k$.

Now we consider a complex two-plane Grassmannian $G_2(\mathbb{C}^{m+2})$ consisting of the set of all complex 2-dimensional linear subspaces in $\mathbb{C}^{m+2}$. Then the situation mentioned above is not so simple if we consider a real hypersurface in such a complex two-plane Grassmannian $G_2(\mathbb{C}^{m+2})$.

In the proof of Theorem 1 we have shown that the one-dimensional distribution $[\xi]$ is contained in either the 3-dimensional distribution $\mathfrak{D}^\perp$ or in the orthogonal complement $\mathfrak{D}$ such that $T_x M = \mathfrak{D} \oplus \mathfrak{D}^\perp$. The case (1) in Theorem 1, which is said to be of type $A$, is just the case that the one dimensional distribution $[\xi]$ is contained in $\mathfrak{D}^\perp$. Of course it is not difficult to check that the second fundamental form of any real hypersurfaces given in Theorem 1 is not parallel. Then it must be natural question to know that whether real hypersurfaces in $G_2(\mathbb{C}^{m+2})$ whose second fundamental form is parallel can exist or not?

According to such a question we are going to show another result due to the present author [11] for the non-existence of all real hypersurfaces in $G_2(\mathbb{C}^{m+2})$ with parallel second fundamental tensor as follows:

**Theorem 4.** *There do not exist any real hypersurface $M$ in $G_2(\mathbb{C}^{m+2})$ with parallel second fundamental form, that is, $\nabla A = 0$.*

The proof of Theorem 4 also works for the more general case of Codazzi hypersurfaces, that is, for hypersurfaces whose shape operator is a Codazzi tensor, which means that the covariant derivative of the shape operator is symmetric in all its variables. Recently, more generalized situations better than Theorem 4 are also given in the works due to the present author [12] and [13].

In Theorem 1 the vector $\xi$ contained in the one-dimensional distribution $[\xi]$ is said to be a *Hopf* vector when it becomes a principal vector for the shape operator $A$ of $M$ in $G_2(\mathbb{C}^{m+2})$. Moreover in such a situation $M$ is said to be a *Hopf* hypersurface. Besides of this, a real hypersurface $M$ in $G_2(\mathbb{C}^{m+2})$ also admits the 3-dimensional distribution $\mathfrak{D}^\perp$, which are spanned by almost contact 3-structure vector fields $\{\xi_1, \xi_2, \xi_3\}$, such that $T_xM = \mathfrak{D} \oplus \mathfrak{D}^\perp$.

Now in this survey let us consider a distribution $\mathfrak{F}$ combined by $[\xi]$ and $\mathfrak{D}^\perp$ in such a way that $\mathfrak{F} = [\xi] \cup \mathfrak{D}^\perp$ for a real hypersurface $M$ in $G_2(\mathbb{C}^{m+1})$. Then naturally as in above classifications we can restrict the condition $\nabla A = 0$ to $\mathfrak{F}$, which means that the second fundamental tensor of $M$ in $G_2(\mathbb{C}^{m+2})$ is parallel on $\mathfrak{F}$. Of course, this condition is much more weaker than $\nabla A = 0$ as in Theorem 4. Then by virtue of Theorem 1 we assert the following remarkable facts :

**Theorem 5.** *There do not exist any real hypersurface $M$ in $G_2(\mathbb{C}^{m+2})$ with the parallel second fundamental tensor on $\mathfrak{F}$ when $\xi$ is invariant by the shape operator of $M$.*

When the structure vector field $[\xi]$ of a real hypersurface $M$ in $G_2(\mathbb{C}^{m+2})$ is invariant by the shape operator, that is, $A\xi = \alpha\xi$, a real hypersurface $M$ is said to be a *Hopf* hypersurface. When the distribution $\mathfrak{D}^\perp$ of a hypersurface $M$ in $G_2(\mathbb{C}^{m+2})$ is invariant by the shape operator, we say $M$ a $\mathfrak{D}^\perp$-*invariant* hypersurface.

Theorem 1 means that in the class of *Hopf* real hypersurfaces in $G_2(\mathbb{C}^{m+2})$ there do not exist any real hypersurface with parallel second fundamental tensor on $\mathfrak{F}$. Now we also assert the following.

**Theorem 6.** *There do not exist any real hypersurface $M$ in $G_2(\mathbb{C}^{m+2})$ with the parallel second fundamental tensor on $\mathfrak{F}$ when $\mathfrak{D}^\perp$ is invariant by the shape operator of $M$.*

Real hypersurfaces of type $B$ in (2) of Theorem 1 are just the case that the one dimensional distribution $[\xi]$ is contained in $\mathfrak{D}$. It can be easily verified in [12] that the tube of type $B$ satisfies the following formula on the orthogonal complement of the one-dimensional distribution $[\xi]$

$$(*) \qquad A\phi_\nu - \phi_\nu A = 0, \quad \nu = 1, 2, 3.$$

Also in Theorem 2 we have given a characterization of real hypersurfaces of type $A$ when the shape operator $A$ of $M$ in $G_2(\mathbb{C}^{m+2})$ commutes with the structure tensor $\phi$, which is equivalent to the condition that the Reeb flow on $M$ is isometric.

Now in this survey we consider another condition that the almost contact 3-structure tensors $\{\phi_1, \phi_2, \phi_3\}$ and the shape operator $A$ of a real hypersurface $M$ in $G_2(\mathbb{C}^{m+2})$ commute with each other. Then we are able to assert the following:

**Theorem 7.** *There do not exist any real hypersurfaces of $M$ in $G_2(\mathbb{C}^{m+2})$ satisfying the condition that the shape operator $A$ and the structure tensors $\{\phi_1, \phi_2, \phi_3\}$ commute with each other.*

From this view point in Theorem 7 and the argument mentioned above we give a characterization of real hypersurfaces of type $B$ in $G_2(\mathbb{C}^{m+2})$. Namely, we want to give a characterization of type $B$ in Theorem 1 as follows:

**Theorem 8.** *Let $M$ be a Hopf real hypersurface in $G_2(\mathbb{C}^{m+2})$ satisfying $(*)$ on the orthogonal complement of the one-dimensional disribution $[\xi]$. Then $M$ is locally congruent to an open part of a tube around a totally geodesic $\mathbb{H}P^n$ in $G_2(\mathbb{C}^{m+2})$, where $m = 2n$.*

Let $M$ be a $(4m-1)$-dimensional Riemannian manifold with an almost contact structure $(\phi, \xi, \eta)$ and an associated Riemannian metric $g$. We put

$$\omega(X, Y) = g(\phi X, Y),$$

where $\omega$ defines a 2-form on $M$ and $\operatorname{rank} \omega = \operatorname{rank} \phi = 4m-2$.

If there is a non-zero valued function $\rho$ such that

$$(**) \qquad \rho g(\phi X, Y) = \rho\omega(X, Y) = d\eta(X, Y),$$

the rank of the matrix $(\omega)$ being $4m-2$, we have

$$\eta \wedge \overbrace{\omega \wedge \cdots \wedge \omega}^{2m-1 \text{ times}} = \eta \wedge \rho^{-(2m-1)} \overbrace{d\eta \wedge \cdots \wedge d\eta}^{2m-1 \text{ times}} \neq 0.$$

Now we consider a $(4m-1)$-dimensional real hypersurface $M$ in complex two-plane Grassmannians $G_2(\mathbb{C}^{m+2})$. Then from the Kähler structure of $G_2(\mathbb{C}^{m+2})$ there exist an almost contact structure $\phi$ on $M$. If the non-zero function $\rho$ satisfies the formula $(**)$, we call $M$ a *contact* hypersurface of the Kähler manifold. Moreover, it can be easily verified that a real hypersurface $M$ in $G_2(\mathbb{C}^{m+2})$ is *contact* if and only if there exists a non-zero constant function $\rho$ defined on $M$ such that

$$(***) \qquad \phi A + A\phi = k\phi, \quad k = 2\rho.$$

The formula (***) means that

$$g((\phi A + A\phi)X, Y) = 2d\eta(X, Y),$$

where the exterior derivative $d\eta$ of the 1-form $\eta$ is defined by

$$d\eta(X, Y) = (\nabla_X \eta)Y - (\nabla_Y \eta)X$$

for any vector fields $X, Y$ on $M$ in $G_2(\mathbb{C}^{m+2})$.

Then finally as a new characterization of type $B$ in (2) of Theorem 1 we introduce the following due to Suh [15].

**Theorem 9.** *Let $M$ be a real hypersurface in $G_2(\mathbb{C}^{m+2})$ satisfying the formula*

$$A\phi + \phi A = k\phi,$$

*where the function $k$ is non-zero and constant. Then the following are equivalent to each other*
(1) *the mean curvature is constant,*
(2) *the function $\alpha$ is constant,*
(3) $\xi \in \mathfrak{D}$ *or* $\xi \in \mathfrak{D}^\perp$.

By virtue of this theorem we also assert the following

**Theorem 10.** *Let $M$ be a real hypersurface in $G_2(\mathbb{C}^{m+2})$ with constant mean curvature satisfying the formula*

$$A\phi + \phi A = k\phi,$$

*where the function $k$ is non-zero and constant. Then we have the following*
(1) *The structure vector field $\xi$ is principal,*
(2) *The function $\alpha$ is constant,*
(3) $\xi \in \mathfrak{D}$ *or* $\xi \in \mathfrak{D}^\perp$.

Now in this survey, by using Theorems 9 and 10 we give another characterization of real hypersurfaces of type $B$ in complex two-plane Grassmannians $G_2(\mathbb{C}^{m+2})$ in terms of the *contact* hypersurface as follows:

**Theorem 11.** *Let $M$ be a real hypersurface in $G_2(\mathbb{C}^{m+2})$ with constant mean curvature satisfying*

$$A\phi + \phi A = k\phi,$$

*where the function $k$ is non-zero and constant. Then $M$ is congruent to an open part of a tube around a totally geodesic $\mathbb{H}P^n$ in $G_2(\mathbb{C}^{m+2})$, where $m = 2n$.*

## References


1. D.V. Alekseevskii, *Compact quaternion spaces*, Funct. Anal. Appl. **2** (1968), 106–114.
2. J. Berndt, *Riemannian geometry of complex two-plane Grassmannians*, Rend. Sem Mat. Univ. Politec Torino **55** (1997), 19–83.
3. J. Berndt and Y.J. Suh, *Real hypersurfaces in complex two-plane Grassmannians*, Monatshefte für Math. **127** (1999), 1–14.
4. J. Berndt and Y.J. Suh, *Isometric flows on real hypersurfaces in complex two-plane Grassmannians*, Monatshefte für Math. **137** (2002), 87–98.
5. T.E. Cecil and P.J. Ryan, *Focal sets and real hypersurfaces in complex projective space*, Trans. Amer. Math. Soc. **269** (1982), 481–499.
6. M. Kimura and S. Maeda, *On real hypersurfaces of a complex projective space II*, Tsukuba J. Math. **15** (1991), 547–561.
7. S. Montiel and A. Romero, *On some real hypersurfaces of a complex hyperbolic space*, Geometriae Dedicata **20** (1986), 245–261.
8. M. Okumura, *On some real hypersurfaces of a complex projective space*, Trans. Amer. Math. Soc. **212** (1975), 355–364.
9. J.D. Pérez, *Real hypersurfaces of quaternionic projective space satisfying* $\nabla_{U_i} A = 0$, J. Geom. **49** (1994), 166–177.
10. J.D. Pérez and Y.J. Suh, *Real hypersurfaces of quaternionic projective space satisfying* $\nabla_{U_i} R = 0$, Diff. Geom. and Its Appl. **7** (1997), 211–217.
11. Y.J. Suh, *Real hypersurfaces in complex two-plane Grassmannians with parallel shape operator*, Bull. of Austral. Math. Soc. **68** (2003), 493–502.
12. Y.J. Suh, *Real hypersurfaces in complex two-plane Grassmannians with commuting shape operator*, Bull. of Austral. Math. Soc. **69** (2003), 379–393.
13. Y.J. Suh, *Real hypersurfaces in complex two-plane Grassmannians with parallel shape operator II*, J. of Korean Math. Soc. **41** (2004), 535–565.
14. Y.J. Suh, *Real hypersurfaces of type A in complex two-plane Grassmannians with vanishing Lie derivatives*, to appear in Canadian Math. Bull. **48** (2004).
15. Y.J. Suh, *Real hypersurfaces of type B in complex two-plane Grassmannians*, Submitted.

# CURVATURE OF THE DETERMINANT LINE BUNDLE ON THE MODULI SPACE OF NULL-CORRELATION BUNDLES

TADASHI TANIGUCHI

*Sendai National College of Technology,*
*Aobaku, Sendai 989-3128, Japan*
*E-mail: ttani@cc.sendai-ct.ac.jp*

The moduli space $\mathcal{N}(\mathbb{C}P^{2n+1})$ of null-correlation bundles over $\mathbb{C}P^{2n+1}$ has a natural $L_2$-Kähler metric. We describe explicitly the $L_2$-Kähler metrics of the moduli spaces $\mathcal{N}(\mathbb{C}P^3)$ and $\mathcal{N}(\mathbb{C}P^5)$ respectively. We also obtain that the $L_2$-Kähler metrics of these moduli spaces can be written down in terms of the Gaussian hypergeometric functions and the Appell's hypergeometric functions of type $F_1$ respectively.

## 1. Introduction

Let $M$ be an $n$-dimensional compact Kähler manifold with Kähler form $\omega$ and $E$ a vector bundle of rank $r$ on $M$. We denote by $\mathcal{N}(M)$ the moduli space of irreducible Einstein-Hermitian connections $\nabla$ on $E$. Then $\mathcal{N}(M)$ also has a Kähler form $\Omega$. For each $\nabla \in \mathcal{N}(M)$ we have an elliptic complex. We denote by $H^i_\nabla$ the $i$-th cohomology group of the elliptic complex. Then we consider the determinant line bundle:

$$\mathcal{L}_\nabla = \bigotimes_i \Big( \bigwedge^{\max} H^i_\nabla \Big)^{(-1)^{i+1}}.$$

Bismut and Freed [B-F] showed that $\mathcal{L}_\nabla$ forms a line bundle as $\nabla$ varies and defined a metric and connection, so-called Quillen metric and Bismut-Freed connection, on it using the zêta function regularization. It is not at all obvious that $\mathcal{L}_\nabla$ forms a line bundle since the dimensions of $H^i_\nabla$ may change as $\nabla$varies. Let $p : \mathcal{N}(M) \times M \to M$ be the projection to the second factor and $\mathcal{E} = p^*(E)$ the universal bundle on $\mathcal{N}(M) \times M$. Then the curvature form of the Bismut-Freed connection on $\mathcal{L}$ is given by

$$2\pi\sqrt{-1} \left\{ \int_M ch(\mathcal{E}) Td(M) \right\}^{(2)},$$

where $(\cdot)^{(2)}$ means the 2-form component.

**Theorem 1.1.** ([B-F]) *The curvature form of the Bismut-Freed connection on the determinant line bundle* $\mathcal{L}$ *is given by* $-2\pi\sqrt{-1}\Omega$, *where* $\Omega(\alpha,\beta) = -2^{n-2}\pi^{-2}c_2(E)\int_M \mathrm{tr}(\alpha\wedge\beta)\wedge\omega^{n-1}$ *for* $\alpha,\beta \in T_{[\nabla]}\mathcal{N}(M)$.

On the other hand, it is known that $\mathcal{N}(M)$ has a Kähler structure defined by $L_2$-metric and the complex structure on the base manifold ([I], [K]). Uhlenbeck and Yau [U-Y] also showed that $\mathcal{N}(M)$ is isomorphic to the moduli space of stable holomorphic vector bundle over compact Kähler manifold $M$. Fundamental example of the stable holomorphic vector bundles is null-correlation bundles on $\mathbb{C}P^{2n+1}$ [Sp]. A holomorphic vector bundle $N$ is called the *null-correlation bundles* if it has an exact sequence

$$0 \longrightarrow \mathcal{O}(-1) \longrightarrow T^*\mathbb{C}P^{2n+1}\otimes\mathcal{O}(1) \longrightarrow N \longrightarrow 0.$$

We denote by $\mathcal{N}(\mathbb{C}P^{2n+1})$ the moduli space of null-correlation bundles.

In [Ni-T], we have already obtained the algorithm which can wrote down the Kähler metric on $\mathcal{N}(\mathbb{C}P^{2n+1})$. In this paper, we will give the explicit curvature, that is the $L_2$-Kähler metric, of the determinant line bunlde $\mathcal{L}$ on $\mathcal{N}(\mathbb{C}P^3)$ and $\mathcal{N}(\mathbb{C}P^5)$ respectively. We will also describe that the $L_2$-Kähler metrics of these moduli spaces can be written down in terms of the Appell's hypergeometric functions of type $F_1$.

## 2. Preliminaries

When a Quaternion-Kähler manifold $X$ has a positive scalar curvature, the twistor space is naturally a Kähler-Einstein manifold. An anti-self-dual connection on $F$ can be pulled back to an Einstein-Hermitian connection on the pulled back vector bundle over the twistor space. The mapping induces a natural injection from the moduli space of anti-self-dual connections over $X$ to the moduli space of Einstein-Hermitian connections of the twistor space. It is known that the embedding is totally real (cf. [Ni.2]).

Let $\mathbb{H}P^2$ be an real eight dimensional quaternionic projective space. Then the twistor space is a complex projective space $\mathbb{C}P^5$ as a Hopf fibration $p : \mathbb{C}P^5 \to \mathbb{H}P^2$.

From now on, we set $F = Sp(3)\times_\rho \mathbb{H}^2$ for the projection $\rho : Sp(1)\times Sp(2) \to Sp(2)$. We call anti-self-dual connections on $F$ *1-instantons* and write the moduli space of 1-instantons as $\mathcal{M}(\mathbb{H}P^2)$ (cf. [Ni.1], [MC-S]). Let $\nabla_0$ be the unique invariant connection on the homogeneous vector bundle $\mathbb{E}$. Then $\nabla_0$ is anti-self-dual and called a *basic* 1-*instanton*. The action of $PGL(3,\mathbb{H})$ on $\mathbb{H}P^2$ preserves the $GL(2,\mathbb{H})\cdot Sp(1)$-structure. Thus we have an anti-self-dual connection $g^*\nabla_0$ on $g^*F$ for $g \in PGL(3,\mathbb{H})$. Using an $Sp(2)$-bundle equivalence $\sigma_g : F \to g^*F$, we obtain a 1-instanton $\sigma_g^* g^*\nabla_0 = \nabla_0 \cdot g$, which is unique up to $Sp(2)$-gauge transformations on $F$. Doi-Okai [D-O.1] and Nitta [Ni.3] proved that the action of $PGL(3,\mathbb{H})$ is transitive on $\mathcal{M}(\mathbb{H}P^2)$, that is $\mathcal{M}(\mathbb{H}P^2) = PSp(3)\setminus PGL(3,\mathbb{H})$.

Let $\wedge_+ = \{\lambda = \mathrm{diag}(1,\lambda_1,\lambda_2) : 1 \geq \lambda_1 \geq \lambda_2 > 0\}$. Let $K(\lambda)$ denote the centralizer of $\lambda \in \wedge_+$ in $Sp(3)$. Setting $h(\lambda, K(\lambda)g) = PSp(3)\lambda g$ for $\lambda \in \wedge_+$ and $g \in PSp(3)$, we have a homeomorphism $h : \wedge_+ \times (K(\lambda)\setminus Sp(3)) \to PSp(3)\setminus PGL(3,\mathbb{H}) = \mathcal{M}(\mathbb{H}P^2)$ ([D-O.2]).

## 3. The moduli space of null-correlation bundles over $\mathbb{C}P^3$

Groisser-Parker [G-P.1], Doi-Matsumoto-Matumoto [D-M-M] and Habermann [Ha] obtain a metric on the moduli space $\mathcal{M}(S^4)$ of 1-instantons on $S^4$, explicitly. The moduli space $\mathcal{M}(S^4)$ is generically a product space $(0,1)\times S^4$ and it has of the $Sp(2)$-isometric action. The metric formula is $h(\lambda)g_{\mathbb{R}} + f(\lambda)g_{S^4}$, where $\lambda \in (0,1)$ is the scale of instantons. Moreover, $h(\lambda)$ and $f(\lambda)$ can be written down in terms of the Gaussian hypergeometric functions.

**Theorem 3.1.** *The Kähler metric $g_{\mathcal{N}(\mathbb{C}P^3)}$ on the moduli space of the null-correlation bundles over $\mathbb{C}P^3$ is explicitly written as follows.*

$$g_{\mathcal{N}(\mathbb{C}P^3)} = h(\lambda)|d\mu|^2 + f(\lambda)g_{G_2(\mathbb{C}^4)},$$

*where*

$$\mu = \lambda + i\nu, \quad \lambda \in (0,1) \quad \nu \in U(1),$$

$$h(\lambda) = \frac{16}{5}\pi^2\lambda^2 F(4,3,6;1-\lambda^2),$$

$$f(\lambda) = 16\pi^2\left(\frac{1-\lambda^2}{1+\lambda^2}\right)^2\left\{1 - \frac{\lambda^4}{5}F(4,3,6;1-\lambda^2)\right\},$$

$$F(\alpha,\beta,\gamma;x) = \frac{\Gamma(\gamma)}{\Gamma(\beta)\Gamma(\gamma-\beta)}\int_0^1 u^{\beta-1}(1-u)^{\gamma-\beta-1}(1-xu)^{-\alpha}du.$$

**Proof.** Since the moduli space $\mathcal{M}(S^4)$ of 1-instantons over $S^4$ is totally real embedded in $\mathcal{N}(\mathbb{C}P^3)$ [Ni.2], the tangent space of $\mathcal{N}(\mathbb{C}P^3)$ is a complexification of the tangent space of $\mathcal{M}(S^4)$. The Kähler metric $g_{\mathcal{N}(\mathbb{C}P^3)}$ is a complexification of $g_{\mathcal{M}(S^4)} = h(\lambda)d\lambda^2 + f(\lambda)g_{S^4}$ on $\wedge_+ = (0,1) \ni \lambda$. Hence $g_{\mathcal{N}(\mathbb{C}P^3)} = h(\lambda)|d\mu|^2 + f(\lambda)g_{G_2(\mathbb{C}^4)}$ on $\wedge_+ \ni \lambda$, $\mu \in \mathbb{C}$, where

$G_2(\mathbb{C}^4)$ is the Grassman manifolds of 2-planes in $\mathbb{C}^4$. It is known that $\mathcal{N}(\mathbb{C}P^3)$ is diffeomorphic to $PSp(2,\mathbb{C}) \setminus PGL(4,\mathbb{C})$ (cf. [Ni.3]). We put $G = P(SU(2) \times SU(2)) \setminus PU(4)$. Then we can find the diffeomorphism map

$$\begin{aligned} \wedge_+ \times G \ni (\lambda, (P(SU(2) \times SU(2)))k) \\ \mapsto PSp(2,\mathbb{C})\lambda k \in PSp(2,\mathbb{C}) \setminus PGL(4,\mathbb{C}), \end{aligned}$$

where $k \in PU(4)$. Moreover, we see that $G$ is the $\frac{P(U(2)\times U(2))}{P(SU(2)\times SU(2))} \cong U(1)$-fiber bundle over $G_2(\mathbb{C}^4) \cong \frac{PU(4)}{P(U(2)\times U(2))}$. Therefore, $\mathcal{N}(\mathbb{C}P^3)$ is the $\wedge_+ \times U(1)$-fiber bundle over $G_2(\mathbb{C}^4)$. $PU(4)$ is the isometry group on $\mathbb{C}P^3$. Hence $g_{\mathcal{N}(\mathbb{C}P^3)}$ is $PU(4)$-invariant by the definition of the $L_2$-metric. In particular the metric is $U(1)$-invariant. Thus the metric decomposition is preserved globally on $\mathcal{N}(\mathbb{C}P^3)$ and we find that $\mu = \lambda + \sqrt{-1}\nu$, $\nu \in U(1)$. □

**Corollary 3.1.** *The moduli space* $(\mathcal{N}(\mathbb{C}P^3), g_{\mathcal{N}(\mathbb{C}P^3)})$ *has finite diameter and hence is incomplete. Moreover it has finite volume.*

**Proof.** Let $L$ be the length from end $(\lambda \to 0)$ to center $(\lambda \to 1)$:

$$L = \int_0^1 \sqrt{h(\lambda)}d\lambda.$$

Using the Stirling formula $\frac{(\alpha)_n(\beta)_n}{(\gamma)_n(1)_n} = \frac{\Gamma(\gamma)\Gamma(1)}{\Gamma(\alpha)\Gamma(\beta)} n^{\alpha+\beta-\gamma-1}(1+o(\frac{1}{n}))$, we have

$$\begin{aligned} |\lambda^2 F(4,3,6;1-\lambda^2)| &= \lambda^2 \left| \sum_{n=0}^{\infty} \frac{(4)_n(3)_n}{(6)_n(1)_n}(1-\lambda^2)^n \right| \\ &\leq \lambda^2 \sum \left| \frac{(4)_n(3)_n}{(6)_n(1)_n} \right| |(1-\lambda^2)^n| \\ &= \lambda^2 \sum \left| \frac{\Gamma(6)\Gamma(1)}{\Gamma(4)\Gamma(3)} n^{4+3-6-1}\left(1+o\left(\tfrac{1}{n}\right)\right) \right| |(1-\lambda^2)^n| \\ &\leq C\lambda^2 \sum (1-\lambda^2)^n = C\frac{\lambda^2}{1-(1-\lambda^2)} = C \end{aligned}$$

where $C$ is constant. Similary, we see that

$$\text{volume} = \int_0^1 \sqrt{h(\lambda)} f^2(\lambda)d\lambda \times (\text{vol}(U(1))) \times (\text{vol}(G_2(\mathbb{C}^4)))$$

is finite. □

One of the motivations for studying the explicit metric arises in physics. Path integrals over the infinite dimensional space of connections modulo gauge transformations arising in quantum Yang-Mills theory are thought to be well-approximated by integrals over the finite dimensional moduli spaces with respect to an appropriate quantum measure. The evaluation of such integrals requires a detailed description of the metric and the volume form of the moduli spaces. Our conjecture is that the integrals over the moduli space of null-correlation bundles on twistor space $\mathbb{C}P^3$ in the semiclassical approximation are finite and calculable.

## 4. Diffeo-type and tangent space of null-correlation bundles

In this section, we write down an explicit metric of the moduli space of 1-instantons on $\mathbb{H}P^2$. The homogeneous manifold $(Sp(1) \times Sp(1) \times Sp(1)) \setminus Sp(3)$ is denoted by $F_{1,2}(\mathbb{H}^3)$ and $\wedge_+$ is decomposed into $\{(1, \lambda_1, \lambda_2) : 1 > \lambda_1 > \lambda_2 > 0\}$, $\{(1, \lambda_1, \lambda_2) : 1 = \lambda_1 > \lambda_2 > 0\}$, $\{(1, \lambda_1, \lambda_2) : 1 > \lambda_1 = \lambda_2 > 0\}$ denoted by $\wedge_+^0$, $\wedge_+'$, $\wedge_+''$. According to the decomposition, $\mathcal{M}(\mathbb{H}P^2)$ is divided into $\mathcal{M}^0$, $\mathcal{M}'$, $\mathcal{M}''$ and a basic instanton.

**Proposition 4.1.**

$$\mathcal{M}^0 \cong \wedge_+^0 \times F_{1,2}(\mathbb{H}^3), \quad \mathcal{M}' \cong \wedge_+' \times \mathbb{H}P^2, \quad \mathcal{M}'' \cong \wedge_+'' \times \mathbb{H}P^2.$$

**Proof.** $\mathcal{M}(\mathbb{H}P^2)$ is a set of orbit spaces $\{{}^t\bar{g}\,(\mathrm{diag}\,(1, \lambda_1, \lambda_2))\, g \ : \ g \in Sp(3),\ (1, \lambda_1, \lambda_2) \in \wedge_+\}$. For a natural projection $p : \mathcal{M}(\mathbb{H}P^2) \to \wedge_+$, the fiber is the orbit space $(Sp(1) \times Sp(1) \times Sp(1)) \setminus Sp(3)$ in the case $1 > \lambda_1 > \lambda_2 > 0$ and $(Sp(1) \times Sp(2)) \setminus Sp(3)$ in the case $1 = \lambda_1 > \lambda_2 > 0$ or $1 > \lambda_1 = \lambda_2 > 0$. Since $\wedge_+^0$, $\wedge_+'$ and $\wedge_+''$ are contractible, $\mathcal{M}^0$, $\mathcal{M}'$ and $\mathcal{M}''$ are product spaces. □

Let $T_{[\nabla]}\mathcal{M}(\mathbb{H}P^2)$ be a tangent space of $\mathcal{M}(\mathbb{H}P^2)$ at $[\nabla] \in \mathcal{M}(\mathbb{H}P^2)$. The second cohomology of (**) with respect to the basic instanton $\nabla_0$ is equal to zero (cf. [Ni.3, lemma 4.1.2]). If $\nabla$ is $\nabla_0 \cdot g$, then the action of $g$ induces a group isomorphism between second cohomology of (**) with respect to $\nabla_0$ and $\nabla$. By the reason, second cohomology for any $\nabla$ vanishes. On the other hand, the tangent space $T_{[\nabla]}\mathcal{M}(\mathbb{H}P^2)$ is identified with the first cohomology and we obtain the following:

**Proposition 4.2.** ([Ni-T] Proposition 3.2)

$$T_{[\nabla]}\mathcal{M}(\mathbb{H}P^2) = \left\{ \iota_V R(\nabla) \;\middle|\; v \in \mathcal{H}_0,\ V = \frac{d}{dt}(\exp tv)|_{t=0} \right\},$$

*where* $\mathcal{H}_0 = \{\, A \in M(3, \mathbb{H}) \mid {}^t\bar{A} = A,\ \mathrm{tr} A = 0 \,\}$.

Let $\mathcal{N}(\mathbb{C}P^5)$ be a moduli space of null-correlation bundles on $\mathbb{C}P^5$. Then $\mathcal{N}(\mathbb{C}P^5) \cong \{B \in GL(6,\mathbb{C}) : {}^tB = -B\}/\mathbb{C}^* \cong PSp(3,\mathbb{C}) \setminus PGL(6,\mathbb{C})$ ([Ni.3], [Sp]).

The Lie group $PGL(3,\mathbb{H})$ is naturally embedded in $PGL(6,\mathbb{C})$. In particular, we regard $\wedge_+$ as a subset of $GL(6,\mathbb{C})$. Then we have the Cartan decompositions $GL(3,\mathbb{H}) = \mathbb{R}^* \cdot Sp(3) \cdot \wedge_+ \cdot Sp(3)$ and $GL(6,\mathbb{C}) = \mathbb{C}^* \cdot Sp(3,\mathbb{C}) \cdot \wedge_+ \cdot U(6)$. If a null correlation bundle $N$ is corresponding to $\lambda g$ ($\lambda \in \wedge_+$, $g \in U(6)$), then $g^*p^*\lambda^*\nabla$ is an Einstein-Hermitian connection on $N$. First, we describe diffeomorphism type of the moduli space of null-correlation bundles. According to $\wedge_+^0$, $\wedge_+'$ and $\wedge_+''$, the moduli space is decomposed into $\mathcal{N}^0$, $\mathcal{N}'$, $\mathcal{N}''$ and a basic null-correlation bundle. We denote $P(SU(2) \times SU(2) \times SU(2)) \setminus PU(6)$, $P(SU(4) \times SU(2)) \setminus PU(6)$, and $P(SU(2) \times SU(4)) \setminus PU(6)$ by $K_1$, $K_2$ and $K_3$, respectively.

**Proposition 4.3.**

$$\mathcal{N}^0 \cong \wedge_+^0 \times K_1, \quad \mathcal{N}' \cong \wedge_+' \times K_2, \quad \mathcal{N}'' \cong \wedge_+'' \times K_3.$$

**Proof.** We denote

$$\begin{pmatrix} 0 & 1 & & & & \\ -1 & 0 & & & & \\ & & 0 & \lambda_1 & & \\ & & -\lambda_1 & 0 & & \\ & & & & 0 & \lambda_2 \\ & & & & -\lambda_2 & 0 \end{pmatrix}$$

by $\mathrm{skew}(1,\lambda_1,\lambda_2)$. $\mathcal{N}(\mathbb{C}P^5)$ is a set of orbit spaces $\{{}^tg(\mathrm{skew}(1,\lambda_1,\lambda_2))g : g \in PU(6),\ (1,\lambda_1,\lambda_2) \in \wedge_+\}$. There is a natural projection $q : \mathcal{N}(\mathbb{C}P^5) \to \wedge_+$, where the fiber is the orbit space $K_1$, $K_2$ or $K_3$ in the case $1 > \lambda_1 > \lambda_2 > 0$, $1 = \lambda_1 > \lambda_2 > 0$ or $1 > \lambda_1 = \lambda_2 > 0$, respectively. Since $\wedge_+^0$, $\wedge_+'$ and $\wedge_+''$ are contractible, $\mathcal{N}^0$, $\mathcal{N}'$ and $\mathcal{N}''$ are product spaces. □

According to a natural embedding $\mathfrak{sl}(3,\mathbb{H}) \subset \mathfrak{sl}(6,\mathbb{C})$, $\wedge_+$, $\mathcal{H}$ and $\mathcal{H}_0$ are considered as subspaces of $\mathfrak{sl}(6,\mathbb{C})$. A complexification $\mathcal{H}_0 \oplus i\mathcal{H}_0$ is a space $\{b \in \mathfrak{sl}(6,\mathbb{C}) : {}^tb = -b\}$. The moduli space $\mathcal{M}(\mathbb{H}P^2)$ is embedded in $\mathcal{N}(\mathbb{C}P^5)$. From now on we consider $\mathcal{M}(\mathbb{H}P^2)$ is a submanifold of $\mathcal{N}(\mathbb{C}P^5)$.

**Proposition 4.4.** ([Ni-T] Proposition 3.4) *For* $[\nabla] \in \mathcal{M}(\mathbb{H}P^2) \subset \mathcal{N}(\mathbb{C}P^5)$,

$$T_{[\nabla]}\mathcal{N}(\mathbb{C}P^5) = \left\{ \iota_V R(\nabla) \;\middle|\; V = \frac{d}{dt}(\exp tv)|_{t=0}, v \in \mathcal{H}_0 \oplus i\mathcal{H}_0 \right\}.$$

## 5. Metric on $\mathcal{M}(\mathbb{H}P^2)$

$\mathbb{H}P^2$ is considered as the quotient space of the unit 11-dimensional sphere $S^{11} = \{ z \in \mathbb{H}^3 : {}^t\bar{z}z = 1 \}$ identifying $z$ with $za$ where $a \in \mathbb{H}$ such that $|a| = 1$. The Fubini-Study metric $\langle \ , \ \rangle_{\mathbb{H}P^2}$ is an invariant metric such that the fibering $p : S^{11} \to \mathbb{H}P^2$ is a Riemannian submersion. Define a map $\tilde{f} : S^{11} \to \mathcal{H} = \{ A \in M(3, \mathbb{H}) : {}^t\bar{A} = A \}$ as follows:

$$\tilde{f}(z) = z^t\bar{z} = \begin{pmatrix} |z_0| & z_0\bar{z}_1 & z_0\bar{z}_2 \\ z_1\bar{z}_0 & |z_1| & z_1\bar{z}_2 \\ z_2\bar{z}_0 & z_2\bar{z}_1 & |z_2| \end{pmatrix}$$

for $z = (z_i) \in S^{11}$. Then $\tilde{f}$ gives a map $f : \mathbb{H}P^2 \to \mathcal{H}$ such that $\tilde{f} = f \circ p$ (cf. [Sak, (2.1)]). Let $\mathcal{H}_0$ be $\{ A \in M(3, \mathbb{H}) : {}^t\bar{A} = A, \ \mathrm{tr}A = 0 \}$ and $V$ be an induced vector field on $\mathbb{H}P^2$ for $v \in \mathcal{H}_0$. Then the vector field $V$ is written on the embedded image of $\mathbb{H}P^2$ in $\mathcal{H}$ as the following.

$$\begin{aligned} V &= \frac{d}{dt}(\exp tv)|_{t=0}(z^t\bar{z}) \\ &= \frac{d}{dt}\Big|_{t=0}\big((\exp tv)z \cdot |\exp tv|^{-2} \cdot {}^t\overline{(\exp tv)z}\big) \\ &= v(z^t\bar{z}) + z\frac{d}{dt}\Big|_{t=0}(\exp tv|^{-2})^t\bar{z} + (z^t\bar{z})v \\ &= v(z^t\bar{z}) + (z^t\bar{z})v - 2(z^t\bar{z})v(z^t\bar{z}). \end{aligned}$$

We denote $Pr$ by the projection from $\mathcal{H}_0$ to vector fields on $\mathbb{H}P^2$, that is $Pr_{(z^t\bar{z})} = (z^t\bar{z})v + v(z^t\bar{z}) - 2(z^t\bar{z})v(z^t\bar{z})$ at $z^t\bar{z}$. Hence the metric $\langle V, W \rangle_{\mathbb{H}P^2}$ is written down as follows.

$$\begin{aligned} \langle V, W \rangle &= \langle Pr_{(z^t\bar{z})}v, Pr_{(z^t\bar{z})}w \rangle \\ &= \langle v, Pr^2_{(z^t\bar{z})}w \rangle = \langle v, Pr_{(z^t\bar{z})}w \rangle \\ &= \langle v, (z^t\bar{z})w + w(z^t\bar{z}) - 2(z^t\bar{z})w(z^t\bar{z}) \rangle \\ &= \mathrm{tr}((vw + wv)(z^t\bar{z})) - 2\mathrm{tr}(v(z^t\bar{z})w(z^t\bar{z})). \end{aligned} \tag{5.1}$$

We prepare the following lemmas to prove the theorems. We denote the $L_2$-metric on $\mathcal{M}(\mathbb{H}P^2)$ by $\langle \ , \ \rangle_{\mathcal{M}(\mathbb{H}P^2)}$.

**Lemma 5.1.** ([Ni-T] Lemma 4.1)

$$\begin{aligned} \langle \iota_V R(\nabla), \iota_W R(\nabla) \rangle_{\mathcal{M}(\mathbb{H}P^2)} &= \int_{\mathbb{H}P^2} g^*g^*\langle V, W \rangle dv_{\mathbb{H}P^2} \\ &\quad + 2\int_{\mathbb{H}P^2} \mathrm{tr}({}^t\overline{(g^*)}g^*)g^*\langle V, W \rangle dv_{\mathbb{H}P^2}, \end{aligned}$$

*where* $\nabla = \nabla_0 \cdot g$.

**Lemma 5.2.** ([Ni-T] Lemma 4.2)

$$
\begin{aligned}
&\langle \iota_V R(\nabla), \iota_W R(\nabla)\rangle_{\mathcal{M}(\mathbb{H}P^2)} \\
&= \mathrm{tr}\int_{S^{11}} (vw+wv)g^*g^*A dv_{S^{11}} - 2\mathrm{tr}\int_{S^{11}} vg^*g^*Awg^*g^*A dv_{S^{11}} \\
&\quad + 2\mathrm{tr}\int_{S^{11}} \mathrm{tr}\big({}^t\overline{(g^*)}g^*\big)(vw+wv)g^*A dv_{S^{11}} \\
&\quad - 4\mathrm{tr}\int_{S^{11}} \mathrm{tr}\big({}^t\overline{(g^*)}g^*\big)vg^*Awg^*A dv_{S^{11}},
\end{aligned}
\tag{5.3}
$$

*where* $A = z\,{}^t\bar{z}$, $z = {}^t(z_0, z_1, z_2) \in S^{11} \subset \mathbb{H}^2$, $V = \frac{d}{dt}(\exp tv)|_{t=0}$, $W = \frac{d}{dt}(\exp tw)|_{t=0}$ *and* $g = \mathrm{diag}(1, \lambda_1, \lambda_2)$.

**Lemma 5.3.** ([Ni-T] Lemma 6.1)

$$
\mathrm{tr}\big({}^t\overline{(g^*)}g^*\big)_{gz} = \frac{1}{|gz|^4}\{|gz|^2 - (1-\lambda_1^2)(1-\lambda_2^2)|z_0|^2 + \lambda_1^2\lambda_2^2\}, \tag{5.4}
$$

*where* $g \in \mathrm{diag}(1, \lambda_1, \lambda_2)Sp(3)$ *and* $\lambda = (\lambda_1, \lambda_2)$.

**Theorem 5.1.** *The Riemannian metrics* $g_{\mathcal{M}^0}$, $g_{\mathcal{M}'}$ *and* $g_{\mathcal{M}''}$ *are of the form as follows:*

$$
\begin{aligned}
g_{\mathcal{M}^0} &= \varphi_1(\lambda_1,\lambda_2)d\lambda_1^2 + \varphi_2(\lambda_1,\lambda_2)d\lambda_2^2 \\
&\qquad + \varphi_3(\lambda_1,\lambda_2)g_{\mathbb{H}}^1 + \varphi_4(\lambda_1,\lambda_2)g_{\mathbb{H}}^2 + \varphi_5(\lambda_1,\lambda_2)g_{\mathbb{H}}^3, \\
g_{\mathcal{M}'} &= \psi_1(\lambda_2)d\lambda_2^2 + \psi_2(\lambda_2)g_{\mathbb{H}P^2}, \\
g_{\mathcal{M}''} &= \xi_1(\lambda_1)d\lambda_1^2 + \xi_2(\lambda_1)g_{\mathbb{H}P^2},
\end{aligned}
$$

**Proof.** We choose a real basis of $\mathcal{H}_0$ as follows:

$$
\left\{ e_1 = \begin{pmatrix} 1 & & \\ & -1 & \\ & & 0 \end{pmatrix}, e_2 = \frac{1}{\sqrt{3}}\begin{pmatrix} 1 & & \\ & 1 & \\ & & -2 \end{pmatrix}, \right.
$$

$$
e_3 = \left(\begin{array}{cc|c} & 1 & \\ 1 & & \\ \hline & & \end{array}\right), e_4 = \left(\begin{array}{cc|c} & i & \\ -i & & \\ \hline & & \end{array}\right), e_5 = \left(\begin{array}{cc|c} & j & \\ -j & & \\ \hline & & \end{array}\right),
$$

$$
e_6 = \left(\begin{array}{cc|c} & k & \\ -k & & \\ \hline & & \end{array}\right), e_7 = \begin{pmatrix} & & 1 \\ & & \\ 1 & & \end{pmatrix}, e_8 = \begin{pmatrix} & & i \\ & & \\ -i & & \end{pmatrix},
$$

$$
e_9 = \begin{pmatrix} & & j \\ & & \\ -j & & \end{pmatrix}, e_{10} = \begin{pmatrix} & & k \\ & & \\ -k & & \end{pmatrix}, e_{11} = \left(\begin{array}{c|cc} & & \\ \hline & & 1 \\ & 1 & \end{array}\right),
$$

$$e_{12} = \left(\begin{array}{c|cc} & & \\ \hline & & i \\ & -i & \end{array}\right), e_{13} = \left(\begin{array}{c|cc} & & \\ \hline & & j \\ & -j & \end{array}\right), e_{14} = \left(\begin{array}{c|cc} & & \\ \hline & & k \\ & -k & \end{array}\right)\Bigg\}.$$

Then $\{e_i; 3 \le i \le 6\}$, $\{e_i; 7 \le i \le 10\}$ and $\{e_i; 11 \le i \le 14\}$ span $\mathbb{H} \oplus \mathbb{H} \oplus \mathbb{H}$ naturally. We identify $\mathcal{H}_0 \simeq \mathbb{R} \oplus \mathbb{R} \oplus \mathbb{H} \oplus \mathbb{H} \oplus \mathbb{H}$. Substituting the following equations

$$g^* A = \frac{gz \cdot {}^t\overline{gz}}{|gz|^2}, \quad g^* g^* A = \frac{g^2 z \cdot {}^t\overline{g^2 z}}{|g^2 z|^2}, \quad \text{and} \quad (5.4)$$

into (5.3) and using a property that the integral of any odd function vanishes over $S^{11}$, we have

$$\int_{\mathbb{H}P^2} \langle \iota_{E_i} R(\nabla), \iota_{E_j} R(\nabla) \rangle = 0 \quad \text{for} \quad i \neq j,$$

where $E_i = \partial_t (\exp te_i)|_{t=0}$, $E_j = \partial_t (\exp te_j)|_{t=0}$. Therefore we calculate it when $i = j$. The metric on $\mathcal{M}(\mathbb{H}P^2)$ is divided into five orthogonal components by Lemma 5.2. □

## 6. Main theorem

**Theorem 6.1.** *The Kähler metrics on $\mathcal{N}(\mathbb{C}P^5)$ are explicitly represented.*

$$\begin{aligned} g_{\mathcal{N}^0} &= \varphi_1(\lambda_1, \lambda_2)|d\mu|^2 + \varphi_2(\lambda_1, \lambda_2)|d\mu|^2 \\ &\qquad + \varphi_3(\lambda_1, \lambda_2) g^1_{\mathbb{C}^4} + \varphi_4(\lambda_1, \lambda_2) g^2_{\mathbb{C}^4} + \varphi_5(\lambda_1, \lambda_2) g^3_{\mathbb{C}^4}, \\ g_{\mathcal{N}'} &= \psi_1(\lambda_2)|d\mu_2|^2 + \psi_2(\lambda_2) g_{G_2(\mathbb{C}^6)}, \\ g_{\mathcal{N}''} &= \xi_1(\lambda_1)|d\mu_1|^2 + \xi_2(\lambda_1) g_{G_2(\mathbb{C}^6)}. \end{aligned}$$

**Proof.** Since $T_{[\nabla]}\mathcal{N}(\mathbb{C}P^5)$ is a complexification of $T_{[\nabla]}\mathcal{M}(\mathbb{H}P^2)$ for $[\nabla] \in \mathcal{M}(\mathbb{H}P^2)$, the Kähler metric $g_{\mathcal{N}(\mathbb{C}P^5)}(x_1 + Jy_1, x_2 + Jy_2)$, $(x_1, x_2, y_1, y_2 \in T_{[\nabla]}\mathcal{M}(\mathbb{H}P^2))$ is equal to $g_{\mathcal{M}(\mathbb{H}P^2)}(x_1, x_2) + g_{\mathcal{M}(\mathbb{H}P^2)}(y_1, y_2)$, where $J$ is the complex structure on $\mathcal{N}(\mathbb{C}P^5)$. Hence $g_{\mathcal{N}(\mathbb{P}^5)} = \varphi_1(\lambda_1, \lambda_2)|d\mu_1|^2 + \varphi_2(\lambda_1, \lambda_2)|d\mu_2|^2 + \varphi_3(\lambda_1, \lambda_2) g^1_{\mathbb{C}^4} + \varphi_4(\lambda_1, \lambda_2) g^2_{\mathbb{C}^4} + \varphi_5(\lambda_1, \lambda_2) g^3_{\mathbb{C}^4}$ on $\wedge^0_+$. The fiber is an orbit space of $PU(6)$ and the action is isometric on $\mathcal{N}^0$ by the definition of the $L_2$-metric. Especially the metric is $U(1) \times U(1)$-invariant whose orbit spaces are $U(1) \times U(1) \times \wedge^0_+$. Thus the metric decomposition is preserved globally on $\mathcal{N}^0$ and we find that $\mu_1 = \lambda_1 + \sqrt{-1}\nu_1$, $\mu_2 = \lambda_2 + \sqrt{-1}\nu_2$, $(\nu_1, \nu_2) \in U(1) \times U(1)$. By Theorem 5.1 and Proposition 4.3, 4.4, the same argument implies explicit formulas of $g_{\mathcal{N}'}$ and $g_{\mathcal{N}''}$. □

We denote by $F_1$ Appell's hypergeometric functions [Sas]:

$$F_1(\alpha, \beta_1, \beta_2, \gamma; x, y)$$
$$= \frac{\Gamma(\gamma)}{\Gamma(\gamma - \beta_1 - \beta_2)\Gamma(\beta_1)\Gamma(\beta_2)} \times$$
$$\int_\Delta u_1^{\beta_1 - 1} u_2^{\beta_2 - 1} (1 - u_1 - u_2)^{\gamma - \beta_1 - \beta_2 - 1} (1 - xu_1 - yu_2)^{-\alpha} du_1 du_2,$$

where $\Delta = \{ (u_1, u_2) : u_1 + u_2 \leq 1, u_1, u_2 \geq 0 \}$ (cf. [Sas]).

Let $\bar{F}_1(\alpha, \beta_1, \beta_2, \gamma; x, y)$ be

$$2^5 \pi^6 \frac{\Gamma(\beta_1)\Gamma(\beta_2)\Gamma(\gamma - \beta_1 - \beta_2)}{\Gamma(\gamma)} F_1(\alpha, \beta_1, \beta_2, \gamma; x, y).$$

**Lemma 6.1.** ([Ni-T] Proposition 6.1)

$$\int_{S^{11}} \frac{|z_0|^{2i_0} |z_1|^{2i_1} |z_2|^{2i_2}}{Z} (|z_0|^2 + a_1 |z_1|^2 + a_2 |z_2|^2)^i dv_{S^{11}} \tag{6.1}$$
$$= 4^{-1} (2\pi^2)^3 \frac{\Gamma(i_1 + 2)\Gamma(i_2 + 2)\Gamma(i_0 + 2)}{\Gamma(i_0 + i_1 + i_2 + 6)} \times$$
$$F_1(i, i_1 + 2, i_2 + 2, i_0 + i_1 + i_2 + 6; 1 - a_1, 1 - a_2).$$

Hence we obtained that the coefficients of the metrics can be written down in terms of the Appell's hypergeometric functions of type $F_1$.

**Theorem 6.2.** *The coefficients $\varphi_i(\lambda_1, \lambda_2)$, $i = 1, \ldots, 5$, are linear combinations of*

$$F_1(\alpha, \beta_1, \beta_2, \gamma; (1 - \lambda_1)^4, (1 - \lambda_2)^4),$$
$$\alpha = 1, 2, \quad \beta_1, \beta_2 = 2, 3, 4, \quad \gamma = 7, 8$$

*and*

$$F_1(\alpha, \beta_1, \beta_2, \gamma; (1 - \lambda_1)^2, (1 - \lambda_2)^2),$$
$$\alpha = 3, 4, \quad \beta_1, \beta_2 = 2, 3, 4, 5, \quad \gamma = 8, 9.$$

## References

[B-F] J.M. Bismut and D.S. Freed, *Metrics and connections o determinant bundles*, Comm. Math. Phys., 106(1986), 159–176.

[D-M-M] H. Doi, Y. Matsumoto and T. Matumoto, *An explicit formula of the metric on the moduli spaces of BPST-instantons over $S^4$*, A Fete of topology, (1988), 543–556.

[D-O.1] H. Doi and T. Okai, *Moduli space of 1-instantons on a quaternionic projective space $\mathbb{H}P^n$*, Hiroshima Math. J., 19(1989), 251–258.

[D-O.2] H. Doi and T. Okai, *Compactifying the moduli space of 1-instantons on* $\mathbb{H}P^n$, Geometry of Manifolds 1989 by Academic Press. Ins.

[G-P.1] D. Groisser and T.H. Parker, *The Riemannian geometry of the Yang-Mills moduli space*, Commun. Math. Phys., 122(1987), 663–689.

[G-P.2] D. Groisser and T.H. Parker, *Semiclassical Yang-Mills theory I: instantons*, Commun. Math. Phys., 135(1990), 101–140.

[Ha] L. Habermann, *on the geometry of the space of* $SP(1)$*-instantons with Pontrjagin index 1 on the 4-sphere*, Ann. Globa Anal. Geom., 6(1988), 3–29.

[I] M. Itoh, *Geometry of anti-self-dual connections and Kuranishi map*, J. Math. Soc. Japan, 40(1988), 9–33.

[K] H.J. Kim, *Moduli of Hermitian-Einstein vector bundles*, Math. Z., 195(1987), 143–150.

[MC-S] M. Mamone Capria and S.M. Salamon, *Yang-Mills fields on quaternionic spaces*, Nonlinearity, 1(1988), 517–530.

[Ni.1] T. Nitta, *Vector bundles over quaternionic Kähler manifolds*, Tohoku Math. J., 40(1988), 425–440.

[Ni.2] T. Nitta, *Moduli spaces of Yang-Mills connections over quaternionic Kähler manifolds*, Osaka J. Math., 26(1989), 287–298.

[Ni.3] T. Nitta, *Compactification of moduli spaces of Einstein-Hermitian connections for null correlation bundles*, Advanced Studies in Pure Math., 18-I(1990), 397–416.

[Ni-T] T. Nitta and T. Taniguchi, *Hypergeometric functions and moduli spaces of Null-Correlation bundles*, J. Math. Physics 41(2001), 1016–1025.

[Sp] H. Spindler, *Holomorphe vectorbündel auf* $\mathbb{P}_n$ *mit* $c_1 = 0$ *und* $c_2 = 1$, Manuscripta Math., 42(1983), 171–198.

[Sak] K. Sakamoto, *Planer geodesic immersions*, Tohoku Math. J., 29(1977), 25–56.

[Sal] S.M. Salamon, *Quaternionic Kähler manifolds*, Invent. Math., 67(1982), 143–171.

[Sas] T. Sasaki, *Contiguity relations of Aomoto-Gel'fand hypergeometric functions and applications to Appell's system* $F_3$ *and Goursat's system* ${}_3F_2$, SIAM J. Math. Anal., 22(1991), 821–846.

[U-Y] K. Uhlenbeck and S.T. Yau, *Hermitian-Einstein connections in stable vector bundles*, Commun. Pure Appl. Math. Suppl., 39(S)(1986), 257–293.

# ISOTROPIC KÄHLER IMMERSIONS INTO A COMPLEX QUADRIC

KAZUMI TSUKADA*

*Department of Mathematics, Ochanomizu University,*
*2-1-1 Otsuka, Bunkyo-ku, Tokyo 112-8610, Japan*
*E-mail: tsukada@math.ocha.ac.jp*

## 1. Introduction

Let $\mathbf{C}P^{n+1}$ be a complex projective space with Fubini-Study metric which has constant holomorphic sectional curvature 4 and $Q^n$ be a complex hyperquadric in $\mathbf{C}P^{n+1}$ defined in terms of the homogeneous coordinate system $z_1, z_2, \ldots, z_{n+2}$ by the following equation

$$z_1^2 + z_2^2 + \cdots + z_{n+2}^2 = 0.$$

We equip $Q^n$ with the induced Kähler metric from the Fubini-Study metric. Then $Q^n$ is a Hermitian symmetric space. Let $\{U_\lambda, \Omega_\lambda\}_{\lambda\in\Lambda}$ be a *complex conformal structure* on $Q^n$ . That is, $\{U_\lambda\}_{\lambda\in\Lambda}$ is an open covering of $Q^n$ and $\Omega_\lambda$ is a holomorphic tensor field of type $(0,2)$ defined on $U_\lambda$ and at each point $p$ of $U_\lambda$, $\Omega_\lambda$ is a non-degenerate symmetric bilinear form $T_pQ^{n+} \times T_pQ^{n+} \to \mathbf{C}$. Moreover on $U_\lambda \cap U_\mu$ there exists a non-zero holomorphic function $f_{\mu\lambda}$ which satisfies

$$\Omega_\mu = f_{\mu\lambda}\Omega_\lambda \quad \text{on} \quad U_\lambda \cap U_\mu.$$

Let $\varphi : M^m \to Q^n$ be a Kähler immersion of an $m$-dimensional Kähler manifold $M^m$ into $Q^n$. We call $\varphi$ an *isotropic Kähler immersion* if at each point $p \in M$ $\varphi^*\Omega_\lambda = 0$, $\varphi(p) \in U_\lambda$.

Our motivation of studying this class of immersions is the following:

**1.** Interesting observation relating to Gauss mappings by G. Ishikawa, M. Kimura and R. Miyaoka [2]. We identify $Q^n$ with the Grassmann manifold $G_2(\mathbf{R}^{n+2})$ of oriented 2-dimensional subspaces in $\mathbf{R}^{n+2}$. Let $V_2(\mathbf{R}^{n+2})$

*Partially supported by Grants-in-Aid for Scientific Research No. 15540065

be the Stiefel manifold of orthonormal 2-vectors in $\mathbf{R}^{n+2}$. Then the natural projection $\pi : V_2(\mathbf{R}^{n+2}) \to G_2(\mathbf{R}^{n+2})$ is a principal circle bundle. We denote by $\eta$ the map of $V_2(\mathbf{R}^{n+2})$ into the unit sphere $S^{n+1}$ defined by $\eta(e_1, e_2) = e_1$ for an orthonormal 2-vector $(e_1, e_2)$. Let $\varphi : M^m \to Q^n = G_2(\mathbf{R}^{n+2})$ be a Kähler immersion and $\tilde{M}$ be the total space of the pullback bundle $\varphi^* V_2(\mathbf{R}^{n+2})$. Then we obtain a map $\tilde{\varphi} : \tilde{M} \to S^{n+1}$ defined by $\tilde{\varphi} = \eta \circ \iota$, where $\iota : \tilde{M} \to V_2(\mathbf{R}^{n+2})$ is the bundle mapping. If $\varphi$ is an isotropic Kähler immersion, then $\tilde{\varphi} : \tilde{M} \to S^{n+1}$ is an immersion and has remarkable properties: its Gauss mapping $\tilde{M} \to G_{2m+2}(\mathbf{R}^{n+2})$ is degenerate and $\tilde{\varphi}$ is *austere* ([2] Proposition 8.2 and Theorem 8.3). Here we say an immersion to be *austere* if for each normal vector $\xi$, the set of eigenvalues of the shape operator $A_\xi$ is invariant under multiplication of $-1$.

**2.** The theory of isotropic Kähler immersions is formally analogous to that of totally complex immersions into a quaternionic Kähler manifold (for the theory of totally complex immersions into a quaternionic Kähler manifold, see [7]).

**3.** A generalization of the theory of holomorphic curves in $Q^n$ by G.R. Jensen, M. Rigoli and K. Yang ([3]). In particular, it is known that there exists a one-to-one correspondence between the set of generalized minimal immersions $S^2 \to S^{2m}$ and the set of "totally isotropic" holomorphic curves $S^2 \to Q^{2m-1}$.

In this note, first we give another definition of a complex conformal structure on $Q^n$ and introduce a (local) tensor field $J$ which satisfies $J^2 = Id$ (the identity map). We consider isotropic Kähler immersions with low codimension in §2. In §3 we construct isotropic Kähler immersions of Kähler C-spaces using orthogonal representations and study the higher normal spaces and the osculating degrees of isotropic Kähler immersions of Hermitian symmetric spaces. We state our results without detailed proofs. We will show the detailed arguments in the forthcoming paper.

## 2. The complex conformal structure on a complex quadric and isotropic Kähler immersions

We give another definition of a complex conformal structure on $Q^n$. We denote by $I$ and $\langle\ ,\ \rangle$ the complex structure and the Kähler metric on $Q^n$, respectively. We view the tangent space $T_pQ^n$ with $I$ as a complex vector space and denote by $S^2(T_pQ^n)$ the complex vector space of complex symmetric bilinear forms on $T_pQ^n$. Then $S^2(TQ^n) = \cup_{p \in Q^n} S^2(T_pQ^n)$ is a complex vector bundle over $Q^n$. We denote by $\tilde{\nabla}$ the Riemannian connection associated with the metric $\langle\ ,\ \rangle$ and by the same notation $\tilde{\nabla}$ the induced connection on $S^2(TQ^n)$.

**Proposition 2.1.** *There exists a unique complex line subbundle $L$ of $S^2(TQ^n)$ which satisfies the following conditions:*

(a) *For each $p \in Q^n$, there is a neighborhood $U$ of $p$ over which there exists a local section $\Omega$ of $L$ satisfying the following: the tensor field $J$ on $U$ defined by the equation*

$$\langle JX, Y\rangle = \text{the real part of } \Omega(X,Y) \quad X, Y \in T_qQ^n,\ q \in U$$

*satisfies $J^2 = Id$ (the identity map).*

(b) *The complex line bundle $L$ is a parallel subbundle of $S^2(TQ^n)$ with respect to the connection $\tilde{\nabla}$ .*

We call the complex line bundle $L$ *a complex conformal structure* on $Q^n$. Let $\Omega$, $J$ be local tensor fields on $U$ in Proposition 2.1 (a). Then we have

(i) $\langle JX, Y\rangle = \langle X, JY\rangle, \quad JI = -IJ$

(ii) there exists a $\mathbf{R}$-valued 1-form $\alpha$ on $U$ such that

$$\tilde{\nabla}_X J = \alpha(X) IJ.$$

The curvature tensor $\tilde{R}$ of $Q^n$ is of the form

$$\begin{aligned}(2.1)\qquad \tilde{R}(X,Y)Z = {} & \langle Y,Z\rangle X - \langle X,Z\rangle Y + \langle IY,Z\rangle IX - \langle IX,Z\rangle IY \\ & - 2\langle IX,Y\rangle IZ + \langle JY,Z\rangle JX - \langle JX,Z\rangle JY \\ & + \langle IJY,Z\rangle IJX - \langle IJX,Z\rangle IJY\end{aligned}$$

where $X, Y, Z$ are tangent vectors.

A complex subspace $W$ of $T_pQ^n$ is called *an isotropic complex subspace* if $\Omega(X,Y) = 0$ for any $X, Y \in W$. A complex subspace $W$ is isotropic if and only if $JW$ is orthogonal to $W$. An $m$-dimensional isotropic complex subspace $W$ satisfies $\tilde{R}(W,W)W \subset W$. Therefore there exists a totally geodesic Kähler submanifold $M$ with $p \in M$ and $T_pM = W$. Moreover $M$ is isometric to $\mathbf{C}P^m$ and it is also totally geodesic in $\mathbf{C}P^{n+1}$.

**Definition 2.2.** Let $\varphi : M^m \to Q^n$ be a Kähler immersion of an $m$-dimensional Kähler manifold $M^m$ into a complex quadric. Then $\varphi$ is said to be *an isotropic Kähler immersion* if for an arbitrary point $p \in M$, $\varphi_*(T_pM)$ is an isotropic complex subspace in $T_{\varphi(p)}Q^n$.

The inclusion map of the totally geodesic submanifold $\mathbf{C}P^m$ in the above is an isotropic Kähler imbedding.

Let $\varphi : M^m \to Q^n$ be an isotropic Kähler immersion and denote by $\sigma$ its second fundamental form. By straightforward calculation, we have the following.

**Lemma 2.3.** *Let $\varphi : M^m \to Q^n$ be an isotropic Kähler immersion. Then for tangent vectors $X, Y, Z, V$ we have*

(1) $\langle \sigma(X,Y), JZ \rangle = 0$,
(2) $\langle (\bar{\nabla}_X \sigma)(Y,Z), JV \rangle = -\langle \sigma(Y,Z), J\sigma(X,V) \rangle$.

By Lemma 2.3 (1), we have the following.

**Corollary 2.4.** *Let $\varphi : M^m \to Q^{2m}$ be an isotropic Kähler immersion. Then $\varphi$ is totally geodesic and $M^m$ is isometric to an open set of $\mathbf{C}P^m$.*

By Lemma 2.3 (2) and the equation of Codazzi, we have $\langle \sigma(X,Y), J\sigma(Z,W) \rangle = \langle \sigma(Z,Y), J\sigma(X,W) \rangle$. Applying this identity, we obtain the following.

**Proposition 2.5.** *Let $\varphi : M^m \to Q^{2m+1}$ be an isotropic Kähler immersion. Then for an arbitrary point $p \in M$, we have $\dim_{\mathbf{C}} RN_p \geq m-1$, where*

$$RN_p = \{\, X \in T_pM \mid \sigma(X,Y) = 0 \textit{ for all } Y \in T_pM \,\}.$$

This, together with Corollary 5 in [1], implies the following.

**Corollary 2.6.** *Let $\varphi : M^m \to Q^{2m+1}$ be an isotropic Kähler immersion of an $m(\geq 2)$-dimensional complete Kähler manifold $M$. Then $\varphi$ is totally geodesic and $M^m$ is isometric to $\mathbf{C}P^m$.*

In addition of the assumption of Proposition 2.5, suppose that the second fundamental form $\sigma$ does not vanish on $M$. At each point $p$, we put

$$\mathcal{D}_p = \{\, X \in T_pM \mid \sigma(X,Y) = 0 \quad \text{for all } Y \in T_pM \,\}.$$

Then $\dim_{\mathbf{C}} \mathcal{D}_p = m-1$. The distribution $\mathcal{D}$ is completely integrable and each leaf of $\mathcal{D}$ is a totally geodesic Kähler submanifold of $M$ and also totally geodesic in $Q^{2m+1}$ and $\mathbf{C}P^{2m+2}$. It is a so-called ruled submanifold of $Q^{2m+1}$ and $\mathbf{C}P^{2m+2}$.

**Example 2.7.** We construct isotropic Kähler immersions $M^m \to Q^{2m+1}$ which are ruled.

We denote by $\tilde{\Omega}$ the canonical non-degenerate complex symmetric bilinear form on $\mathbf{C}^{2(m+1)+1}$ and by $\pi : \mathbf{C}^{2(m+1)+1} - \{0\} \to \mathbf{C}P^{2(m+1)}$ a natural projection. Let $\Xi : U \to \mathbf{C}P^{2(m+1)}$ be an $m$-th order isotropic curve of a domain $U$ of $\mathbf{C}$ (cf. [3]), i.e., if $\Xi = \pi \circ \xi$, where $\xi : U \to \mathbf{C}^{2(m+1)+1}$, $\xi$ satisfies

- $\xi$, $\xi'$, $\xi''$, $\ldots$, $\xi^m$, $\xi^{m+1}$ are linearly independent in $\mathbf{C}^{2(m+1)+1}$,
- $\tilde{\Omega}(\xi, \xi) = \tilde{\Omega}(\xi', \xi') = \tilde{\Omega}(\xi'', \xi'') = \cdots = \tilde{\Omega}(\xi^m, \xi^m) = 0$, $\tilde{\Omega}(\xi^{m+1}, \xi^{m+1}) \neq 0$,

where $\xi^i = \partial^i \xi / \partial z^i$ . Then we note that $\tilde{\Omega}(\xi^i, \xi^j) = 0$ for $i + j \leq 2m + 1$. We define a holomorphic map $\phi : M = U \times \mathbf{C} \times \cdots \times \mathbf{C} \times (\mathbf{C} - \{0\}) \to \mathbf{C}^{2(m+1)+1} - \{0\}$ by

$$\phi(z, t_1, \cdots, t_{m-2}, t_{m-1}) = \xi(z) + t_1 \xi'(z) + \cdots + t_{m-2} \xi^{m-2}(z) + t_{m-1} \xi^{m-1}(z).$$

Then we obtain a holomorphic immersion $\Phi = \pi \circ \phi : M \to \mathbf{C}P^{2(m+1)}$. Moreover the image of $\Phi$ is contained in $Q^{2m+1}$ and $\Phi$ is an isotropic Kähler immersion into $Q^{2m+1}$.

## 3. Isotropic Kähler immersions of Hermitian symmetric spaces

First we recall the construction of Kähler C-spaces $M$ with $\dim H^2(M, \mathbf{R}) = 1$ and the canonical imbeddings into a complex projective space (cf. [4], [5]). Let $\bar{\mathfrak{g}}$ be a complex simple Lie algebra of rank $l$. Let $\{\alpha_1, \ldots, \alpha_l\}$ and $\{\Lambda_1, \ldots, \Lambda_l\}$ be a fundamental root system of $\bar{\mathfrak{g}}$ and the fundamental weight system associated with $\{\alpha_1, \ldots, \alpha_l\}$,respectively. Let $\mathfrak{g}$ be a compact real form of $\bar{\mathfrak{g}}$ and $G$ be a compact simply connected Lie group which corresponds to $\mathfrak{g}$. It is known that every Kähler C-space $M$ with $\dim H^2(M, \mathbf{R}) = 1$ can be obtained from the pair $(\bar{\mathfrak{g}}, \alpha_i)$ of a complex simple Lie algebra $\bar{\mathfrak{g}}$ and a simple root $\alpha_i$. We fix a pair $(\bar{\mathfrak{g}}, \alpha_i)$. Let $\rho$ be an irreducible complex representation of $\bar{\mathfrak{g}}$ with the highest weight $p\Lambda_i$ for a positive integer $p$. Then we obtain a holomorphic imbedding of $M$ into a complex projective space by the $\rho(G)$-orbit through the highest weight vector. It is called the *p-th canonical imbedding* of $M$.

Next we construct Kähler imbeddings of the product manifold of a number of Kähler C-spaces $M$ with $\dim H^2(M, \mathbf{R}) = 1$ into a complex projective space. Let $M_k$ $(1 \leq k \leq s)$ be Kähler C-spaces obtained from the pairs $(\bar{\mathfrak{g}}_k, \gamma_k)$ of complex simple Lie algebras $\bar{\mathfrak{g}}_k$ and simple roots $\gamma_k$, where $\gamma_k$ denotes one of simple roots $\alpha_1, \ldots, \alpha_l$ of $\bar{\mathfrak{g}}_k$. We have the direct sum $\bar{\mathfrak{g}} = \bar{\mathfrak{g}}_1 \oplus \cdots \oplus \bar{\mathfrak{g}}_s$ and its compact real form $\mathfrak{g} = \mathfrak{g}_1 \oplus \cdots \oplus \mathfrak{g}_s$. Let $G_k$

$(1 \leq k \leq s)$ be compact simply connected Lie groups which correspond to $\mathfrak{g}_k$ and $G = G_1 \times \cdots \times G_s$ be the product Lie group which corresponds to $\mathfrak{g}$. Let $f_k : M_k \to \mathbf{C}P^{n_k}$ $(1 \leq k \leq s)$ be the $p_k$-th canonical imbeddings of $M_k$ constructed by the representations $\rho_k$ of $\bar{\mathfrak{g}}_k$. We consider the external tensor product $\rho = \rho_1 \otimes \cdots \otimes \rho_s$ of the representations $\rho_k$ of $\bar{\mathfrak{g}}_k$. It is a representation of $\bar{\mathfrak{g}} = \bar{\mathfrak{g}}_1 \oplus \cdots \oplus \bar{\mathfrak{g}}_s$. Let $v_k$ be the highest weight vectors with the highest weights $p_k\Lambda_{\gamma_k}$. Then $v = v_1 \otimes \cdots \otimes v_s$ is the highest weight vector of $\rho$ with the highest weight $\sum_{k=1}^{s} p_k\Lambda_{\gamma_k}$. Then we obtain the Kähler imbedding of the product Kähler manifold $M = M_1 \times \cdots \times M_s$ into $\mathbf{C}P^n$ by the $\rho(G)$-orbit through the highest weight vector $v$, where $n = \Pi_{k=1}^{s}(n_k + 1) - 1$. We call this Kähler imbedding *the tensor product of* $f_1, \ldots, f_s$.

The representation $\rho$ of a complex semi-simple Lie algebra $\bar{\mathfrak{g}}$ is called *orthogonal* or *symplectic* according as it has an invariant symmetric or skew-symmetric bilinear form. Orthogonal or symplectic representations of a complex simple Lie algebra have been determined in Tits ([6]). Let $\{\Lambda_1, \cdots, \Lambda_l\}$ be the fundamental weight system of a complex simple Lie algebra $\bar{\mathfrak{g}}$ and $\rho_j$ be the irreducible complex representation of $\bar{\mathfrak{g}}$ with the highest weight $\Lambda_j$. The members of orthogonal or symplectic representations $\rho_j$ have been determined in [6]. If $\rho_j$ is an orthogonal representation, then the irreducible representation $\rho$ with the highest weight $p\Lambda_j$ is also an orthogonal representation. If $\rho_j$ is a symplectic representation, then the irreducible representation $\rho$ with the highest weight $p\Lambda_j$ is orthogonal or symplectic according as $p$ is even or odd. In particular, we give in Table an irreducible Hermitian symmetric space of compact type obtained from the pair $(\bar{\mathfrak{g}}, \alpha_j)$ of a complex simple Lie algebra $\bar{\mathfrak{g}}$ and its simple root $\alpha_j$ such that $\rho_j$ is an orthogonal or symplectic representation.

**Theorem 3.1.** *Let $M_k$ $(1 \leq k \leq s)$ be Kähler C-spaces obtained from the pairs $(\bar{\mathfrak{g}}_k, \gamma_k)$ and $f_k : M_k \to \mathbf{C}P^{n_k}$ $(1 \leq k \leq s)$ be the $p_k$-th canonical imbeddings of $M_k$ constructed by the representations $\rho_k$ of $\bar{\mathfrak{g}}_k$. Let $M$ be the product Kähler manifold of $M_k$ and $f : M \to \mathbf{C}P^{n+1}$ be the tensor product of $f_k$ constructed by the external tensor product $\rho$ of $\rho_k$ $(1 \leq k \leq s)$. We assume that each $\rho_k$ is either orthogonal or symplectic and that the number of symplectic representations is even. Then $\rho$ is an orthogonal representation and $f(M)$ is contained in some complex quadric $Q^n$ in $\mathbf{C}P^{n+1}$ We denote by $\tilde{f}$ the Kähler imbedding of $M$ into $Q^n$ obtained by this way. Except for a few cases, $\tilde{f}$ is isotropic.*

**Remark 3.2.** The first canonical imbedding of an $m$-dimensional complex quadric $Q^m$ is a hypersurface imbedding into $\mathbf{C}P^{m+1}$. This is the exceptional case in Theorem 3.1 for Hermitian symmetric spaces.

**Table.** Hermitian symmetric spaces with orthogonal representations:

| $M$ | $\dim_{\mathbf{C}} M$ | rank $M$ | remark |
|---|---|---|---|
| $G_{2k+2,2k+2}(\mathbf{C})$ $= SU(4(k+1))/S(U(2k+2)\times U(2k+2))$ | $(2k+2)^2$ | $2k+2$ | $k \geq 0$ |
| $Q^{2l-1} = SO(2l+1)/SO(2l-1)\times SO(2)$ | $2l-1$ | 2 | $l \geq 2$ |
| $Sp(2k)/U(2k)$ | $k(2k+1)$ | $2k$ | $k \geq 2$ |
| $Q^{2l-2} = SO(2l)/SO(2l-2)\times SO(2)$ | $2l-2$ | 2 | $l \geq 4$ |
| $SO(8k)/U(4k)$ | $2k(4k-1)$ | $2k$ | $k \geq 2$ |

Hermitian symmetric spaces with symplectic representations:

| $M$ | $\dim_{\mathbf{C}} M$ | rank $M$ | remark |
|---|---|---|---|
| $\mathbf{C}P^1$ | 1 | 1 | |
| $G_{2k+1,2k+1}(\mathbf{C})$ $= SU(4k+2)/S(U(2k+1)\times U(2k+1))$ | $(2k+1)^2$ | $2k+1$ | $k \geq 1$ |
| $Sp(2k+1)/U(2k+1)$ | $(2k+1)(k+1)$ | $2k+1$ | $k \geq 1$ |
| $SO(2(4k+2))/U(4k+2)$ | $(2k+1)(4k+1)$ | $2k+1$ | $k \geq 1$ |
| $E_7/E_6\cdot T^1$ | 27 | 3 | |

We show interesting properties of those isotropic Kähler immersions of Hermitian symmetric spaces constructed in Theorem 3.1. First we define higher fundamental forms inductively. Let $M$ and $\tilde{M}$ be Kähler manifolds and $\varphi : M \to \tilde{M}$ be a Kähler immersion of $M$ into $\tilde{M}$. Then we have the following orthogonal decomposition:

$$\varphi^* T\tilde{M} = TM + T^{\perp}M.$$

We denote by $\tilde{\nabla}$ the induced connection of $\varphi^* T\tilde{M}$ from the Riemannian connection of $\tilde{M}$. For $j \geq 2$, we define the *$j$-th fundamental forms* $\sigma_j$, *$j$-th normal spaces* $N^j$ and *$j$-th osculating spaces* $\mathcal{O}^j$ inductively. In particular,

$\sigma_j$ is defined as a smooth section of the vector bundle $\mathrm{Hom}(\otimes^j TM, T^\perp M)$ which satisfies the following identity:

$$\sigma_j(X_1,\cdots,IX_{j-1},X_j) = \sigma_j(X_1,\cdots,X_{j-1},IX_j) = I\sigma_j(X_1,\cdots,X_{j-1},X_j), \tag{3.1}$$

where $I$ denotes the complex structure tensor on $\varphi^*T\tilde{M}$. Let $\sigma$ be the second fundamental form of $\varphi$. It is well-known that $\sigma$ satisfies (3.1). We put $\sigma_2 = \sigma$. At a point $p \in M$, we define $N_p^2$ by the subspace of $T^\perp M$ spanned by $\sigma_2(X,Y)$, $X,Y \in T_pM$. By (3.1), $N_p^2$ is a complex subspace of $T^\perp M$. We put

$$\mathcal{O}_p^2 = T_pM + N_p^2,$$

which is called *the second osculating space*. We define

$$\mathcal{R}_2 = \left\{\, p \in M \;\middle|\; \dim_{\mathbf{C}} N_p^2 = \max\{\, \dim_{\mathbf{C}} N_{p'}^2 \mid p' \in M \,\} \right\}.$$

Then $\mathcal{R}_2$ is open in $M$ and $\mathcal{O}^2$ is the subbundle of $\varphi^*T\tilde{M}$ on $\mathcal{R}_2$. At $p \in \mathcal{R}_2$, we define *the third fundamental form* $\sigma_3$ by

$$\sigma_3(X_1,X_2,X_3) = \left(\tilde{\nabla}_{X_1}\sigma_2(X_2,X_3)\right)_{(\mathcal{O}_p^2)^\perp},$$

for $X_1, X_2, X_3 \in T_pM$ arbitrarily extended to the vector fields on $\mathcal{R}_2$, where $(\mathcal{O}_p^2)^\perp$ denotes the orthogonal complement of $\mathcal{O}_p^2$ in $T_{\varphi(p)}\tilde{M}$ and $(*)_{(\mathcal{O}_p^2)^\perp}$ denotes the $(\mathcal{O}_p^2)^\perp$-component of $*$ with respect to the orthogonal decomposition

$$T_{\varphi(p)}\tilde{M} = \mathcal{O}_p^2 + (\mathcal{O}_p^2)^\perp.$$

Then $\sigma_3$ is a smooth section of $\mathrm{Hom}(\otimes^3 TM, T^\perp M)$ defined on $\mathcal{R}_2$ and $\sigma_3$ satisfies (3.1). At a point $p \in \mathcal{R}_2$ we define $N_p^3$ by the subspace of $T^\perp M$ spanned by $\sigma_3(X_1,X_2,X_3)$, $X_1,X_2,X_3 \in T_pM$. It is a complex subspace of $T^\perp M$. We put $\mathcal{O}_p^3 = T_pM + N_p^2 + N_p^3$ and define

$$\mathcal{R}_3 = \left\{\, p \in \mathcal{R}_2 \;\middle|\; \dim_{\mathbf{C}} N_p^3 = \max\{\, \dim_{\mathbf{C}} N_{p'}^3 \mid p' \in \mathcal{R}_2 \,\} \right\}.$$

We define $\sigma_j$, $N^j$ and $\mathcal{O}^j$ for $j = 4, 5, \ldots$ inductively on the open set $\mathcal{R}_{j-1}$ of $M$ and $\sigma_j$ satisfies (3.1) and $N_p^j$ is a complex subspace of $T_p^\perp M$, $p \in \mathcal{R}_{j-1}$. Clearly there exists a unique integer $d$ such that $\sigma_d \neq 0$ but $\sigma_{d+1} \equiv 0$. The integer $d$ is called *the degree* of $\varphi$ and denoted by $d(\varphi)$. A Kähler immersion $\varphi : M \to \tilde{M}$ is said to be *osculating full* if $\dim_{\mathbf{C}} \mathcal{O}_p^{d(\varphi)} = \dim_{\mathbf{C}} \tilde{M}$ at $p \in \mathcal{R}_{d(\varphi)}$.

**Theorem 3.3.** *In addition to the assumption in Theorem 3.1, we assume that each pair* $(\bar{\mathfrak{g}}_k, \gamma_k)$ *defines an orthogonal symmetric Lie algebra of Hermitian type. If* $\tilde{f}$ *is an isotropic Kähler imbedding of* $M$ *into* $Q^n$ *constructed in Theorem 3.1, then* $\tilde{f}$ *is osculating full and the degree* $d(\tilde{f})$ *of* $\tilde{f}$ *is given by*

$$d(\tilde{f}) = d(f) - 1 = \sum_{k=1}^{s} r_k p_k - 1,$$

*where* $r_k$ *denotes the rank of* $M_k$ *as an Hermitian symmetric space* $(1 \leq k \leq s)$. *In particular* $d(\tilde{f})$ *is odd. Moreover for the normal spaces* $N^j$, *we have*

$$JN^j = N^{d(\tilde{f})-j+1} \quad \textit{for} \quad j = 1, \ldots, d(\tilde{f}),$$

*where we put* $N^1 = TM$.

## References

1. K. Abe: *Applications of a Riccati type differential equation to Riemannian manifolds with totally geodesic distributions*, Tôhoku Math. 25(1973), 425–444.
2. G. Ishikawa, M. Kimura and R. Miyaoka: *Submanifolds with degenerate Gauss mappings in spheres*, Adv. Studies in Pure Math. 37(2002), 115–149
3. G.R. Jensen, M. Rigoli and K. Yang: *Holomorphic curves in the complex quadric*, Bull. Austral. Math. Soc., 35(1987), 125–148.
4. H. Nakagawa and R. Takagi: *On locally symmetric Kaehler submanifolds in a complex projective space*, J. Math. Soc. Japan, 28(1976), 638–667.
5. R. Takagi and M. Takeuchi: *Degree of symmetric Kählerian submanifolds of a complex projective space*, Osaka J. Math., 14(1977), 501–518.
6. J. Tits: *Tabellen zu den einfachen Lie Gruppen und ihren Darstellungen*, Springer Lecture Notes in Math., 40, Springer-Verlag, 1967.
7. K. Tsukada: *Parallel submanifolds in a quaternion projective space*, Osaka J. Math. 22 (1985), 187–241

# PROFESSOR KOUEI SEKIGAWA AND HIS FRIENDS IN MATHEMATICS

Yasuo Matsushita

On behalf of Editors of the present volume, it is a great honor for me to write about Professor Kouei Sekigawa and his encounter with friends in mathematics.

He was born on March 12, 1944 at a small town of Joetsu city in Niigata prefecture, located at seashore of Japan Sea. When he was 8 years old, his family moved to Yokohama city in Kanagawa prefecture due to transference of his father and stayed there for 8 years. Then, his family was back to Niigata prefecture in March 1960. In April 1962, he entered in the Department of Mathematics of Niigata University and learned from Professors S. Sawaki, H. Mizusawa, K. Aoki, T. Watabe and other professors. He learned Geometry from Professor H. Mizusawa. After graduated the Department of Mathematics, he entered in the Graduate School (Master Course) of Niigata University and studied several topics in Differential Geometry under Professor Sawaki. H. Takagi was a graduate student, a senior by one year. Since then, Takagi is one of his best friends and a good rival in his research life. He was influenced by Takagi so much in Mathematics. After finished the Master Course, he got a position as a Lecturer in Kamo-Gyosei Commercial Junior College in April 1968. He left there next year and got a new position as an Assistant Professor in the Department of Mathematics, Niigata University. Takagi was already a teaching staff in Niigata University. In a private communication, Takagi introduced him the Nomizu conjecture: *An* $n(\geq 3)$*-dimensional complete irreducible Riemannian manifold satisfying the condition* $R(X,Y)\cdot R = 0$ *is locally symmetric* [14], and suggested him to study the conjecture together. In August 1970, he was invited to the 17th Symposium on Geometry held at Nara Women's University, and delivered a lecture on the Nomizu Conjecture. He showed in a joint work with Takagi [29] that the conjecture is true for an $n(\geq 3)$-dimensional conformally flat space. Since then, he obtained several partial affirmative answers to the conjecture. In April 1971, Takagi left Niigata University and got a position in Tohoku University, and he was promoted to be a Lecturer in his Department. On some day in June, Takagi suddenly informed him that a counter example to the conjecture was found. The news gave him a nasty jar. Takagi's example is a hypersurface in a 4-dimensional Euclidean space $E^4$, which is given by a graph of a non-linear function defined on entire 3-dimensional Euclidean space $E^3$ [33]. He was indeed defeated by such counter examples. He could find, however, higher dimensional counter examples in a similar way as in Takagi's construction [20]. He has continued to try to find other counter examples to the conjecture under some additional curvature conditions and succeeded in construction of a 3-dimensional counter example with constant scalar curvature [21], [22]. Later, Takagi pointed out [34] that this example is an answer to the question posed by I.M. Singer [32] at a Meeting on Geometry at Kanazawa University in 1973. Namely, this is the first example of a complete simply connected curvature homogeneous space which is not homogeneous. From such a discovery, the several topics related to curvature homogeneous spaces (including pseudo-Riemannian cases and also higher order cases) have been studied by many geometers, for example, L. Vanhecke and F. Tricerri [35], O. Kowalski [11], P. Gilkey [6], and others.

One of other major research subjects, which interests Professor Sekigawa greatly, was Almost Hermitian Geometry. Professor Sawaki, his supervisor in his master course, is an international expert in Almost Complex Geometry and he learned many things under Sawaki's instruction. He started a research on Almost Hermitian Geometry, especially focusing his attention on the relationships between almost complex structures and Riemannian structures. Parallel with the study of such topics in Almost Hermitian Geometry, he continued the study of the related topics on curvature homogeneous space. In 1973, he was promoted to Associate Professor in the Deapartment. In 1979, he was conferred PhD at Tokyo Institute of Technology under the instruction of the supervisor Professor S. Ishihara. The topic of his thesis was concerned with homogeneity of some almost Hermitian manifolds [23]. In November 1981, he was promoted to Professor in the Department. His research interest then turned into topics concerning submanifolds in a 6-sphere $S^6$, which can be considered as a nearly Kähler manifold. He obtained a classification of $J$-holomorphic curves of constant Gaussian curvature in $S^6$, together with some significant examples, and reported further results concerning the pinching problems of the Gaussian curvatures. He also studied CR-submanifolds in $S^6$, and obtained an example of 3-dimensional compact minimal proper CR-submanifold in $S^6$. These results and examples were quoted by many geometers, in fact, by F. Dillen, L. Verstraelen and L. Vrancken [4], H. Hashimoto, K. Mashimo [7], [8], [9] and others.

In the early spring of 1970, Professor Sekigawa visited S. Tanno at his home. Then, Tanno introduced him a conjecture by S.I. Goldberg: *A compact almost complex structure of an almost Kähler Einstein manifold is integarable (and hence, a Kähler manifold)* [5]. The conjecture greatly interests him, because it is closely related to the generalization of Matsushima's result by his teacher Sawaki [19]. However, he did not start to study the Goldberg Conjecture at that time. In 1983, he began to consider the Goldberg Conjecture and obtained a partial affirmative answer to the conjecture in dimenson 4. Namely, he showed that the conjecture is true if the scalar curvature is non-negative in dimension 4 in 1985 [26], and then in any dimension in 1987 [27]. The conjecture is, however, still open even now in the case where the scalar curvature is negative. Later, in the joint works with T. Oguro et al, he obtained several affirmative partial answers to the conjecture (cf. [15], [16], [17], and etc). He talked about the former result in the invited lecture of the Annual Conference of Mathematical Society of Japan held at Tokyo Metropolitan University in April 1985.

In July 1984, Professor Sekigawa and his family stayed at Leuven for two weeks. During his stay at Leuven, he visited Katholieke University and met L. Vanhecke. They discussed intensively on a problem posed by S. Kobayashi and obtained an interesting result [31], and this result was developed in several directions for generalization of locally symmetric spaces. Vanhecke is one of his best friends and he learned much from Vanhecke's passionate life style in research. During his stay at Leuven, Vanhecke showed him a paper by O. Muskarov (Bulgarian Academy of Sciences) [13]. After returned to Niigata, he sent some positive comments to Muskarov. He was pleased that Muskarov remarked his comments in that paper. By such a good opportunity to contact with Muskarov, he was invited to the Third International Conference on Complex Analysis and Applications at Varna (Bulgaria) in May 1985. It was indeed delightful for him, and he accepted immediately the invitation and attended the Conference together with his wife. He met S. Dimiev (Bulgarian Academy of Sciences) there. The conference was the most memorial one for him. Dimiev is one of his best friends. During his stay in Varna, he got acquainted with R. Parvatham (Ramanujan Institute of Advanced Mathematics). Her research area was Complex Analysis (Univalent Function Theory) at that time. She invited him to the Ramanujan Institute of Advanced Mathematics. He accepted her invitation. Next year (1986), he stayed at the Ramanujan Institute of Advanced Mathematics (Madras University) in Madras during March and delivered a course of lectures. This is one of the most memorial and fruitful stay for him. During his stay in Madras, he met many mathematicians of various fields. The Ramanijan Mathematical Society started from the next year the Ramanujan Centennial Anniversary, and the first volume of the Journal of the Ramanujan Mathematical Society was issued in the same year. He worked as an Editor (Geometry) of J. Ramanujsn Math. Soc. from 1987 till 1996. This was a golden opportunity for him. In July and August 1987, his family stayed again at Leuven for about 6 weeks. In the midst of his stay, he also visited at Krakow (Poland) by the invitation of B. Opozda and he got acquainted with W. Jelonek and P. Kowak who kindly supported him in everything during his stay in Krakow.

In Autumn 1991, Dimiev visited Niigata and stayed ten days. At that time, Dimiev proposed to hold an International Workshop at some place in Bulgaria the next year. He agreed with Dimiev's proposal. Then, in 1992, the first International Workshop on Almost Complex Structure was held at Bulgarian Academy of Sciences, Sofia. He got acquainted with K. Spallek and J. Ławrynowicz. After that, the International Workshop was held every one year in Bulgaria. The second Workshop was held at Plavetz in 1994. He got acquainted with V. Apostolov and T. Draghici there. He has attended the Workshops, together with his wife from the second one. In the same year, he was invited to the fifth Workshop of the Topology and Geometry Research Center held at Kyungpook National University during the first week of July and delivered a course of lectures on some topics in Almost Hermitian Geometry. He was aquainted with K. Grove and also many Korean mathemticians there. Including this opportunity, he and his wife visited Korea several times by the invitations of U-H. Ki, Y.J. Suh and J.T. Cho. They always treated him very warmly during every his stay in Korea. In August 1995, he and his family visited East Lansing (USA) by the invitation of D.E. Blair and stayed there for two weeks. He discussed with T. Draghichi mainly on the topics related to the Goldberg Conjecture. They were given a warm reception there by the families of D.E. Blair and T. Draghici. He indeed recognized that T. Draghici has developed largely his results concerning Goldberg Conjecture in the joint works with V. Apostolov [1], [2], [3] and etc. He is also pleased with a development on the studies concerning the Goldberg Conjecture by M. Itoh [10], H. Sato [18], Y. Matsushita [12] and others.

In the fourth Workshop 1998, he was presented *COMMEMORATIVE DIPLOMA* from the Bulgarian Academy of Sciences in recognition of his contribution to the series of the International Workshops. He is proud of this very much. He is also proud of a fact that in 2000, H. Hashimoto and T. Oguro, both his students, were presented the same Diplomas from the Bulgarian Academy of Sciences in recognition of their contributions to the series of the Workshops. H. Hashimoto has made efforts in the propaganda of the Workshops, and T. Oguro has engaged in the arrangement of the manuscripts contributed to the Proceedings of each Workshop. In the last year, the 7th International Workshop was held at Plovdiv University (Plovdiv, Bulgaria) from August 31 to September 4. Professor Sekigawa and his colleague have published the Proceedings of every International Workshop from the World Scientific Publishing. The Proceedings of the 7th International Workshop, the latest one, will be published soon.

Professor Sekigawa has supervised six PhD students, H. Hashimoto, T. Koda, J.T. Cho, T. Oguro, A. Yamada and R.S. Lemence up to now. All of them grew up to be nice and active mathematicians and have contributed much to the study of Differential Geometry. He is very proud of his students and their works.

Professor Sekigawa expresses his thanks to all of his friends, his students, and especially to his wife and daughter.

## References


1. V. Apostolov and T. Drăghici, *Almost Kähler 4-manifolds with J-invariant Ricci tensor and special Weyl tensor*, Quart. J. Math. **51** (2000), 275–294.
2. V. Apostolov, T. Drăghici and A. Moroianu, *A splitting theorem for Kähler manifolds whose Ricci tensors have constant eigenvalues*, Internat. J. Math. **12** (2001), 769–789.
3. V. Apostolov, J. Armstrong and T. Drăghici, *Local models and integrability of certain almost Kähler 4-manifolds*, Math. Ann. **323** (2002), 633–666.
4. F. Dillen, L. Verstraelen and L. Vrancken, *Classification of totally real 3-dimensional submanifolds of* $S^6(1)$ *with* $K \geq 1/16$, J. Math. Soc. Japan **42** (1990), 565–584.
5. S.I. Goldberg, *Integrability of almost Kaehler manifolds*, Proc. Amer. Math. Soc. **21** (1969), 96–100.
6. P. Gilkey and S. Nikčević, *Complete curvature homogeneous pseudo-Riemannian manifolds*, Classical Quantum Gravity **21** (2004), 3755–3770.
7. H. Hashimoto, *J-holomorphic curves of a 6-dimensional sphere*, Tokyo J. Math. **23** (2000), 137–159.
8. H. Hashimoto, *Deformations of super-minimal J-holomorphic curves of a 6-dimensional sphere*, Tokyo J. Math. **27** (2004), 285–298.

9. H. Hashimoto and K. Mashimo, *On some 3-dimensional CR-submanifolds in* $S^6$, Nagoya Math. J. **156** (1999), 171–185.
10. M. Itoh, *Almost Kähler 4-manifolds,* $L^2$*-scalar curvature functional and Seiberg-Witten equations*, Internat. J. Math. **15** (2004), 573–580.
11. O. Kowalski, F. Tricerri and L. Vanhecke, *Curvature homogeneous Riemannian manifolds*, J. Math. Pures Appl. **71** (1992), 471–501.
12. Y. Matsushia, Private communication.
13. O. Muškarov, *Complex submanifolds in almost complex manifolds*, Simon Stevin **60** (1986), 193–209.
14. K. Nomizu and K. Yano, *Some results related to the equivalence problem in Riemannian geometry*, Proceedings of the U.S.-Japan Seminar in Differential Geometry, Kyoto, Japan, (1965), 95–100.
15. T. Oguro and K. Sekigawa, *Non-existence of almost Kähler structure on hyperbolic space of dimension* $2n(\geq 4)$, Math. Ann. **300** (1994), 317–329.
16. T. Oguro and K. Sekigawa, *Four-dimensional almost Kähler Einstein and* $*$*-Einstein manifolds*, Geometriae Dedicata, **69** (1998), 91–112.
17. T. Oguro, K. Sekigawa and A. Yamada, *Four-dimensional almost Kähler Einstein and weakly* $*$*-Einstein manifolds*, Yokohama Math. J. **47** (1999), 75–92.
18. H. Sato, *Compact almost Kähler manifolds with divergence-free Weyl conformal tensor*, preprint.
19. S. Sawaki, *A generalization of Matsushima's theorem*, Math. Ann. **146** (1962), 279–286.
20. K. Sekigawa, *On some hypersurfaces satisfying* $R(X,Y)\cdot R = 0$, Tensor(NS), **25** (1972), 133–136.
21. K. Sekigawa, *On some 3-dimensional Riemannian manifolds*, Hokkaido Math. J. **2** (1973), 259–270.
22. K. Sekigawa, *On the Riemannian manifolds of the form* $B\times_f F$, Kōdai Math. Sem. Rep. **26** (1975), 343–347.
23. K. Sekigawa, *Almost Hermitian manifolds satisfying some curvature conditions* $B\times_f F$, Kōdai Math. J. **2** (1979), 384–405.
24. K. Sekigawa, *Almost complex submanifolds of a 6-dimensional sphere*, Kodai Math. J. **6** (1983), 174–185.
25. K. Sekigawa, *Some CR-submanifolds in a 6-dimensional sphere*, Tensor (NS) **41** (1984), 13–20.
26. K. Sekigawa, *On some 4-dimensional compact Einstein almost Kähler manifolds*, Math. Ann. **271** (1985), 333–337.
27. K. Sekigawa, *On some compact Einstein almost Kähler manifolds*, J. Math. Soc. Japan. **39** (1987), 677–684.
28. K. Sekigawa, H. Suga and L. Vanhecke, *Curvature homogeneity for four-dimensional manifolds*, J. Korean Math. Soc. **32** (1995), 93–101.
29. K. Sekigawa and H. Takagi, *On conformally flat spaces satisfying a certain condition on the Ricci tensor*, Tôhoku Math. J. **23** (1971), 1–11.
30. K. Sekigawa and S. Tanno, *Sufficient conditions for a Riemannian manifold to be locally symmetric*, Pacific J. Math. **34** (1970), 157–162.
31. K. Sekigawa and L. Vanhecke, *Symplectic geodesic symmetries on Kähler*

*manifolds*, Quart. J. Math. Oxford **37** (1986), 95–103.
32. I.M. Singer, *Infinitesimally homogeneous spaces*, Comm. Pure Appl. Math. **13** (1960), 685–697.
33. H. Takagi, *An example of Riemannian manifolds satisfying* $R(X,Y)\cdot R=0$ *but not* $\nabla R=0$, Tôhoku Math. J. **24** (1972), 105–108.
34. H. Takagi, *On curvature homogeneity of Riemannian manifolds*, Tôhoku Math. J. **26** (1974), 581–585.
35. F. Tricerri and L. Vanhecke, *Curvature homogeneous Riemannian manifolds*, Ann. Sci. École Norm. Sup. **22** (1989), 535–554.
36. K. Tsukada, *Curvature homogeneous hypersurfaces immersed in a real space form*, Tohoku Math. J. **40** (1988), 221–244.